In a world where most people live in cities, charting a sustainable course into the future demands that we take urban environments as seriously as any other. This volume is filled with interesting ideas about what that means—and how it might give us real leverage against the scourge of radical inequality.

Bill McKibben, *author of* The End of Nature

The City Is an Ecosystem artfully demonstrates the sciences and the humanities working together under climate crisis conditions. With integrated knowledge around everyday concerns in urban settings, and aiming for systemic change, this volume will be invaluable to teachers, researchers and activists working on the inequities introduced by climate change.

Elizabeth Mazzolini, *author of* The Everest Effect: Nature, Culture, Ideology

The City Is an Ecosystem: Sustainable Education, Policy, and Practice delves deeply into the interrelated nature of urban climate change, biodiversity loss, and social inequality. It highlights cities as socio-ecological systems with a focus on human rights, equity, and justice and identifies impactful opportunities for enhancing urban sustainability education.

Cynthia Rosenzweig, *Head of Climate Impacts Group, NASA Goddard Institute for Space Studies*

THE CITY IS AN ECOSYSTEM

The City Is an Ecosystem maps an interdisciplinary, community-engaged response to the great ecological crises of our time—climate change, biodiversity loss, and social inequality—which pose particular challenges for cities, where more than half the world's population currently live.

Across more than 20 chapters, the three parts of the book cover historical and scientific perspectives on the city as an ecosystem, human rights to the city in relation to urban sustainability, and the city as a sustainability classroom at all educational levels inside and outside formal classroom spaces. It argues that such efforts must be interdisciplinary and widespread to ensure that an informed public and an educated new generation are equipped to face an uncertain future, particularly relevant in the post-COVID-19 world.

Gathering multiple interdisciplinary and community-engaged perspectives on these environmental crises, with contemporary and historical case study discussions, this timely volume cuts across the humanities and social and health sciences and will be of interest to policy makers, urban ecologists, activists, built environment professionals, educators, and advanced students concerned with the future of our cities.

Deborah Mutnick is Professor of English at Long Island University's Brooklyn Campus, USA.

Margaret Cuonzo is Professor of Philosophy at Long Island University's Brooklyn Campus, USA.

Carole Griffiths is Professor Emeritus of Biology at Long Island University's Brooklyn Campus and Research Associate in Ornithology at the American Museum of Natural History in New York, USA.

Timothy Leslie is Associate Professor of Biology at Long Island University's Brooklyn Campus, USA.

Jay M. Shuttleworth is Assistant Professor of Social Studies at City University of New York, Queens College, USA.

Advances in Urban Sustainability

Urban Sustainability in Theory and Practice
Circles of sustainability
Paul James

Sustainability Citizenship in Cities
Theory and Practice
Edited by Ralph Horne, John Fien, Beau B. Beza, and Anitra Nelson

Street Fights in Copenhagen
Bicycle and Car Politics in a Green Mobility City
Jason Henderson and Natalie Marie Gulsrud

Adaptation Urbanism and Resilient Communities
Transforming Streets to Address Climate Change
Billy Fields and John L. Renne

Performance Metrics for Sustainable Cities
Edited by Sylvie Albert and Manish Pandey

The City Is an Ecosystem
Sustainable Education, Policy, and Practice
Edited by Deborah Mutnick, Margaret Cuonzo, Carole Griffiths, Timothy Leslie, and Jay M. Shuttleworth

For more information about this series, please visit: www.routledge.com/Advances-in-Urban-Sustainability/book-series/AUS

THE CITY IS AN ECOSYSTEM

Sustainable Education, Policy, and Practice

Edited by Deborah Mutnick, Margaret Cuonzo, Carole Griffiths, Timothy Leslie, and Jay M. Shuttleworth

Routledge
Taylor & Francis Group
LONDON AND NEW YORK

earthscan
from Routledge

Cover image: East River Raft (Series), Williamsburg, Brooklyn, New York, 1958

First published 2023
by Routledge
4 Park Square, Milton Park, Abingdon, Oxon OX14 4RN

and by Routledge
605 Third Avenue, New York, NY 10158

Routledge is an imprint of the Taylor & Francis Group, an informa business

British Library Cataloguing-in-Publication Data
A catalogue record for this book is available from the British Library

Library of Congress Cataloging-in-Publication Data
A catalog record for this book has been requested

ISBN: 978-1-032-10864-3 (hbk)
ISBN: 978-1-032-10865-0 (pbk)
ISBN: 978-1-003-21744-2 (ebk)

DOI: 10.4324/9781003217442

Typeset in Bembo
by Apex CoVantage, LLC

CONTENTS

PART III
The City as Classroom 191

FIGURES

ABOUT THE CONTRIBUTORS

Editors

Deborah Mutnick is Professor of English at Long Island University's Brooklyn Campus. She is author of *Writing in an Alien World: Basic Writing and the Struggle for Equality in Higher Education* (Boynton/Cook 1996) and coeditor of *Writing Democracy: The Political Turn in and Beyond the Trump Era* (Routledge 2019), as well as numerous articles and book chapters. Currently, she is at work on a monograph about Richard Wright's enduring relevance to 21st-century America, including his proto-environmental consciousness.

Margaret Cuonzo is Professor of Philosophy at Long Island University's Brooklyn Campus. She is the author of *Paradox* (MIT Press 2014) and numerous articles on the philosophy of language, logic, and philosophy of science. She is currently completing a book on the philosophy of rumor and gossip. A Brooklyn native, Margaret coordinates the LIU Brooklyn Community Garden and grows hazelnuts at Morning Glory & Bumble Bee, her regenerative farm near Lambertville, New Jersey.

Carole Griffiths is Professor Emeritus of Biology at Long Island University's Brooklyn Campus and Research Associate in Ornithology at the American Museum of Natural History in New York. Her research focuses on evolutionary relationships within the families of diurnal birds of prey. She is on the board of Federated Conservationists of Westchester County, a nonprofit working on environmental advocacy, education, and environmental justice.

Timothy Leslie is Associate Professor of Biology at Long Island University's Brooklyn Campus. His research program in insect ecology focuses on conserving biodiversity and ecosystem services in both agricultural and urban settings. Recent work has addressed topics such as integrated pest and pollinator management (IPPM) in tree fruit, nontarget effects of genetically modified crops, and arthropod diversity in urban community gardens. He teaches courses in general biology, ecology, and sustainability and advises the campus Outdoors Club.

Jay M. Shuttleworth is Assistant Professor of Social Studies at City University of New York, Queens College. His research interests include issues-centered education, global citizenship education, social studies teacher education, and sustainable living. Jay started his career as a high school history teacher in Northern California.

All five coeditors also codirect LIU Brooklyn's Campus Community Urban Sustainability Program (CUSP), which was generously funded by a 2017 National Endowment for the Humanities Grant and set the stage for this edited volume.

Contributors

Rob Alexander is Associate Professor in the Master of Public Administration program at James Madison University (JMU), where he teaches courses on public sector management and public policy solutions for sustainable communities. His research focuses on the collaborative governance of socioecological problems and achievement of holistic sustainability goals. Rob spent 15 years as an environmental educator and nonprofit executive before obtaining his PhD from the Maxwell School at Syracuse University. He is also a codirector of the Institute for Constructive Dialogue and Advocacy at JMU.

Loren Byrne is an urban soil ecologist who received his PhD in Ecology from The Pennsylvania State University. He is Professor of Biology and Environmental Science at Roger Williams University (Bristol, Rhode Island, USA), where he teaches diverse environmental and sustainability studies courses including urban ecology and conservation biology. His publications include book chapters about urban soils and their restoration and an edited volume entitled *Learner-Centered Teaching Activities for Environmental and Sustainability Studies*. He serves as an associate editor for the journal *Urban Ecosystems*, for which he created a new manuscript category for advancing the study and practice of ecology with cities.

Scott Carlin is Associate Professor of Geography at Long Island University's Post Campus with expertise in urban sustainability and environmental health. At the United Nations, he is an ECOSOC representative for the International Society of Doctors for Environment. He cochaired the 66th UN Department of Global Communications NGO conference in South Korea in 2016, entitled Education for Global Citizenship: Achieving the Sustainable Development Goals for All. In 2021, as second vice president, he led the strategic planning process for the Global NGO Executive Committee. He coauthored "The Role of Civil Society in Advancing

Global Citizenship" for UN Chronicle. As a consultant for the Long Island chapter of the New York State Centers for Children's Environmental Health Centers, he launched their Children's Advisory Board on environmental health in Wyandanch in 2021.

Melissa Checker is Associate Professor of Urban Studies at Queens College and of Anthropology at the CUNY Graduate Center. Her book, *The Sustainability Myth: Environmental Gentrification and the Politics of Justice* (NYU Press 2020) won the 2020–2021 Delmos Jones and Jagna Sharff Memorial Book Prize from the Society of the Anthropology of North America. She is also the author of *Polluted Promises: Environmental Racism and the Search for Justice in a Southern Town* (NYU Press 2005), which won the 2007 Association for Humanistic Sociology Book Award. In addition, she coedited *Sustainability in the Global City: Myth and Practice* (with Cynthia Isenhour and Gary McDonogh, Cambridge 2015) and *Local Actions: Cultural Activism, Power and Public Life* (with Maggie Fishman, Columbia University Press, 2004).

Nicole K. Ciulla earned a PhD in Rhetoric and Scientific and Technical Communication from the University of Minnesota in 2020 and is currently Assistant Professor of English at Saint Mary's University in Winona, Minnesota. As a graduate student, she became interested in community-engaged research methods while participating in local climate action efforts in the Twin Cities area. She continues to study climate change communication and to serve on multiple environmental committees for the municipal government. Always looking for beneficial overlap between research and teaching, she is currently developing community partnership projects with her capstone students to explore the possibilities of civic engagement with them.

Alice Andrea Cortés is a doctor of philosophy with more than 15 years of experience as professor and researcher in universities in Europe and South America, including Freie Universität Berlin–Germany and Sergio Arboleda University, Bogotá. Her research areas are ontology, hermeneutics, phenomenology, philosophy of technology, and philosophy of art and image. She also works with new technologies and digital humanities. She is a member of the Heidegger-Gesellschaft Freiburg–Germany, the Ernst-Reuter Society alumni of the Freie Universität Berlin, and Dialogos en Mercosur, Brazil, among others. She has national and international publications.

Matthew Crippen, nominated for teaching awards, has held academic posts in Canada, Egypt, and the United States and currently has a professorship at Pusan National University and a research position at Humboldt University's Berlin School of Mind and Brain. His research focuses on history, cognitive science, and cross-cultural value theory informed by one another and orbiting ecological concerns. He has 40 publications, many in ranking journals such as *Synthese*, *Topoi*, the *British Journal of Aesthetics*, *Frontiers in Psychology*, *Environmental Ethics*, and *Transactions of the Charles Peirce Society*, plus a Columbia University Press book, *Mind Ecologies: Body, Brain, and World* (with Jay Schulken, 2020). Outside the academy, he has worked as a musician, mandolin and guitar instructor, and gymnastics coach.

Mariia Ermilova is an ecologist and environmental educator who specializes in biocultural diversity concept–based community design. She has worked as a comanager of the Iwase Neighborhood since 2016, cultivating native flowers and herbal gardens in the area. She is currently a postdoctoral researcher at the Graduate School of Horticulture, Chiba University, Japan. Mariia also draws sketches and keeps nature journals as a member of Urban Sketchers International.

Laura M. Holzman is Associate Professor of Art History and Museum Studies at Indiana University–Purdue University Indianapolis (IUPUI), where she is also Director of the Museum Studies Program and Public Scholar of Curatorial Practices and Visual Art. As a specialist in engaged art history, she activates art, its history, and its institutions to strengthen communities, expand democratic discourse, and pursue equity. She regularly develops exhibitions and public programs in collaboration with students and community partners. Holzman is the author of *Contested Image: Defining Philadelphia for the Twenty-First Century* (Temple University Press 2019). Her work has also been published in outlets such as *Public Art Dialogue*, *Panorama*, and *Newsweek Japan*.

Alice S. Horning is Professor Emerita of Writing and Rhetoric and of Linguistics at Oakland University. Her research over her entire career has focused on the intersection of reading and writing, concentrating lately on the increasing evidence of students' reading difficulties and how to address them in writing courses and across the disciplines. Her work has appeared in major professional journals and in books published by Parlor Press and Hampton Press. Her most recent books include *Teaching Critical Reading and Writing in the Era of Fake News*, coedited with Ellen Carillo (Peter Lang 2021), *Literacy Heroines: Women and the Written Word* (Peter Lang 2021), and *Talking Back: Senior Scholars and Their Colleagues Deliberate the Past, Present, and Future of Writing Studies* (Utah State University Press 2020, coedited with Norbert Elliot). She is the editor of the Studies in Composition and Rhetoric book series for Peter Lang.

Leah Humphrey is a community organizer and activist. She is an organizer with Indy10 Black Lives Matter Indianapolis and a student at IUPUI. She represented Kheprw Institute in the "Interpreting Environmental Justice" class in fall 2019 and is a member of the Climate Justice Alliance Black Caucus. Pronouns: she/her/hers.

Glen Hvenegaard is Professor of Environmental Science at the University of Alberta's Augustana Campus in Camrose, Alberta. His research examines human interactions with nature, with a focus on environmental education, parks, birds, and ecotourism. He is a member of the World Commission on Protected Areas and is a Killam Laureate.

Claudia Joseph, an environmental educator practicing whole systems design, is accredited by the Permaculture Institute of North America and Permaculture Institute USA. She has taught numerous Permaculture Design Consultant courses. As the Environmental Educator for the Old Stone House & Washington Park, she established gardens and programs from 2004–2018 and was invited to teach permaculture workshops or classes for Brooklyn Botanic Garden, Just Food, Farm School New York City, GreenThumb, and the NYC Department of Parks and Recreation, Horticultural Division between 2003 and 2018. From 1996 to 2000, she taught permaculture for the Horticulture Department at Merritt College, modeled permaculture at the Oakland Botanical Demonstration Gardens, and worked and taught at the Berkeley Ecology Center. She currently lives in the Mid-Hudson Valley, where she is designing a suburban permaculture model and continues to work at the New York Permaculture Exchange, which she founded and directs.

Greg King is Assistant Professor of Environmental Science at the University of Alberta Augustana Campus in Camrose, Alberta. His research investigates tree response to environmental drivers, specifically using tree rings and dendrochronology. He has worked in northern and alpine ecosystems and is now applying his research methods to urban forests.

Isami Kinoshita is Professor in the Faculty of Social Information Studies, Otsuma Women's University, and Professor Emeritus at Graduate School of Horticulture, Chiba University. His work focuses on city planning, citizen participation, and landscape management. He serves as the chairman of the UNICEF Child Friendly Cities Project Committee in Japan and previously sat on the editorial board of *Children, Youth and Environments*.

Jerome Krase, Emeritus and Murray Koppelman Professor at Brooklyn College of The City University of New York, is an activist-scholar regarding urban community issues. With an Indiana University BA and a New York University PhD, he researches, lectures, writes, and shoots photographs about urban life and culture globally. Sample books are *Self and Community in the City* (1982), *Ethnicity and Machine Politics* (with Charles LaCerra, 1992), *Race and Ethnicity in New York City* (2005), *Ethnic Landscapes in an Urban World* (edited with Ray Hutchison, 2006), *Seeing Cities Change: Local Culture and Class* (2012), and *Race, Class, and Gentrification in Brooklyn* (with Judith N. DeSena, 2016). He coedits *Urbanities* and serves on the editorial boards of *Visual Studies* and the *Journal of Video Ethnography*. Professor Krase is active in the American, European, and International Sociological Associations, Commission on Urban Anthropology IUAES, International Urban Symposium, H-NET Humanities on Line, and the International Visual Sociology Association.

Elizabeth Kryder-Reid is Chancellor's Professor of Anthropology and Museum Studies in the IU School of Liberal Arts at IUPUI, Director of the Cultural Heritage Research Center, and former Director of the IUPUI Museum Studies Program. With a background in archaeology, art history, and public history, she has been instrumental in creating the interdisciplinary IUPUI Museum Studies program, which is now widely recognized as a leading program in the United States for promoting a civically engaged curriculum as a training path for cultural heritage professionals. Her research investigates how humans appropriate tangible and intangible remnants of the past, particularly the built environment, and how they mobilize them in the constitution of social relationships. She is the author of *California Mission Landscapes: Race, Memory, and the Politics of Heritage* (2016). Her current research investigates toxic heritage and ways in which places of environmental harm are deployed and marginalized in contemporary memory practices.

Aghilah Nadaraj has worked with the Kheprw Institute (KI) since 2016 in the areas of fund development and climate and environmental justice. She represents KI, serving on the board of the Climate Justice Alliance. She is passionate about social justice and liberation especially within the global context, coming from her experience as a diasporic Malaysian Tamil. She believes in the power of community and relationships to achieve a radical vision of a just and whole world.

Leonel Lima Ponce is Acting Academic Director of the MSc in Sustainable Environmental Systems (SES) program at Pratt Institute's Graduate Center for Planning and the Environment (GCPE), in Brooklyn, New York. Leonel is also an instructor in the department classes that focus on participatory action research and client-based projects, including an international planning exchange course in regenerative infrastructures, community client-based interdisciplinary sustainable development and urban planning studios, and the client-based thesis SES Demonstration of Professional Competence course at GCPE. As a Registered Architect in the State of New York with LEED-GA accreditation, he has experience in institutional design projects and restoration and rehabilitation work. Leonel is a founding member of Operation Resilient Living and Innovation Plus (ORLI+), a participatory research and design consultancy dedicated to the development and implementation of inclusive, climate adaptive, and resilient processes and projects.

Devjani Roy is Senior Research Associate and former Postdoctoral Fellow at Harvard Kennedy School. She writes on the literature and culture of the long eighteenth-century, the history of the book, and the economic history of eighteenth- and nineteenth-century Britain.

Anjali Sauthoff is an environmental health scientist who is interested in the translation of scientific research into policy and practice through interdisciplinary engagement. Recently, she has been focused on developing integrated approaches to address climate change mitigation and adaptation through a lens of equity, health, and justice. She is currently working with Westchester County and the New York State government to assess the effects of climate change on public health and to develop software tools to aid in climate impact analysis and communication. She works with universities on capstone courses and with the nonprofit sector on community engagement in the context of climate and health. She received her PhD in Environmental Health Sciences from Columbia University and her master's degree in Neurobiology and Behavior from SUNY Stony Brook.

Mary Steggles received her PhD as a Commonwealth Scholar at the University of Leicester. She focuses on the politics of art and its environmental impact. Recent book chapters include "No Falling Statues. India's Response to the Black Lives Matter Movement," "Imaging the Climate Crisis: The Ceramics of Ayumi Horie, Julia Galloway, Amy Snider and Julianna Zwierciadlowska-Rhymer," "Lessening the Carbon Footprint of Ceramics," "The Vietnam Era Resisters Who Shaped Canada's Ceramic Heritage," "Transnationalism and Resistance. The Border at the 49th parallel, the Vietnam Resister, Then and Now," and "Six Public Art Commissions: The Case of the Aga Khan III and British India." She contributes to a variety of journals, including *Studio Potter, Neue Keramic, Ceramics Art and Perception*, *Ceramics Technical*, *Ceramics Monthly*, *Studio Pottery*, and *History Today*, and is a regular contributor to *Kalspatz*. She retired as Professor of Ceramics and Art History at the University of Manitoba in 2020.

Mitsunari Terada (Charlie), PhD, is an IPA Japan board member and a senior researcher at the Information Research Center for Japan Adventure Playground Association. He is a specially appointed Assistant Professor in the Faculty of Regional Policy of Takasaki City University of Economics and lives in a community center as a comanager of the Iwase Neighborhood. His profession is landscape planning and local governance for child-friendly communities.

Henry Way is a Professor of Geography and Director of the School of Integrated Sciences at James Madison University in Virginia. He teaches courses on urban design and sustainability, city planning, and cultural geography. Following a PhD in political geography, his research examines the patterns and prospects for sustainability urbanism and planning in smaller and midsize cities and how addressing climate change is mediated through smaller-scale understandings of the concept of place. He has participated in urban planning through chairing a city planning commission for a number of years, and he helped lead various design and master planning initiatives around downtown revitalization and greenspace planning.

ACKNOWLEDGMENTS

This edited volume would not exist were it not for the National Endowment for the Humanities' timely introduction of the Humanities Connections Grant. We applied for funding the first year the grant was offered and were fortunate to be part of a remarkable cohort of recipients, many of whom, not surprisingly, were also focusing on environmental issues. Designed to bring together faculty across the disciplines in order to infuse humanities content into the sciences and social sciences and vice versa, the grant enabled the five of us in biology, philosophy, English, and education to pool our knowledge and expertise to create the Campus Urban Sustainability Program (CUSP) at Long Island University's Brooklyn Campus. We designed urban sustainability curricula in our own programs; created a campus garden; held Earth Day Conferences with speakers like Puerto Rican attorney and climate justice leader Elizabeth Yeampierre, landscape ecologist and author Eric Sanderson, student climate activist and founder of Earth Uprising Alexandria Villaseñor, and climatologist and agronomist Cynthia Rosenzweig; offered "sustainability across the curriculum" workshops to our colleagues; and put out a call in 2019 for contributors to *The City Is an Ecosystem.*

Along with our deep gratitude to NEH for having the foresight to develop and offer the Humanities Connections grant opportunity and for awarding us the funds to create CUSP, we are thankful to Long Island University for supporting our efforts. We are especially thankful to our program officer Deborah Hurtt for her guidance throughout the grant period. We also thank our LIU colleagues who participated in the curriculum workshops, attended our events, and answered our call to incorporate assignments, exercises, readings, and discussions about urban sustainability into their courses. In addition, we are grateful to Bob Barry, then chair of the art department, for contributing ceramic bowls made by students, faculty, and friends to our annual Earth Day "empty bowls" fundraisers for which we raised money for victims of Hurricane Maria in Puerto Rico and for greater access to healthy food in New York City. One other delightful LIU connection is the photograph by Jerry Dantzic on the front cover, "East River Raft Series 29." We thank longtime Professor Emerita of Art Cynthia Dantzic and her son Grayson for giving us permission to use the image. We also wish to acknowledge the many contributors to this volume, whose excellent work on urban sustainability across the disciplines reflects and enlarges our vision of the transdisciplinary, interinstitutional, community-engaged work needed to meet the grave challenges we face in what will undoubtedly be a crucial decade for world history.

Of course, when we made our call for proposals in 2019, we could not have anticipated that the entire edited collection would emerge during a global pandemic that could not be more relevant to questions of urban—and planetary—sustainability and that, as we write, is still raging worldwide. This universally transformative experience slowed down everyone involved, including peer reviewers, resulted in a somewhat different book than had been originally proposed, and added to the urgency of the subject matter. Finally, we thank Routledge for giving us a home for this project and Routledge editor Rebecca Brennan, whose interest at a 2019 conference of the Association for the Study of Literature and the Environment (ASLE) planted the first seeds of the book, editor Grace Harrison, editorial assistants Rosie Anderson and Matthew Shobbrook, and our anonymous peer reviewers for shepherding us through the publication process.

1

INTRODUCTION

Margaret Cuonzo, Carole Griffiths, Timothy Leslie, Deborah Mutnick, and Jay M. Shuttleworth

In conceptualizing *The City Is an Ecosystem*, we focused on three major crises of the epoch of the Anthropocene[1]—climate change, biodiversity loss, and social inequality—in relation to urban ecosystems. These ecological and social crises produce devastating consequences for increasingly large numbers of the world's interconnected human, nonhuman animal, and plant populations. As educators, we look to that world and its internal processes to teach us and guide us in our work in classrooms and in communities. As the protagonist Lauren Oya Olamina puts it in Octavia Butler's (1993/2017) prescient novel *Parable of the Sower*:

> Your teachers
> Are all around you.
> All that you perceive,
> All that you experience,
> All that is given to you
> or taken from you,
> All that you love or hate,
> need or fear
> Will teach you . . .

Like Olamina, whose story begins in Los Angeles in 2024, we find ourselves two years away in 2022 in a world of increasing drought, rising temperatures, uncontrollable wildfires, racial injustice, and social breakdown—complex problems that make planetary life as we know it unsustainable. At the end of the novel, having sheltered in Northern California from multiplying ecological and social crises, Olamina's new, much older lover Taylor Franklin Bankole laments, "I don't think we have a hope in hell of succeeding here." She replies: "Let's go back. We have work to do" (Butler 1993/2017).

DOI: 10.4324/9781003217442-1

We have work to do, too. First, like many other educators, we have come to see our focus on urban ecosystems in relation to global sustainability as paramount—across our disciplines and beyond our disciplinary identities, specific areas of expertise, and national borders. Second, shaped by our conviction that a solution to these complex problems demands a combination of disciplinary and interdisciplinary approaches, we believe it is our responsibility to demonstrate a willingness and capacity to apply the lessons of science to the humanities and vice versa, and to cultivate a general transdisciplinary understanding of these complex phenomena in our classrooms and communities. Such understanding, we argue, depends on the collective wisdom of human history in all domains of knowledge and respect for what each discipline contributes to integrative ways of knowing. Third, given the accelerated pace of urbanization worldwide and in our own location in New York City, we are especially concerned about the dynamics of these pressing problems in urban spaces and their implications for global solutions. These are the underlying principles and realities that inform this volume.

Since we began the project, the conditions that motivated it have worsened with hotter temperatures, more wildfires, and more extreme weather events worldwide. We have also, like the rest of the humanity, lived through two years of a devastating global pandemic resulting in close to 5.5 million deaths by the end of 2021. When the United Nations Climate Change Summit, COP26 (Conference of the Parties), met in Glasgow, Scotland, in October 2021, the mood was grim as delegates acknowledged that time is running out to limit global warming to 1.5°C. As COP26 President Alok Sharma put it, "[W]e can say with credibility that we have kept 1.5 degrees within reach, but its pulse is weak" (Ellerbeck 2021). In addition to keeping close track of ecological and political developments that will affect urban spaces worldwide, the effects of climate change have now personally affected one member of our editorial team. Jay M. Shuttleworth's parents lost their family home of 41 years during the summer of 2021 fires in California. Takeaways are elusive amid such grief, but this loss underlines that climate change is happening now in our communities and must be addressed as an immediate threat that demands rapid international action to reduce fossil fuel emissions.

Although each of us was actively engaged in research and teaching related to various aspects of urban environments, we only started working together programmatically in 2017 when we applied for and received a Humanities Connections Grant from the National Endowment for the Humanities (NEH) for the Campus-Community Urban Sustainability Program (CUSP). This three-year, NEH-funded program has allowed us to design new courses on urban sustainability across the disciplines in biology, education, English, and philosophy and to incorporate sustainability themes in dozens of other courses through broader faculty involvement. CUSP has also organized campus speaker series and partnered with community organizations and nonprofits to sponsor public events on sustainability issues. Because no single discipline can address the complex, ecological problems we face in the 21st century, just as no single city or nation can do so, our overarching goal has been to create a cross-disciplinary program on urban sustainability that integrates the humanities with science, social science, and the professions, rooted in the adage "think globally and act locally," a phrase attributed to Scottish biologist, sociologist, and urban planner Patrick Geddes in his 1915 book *Cities in Evolution*. As such, some of CUSP's main learning objectives have been to develop an intergenerational perspective, to promote a systems-thinking approach to the living world, and to cultivate knowledgeable, ethical, and socially just responses to all living things. Following the lead of national organizations like the Association for the Advancement of Sustainability in Higher Education, we set out to build this campus-wide program to raise consciousness and promote civic and ethical behavior among all community members, especially undergraduate students, through engagement in interdisciplinary studies of urban sustainability.

The aim of this volume is to consider the idea and practice of urban sustainability through multiple lenses that echo, contest, and go beyond the Brundtland Commission's 1987 oft cited definition of sustainability as that which "meets the needs of the present without compromising the ability of future generations to meet their own needs." Other common points of reference are Theodore Roosevelt's declaration of the importance of preserving nature for "your children's children" (1903, 132), which ignited debates over what should be cared for and for whose benefit—a key aspect of an intergenerational perspective—and Garrett Hardin's (1968) proposition that if natural resources were free and humans acted self-interestedly without having to pay for the consequences, then those resources would disappear. Critics of Hardin's widely read essay "The Tragedy of the Commons" question his assumption that self-interest is a universal, natural trait rather than a consequence of particular historical conditions.[2] Similarly, urban ecologists and sociologists note the disappointing results of much of the "sustainable development" stimulated by the Brundtland Report and the international consensus it forged.

Nevertheless, these canonical environmental texts led appreciably to a new kind of thinking about production and consumption bound by ethical considerations of others' needs, but in building a broad consensus on the desirability of sustainable development, they also contributed to creating "the illusion of certain outcomes while obscuring the undesirable consequences" (Krueger, Freytag, and Mössner 2019, xx). In other words, failure to address systemic priorities of profit over people and private property over public welfare has undermined global policies for sustainable cities. Together with the acceleration of climate change, the disproportionate impact of environmental degradation on poor, working class, and communities of color worldwide underscores both the illusion of many sustainable development efforts and the long history of ravaged lands and peoples, epitomized, for example, by the work of Leslie Marmon Silko (1977, 2006), chronicling the destruction wreaked by European and American conquest of Laguna land.

As M. H. Austin argues (1903, 2006), an immediate challenge for humans is to reject their anthropocentrism, a worldview that puts us at the center of the universe, to understand our embeddedness in nature, and to undergo the transformative, often traumatic psychological repositioning such knowledge entails. Indeed, unsustainable living choices often derive from the gulf between one's basic needs and an understanding of what is required—or what we in modern societies have come to assume is required—to fulfill them. Gary Snyder (1969) remarked that resources required stewards and custodians. Such champions include writers like Rachel Carson (1962) for wildlife decimated by pesticide use, Lois Marie Gibbs (1982) for the health of families, and Caesar Chavez (1989) for the welfare of migrant workers and consumers. We also look to political economists who make clear the relationship between the accumulation of surplus capital and the degradation of both labor and nature in the continual extraction of resources from the Earth that form the basis for consumerist industrial and postindustrial societies. In a critical review of research on sustainability, George Liodakis (2010) notes, "Important questions have been raised in this literature concerning the content itself of sustainability, and more specifically, whose sustainability, what type of natures are to be sustained, through what processes, and for whose benefit?" (2602). Within such broad considerations of urban sustainability, persistent humanistic questions must be asked. What kind of global and local communities do we want? How do we achieve them? What kind of global and local citizen is needed for a sustainable future? How are these kinds of citizenship related?

As Loren Byrne notes in the afterword to this volume, the increasingly widespread and liberal use of the term *ecosystem* "risks turning the word into an abused cliché without clear meaning" (265).[3] Ideas associated with urban ecosystems may potentially be even more elusive according to ecologists for whom ecosystems are traditionally defined by metabolic interactions between organisms and environments rather than by the anthropogenic and social aspects of cities. However, like Byrne, we conclude that a transdisciplinary approach to the complex dynamics of urban spaces is necessary to integrate humans and their social, economic, and politics systems into conceptions of city life. Although other terminology like "social-ecological system" may describe more precisely how human beings interact with variables by which ecologists have traditionally understood ecosystems, the more straightforward idea of an urban ecosystem can serve to popularize and disseminate knowledge that can, in turn, foster engagement in activities such as green projects and urban land use planning in local communities. Thus, the pedagogical value of an integrative approach to urban ecosystems lies in educating the educator across disciplines. This then positions us to take up the urgent issues of planetary and urban sustainability from discipline-specific and transdisciplinary perspectives, to collaborate in interdisciplinary and interinstitutional forums with colleagues, and to introduce learners in traditional classroom and community settings to the key principles and concepts needed to participate fully in determining our collective future.

CUSP's educative challenge is how to foster public receptivity to scientific and social imperatives about sustainable living, on the one hand, and critical analysis of the underlying causes of unsustainable practices and conditions on the other. Beliefs derived from the observable are often the inverse of evidence-based science; at the same time, given its complexity and global scope, what one knows about sustainability and what one is willing or able to do to achieve it may also be at odds. The case for CUSP's transdisciplinary model rests on the necessity of a collaborative, integrative, ethical approach to sustainable living instruction (e.g., Neeser et al. 2010). Since we launched the program in 2017, our commitment to producing and disseminating knowledge about urban sustainability has redoubled in light of alarming international and national reports on the impact of climate change, biodiversity loss, and gross social inequality.[4] Yet, even as we are distressed by the catastrophic future predicted by these reports, we are heartened by the groundswell of support for global movements demanding a response to climate change and social justice issues, of which many are headed by a new generation of youth leaders. These young people are making the same connections between social injustice and ecological destruction that we make in this book.

While the ecological and social problems we examine here occur in all areas rural, urban, suburban, advanced and developing countries—our focus is on urban spaces like New York City, the location of our universities and a microcosm of many of the issues at the heart of the scholarly, educational, and political work on worldwide urban sustainability that we see as imperative. Shaped by our own positions as researchers and teachers across the disciplines in urban universities in one of the most diverse and vulnerable coastal cities of the world, we believe that educators and scholars have a crucial role to play in understanding these pressing problems, analyzing their root causes, helping to solve them, and integrating them into teaching across levels, disciplines, and communities.

To that end, we are guided by the following standpoints: (1) these issues call all of us to act decisively and to act *now*; (2) unprecedented environmental despoliation and extreme wealth and racial inequality are inextricably linked and must be dealt with in relation to one another; (3) addressing the unique sustainability challenges faced by cities is necessary as more than half the world's population now lives in urban areas; (4) the complexity of the problems requires transdisciplinary, interdisciplinary, international, and community-engaged approaches; and (5) as educators in schools and communities, we can play a vital role in researching, analyzing, and teaching about these complex, diffuse, and "troublesome" concepts (Meyer and Land 2006). Understanding phenomena like climate change, biodiversity loss, and global inequality necessitates a grasp of interrelated abstract concepts from across disciplines, together with respect for disciplinary knowledge outside our own areas of expertise and a capacity for critique based on seemingly opposite qualities of skepticism and open-mindedness. In his book on the Anthropocene, Timothy Clark (2015) writes:

> The newly recognized agent of humanity as a geological force is something indiscernible in any of the individuals or even large groups of which it is composed. It is a power that barely recognizes itself as such and which is not really capable of voluntary action or planning, as it arises from the often unforeseen consequences of the plans and acts of its constituents.
>
> *(15)*

What concerns us as educators is the difficulty for learners, first, of perceiving and grasping complex realities that may be remote from their experience or that clash with their worldview and, second, of integrating ecological and social concepts with theory and practice. Mathematical equations, chemical compounds, biological processes, linguistic meanings and forms, literary texts, discourse-specific vocabulary, historical contexts, and philosophical and ethical problems—all may present overwhelming obstacles that derail comprehension in both disciplinary and interdisciplinary studies and arouse feelings of alienation and resentment in learners. Such impasses result partly from our own institutional failure to facilitate integrative learning, leaving it entirely to learners to connect dots that may not yet even be visible to them. But they are also arguably endemic to any learning process that challenges preexisting beliefs.

In classrooms or communities, resistance to new knowledge and its transformational effects can further complicate the learning process. Learners may experience a sense of losing control over their own fate in paradigmatic shifts of knowledge that threaten to displace self, culture, nation, species, or God from the center of the universe, perhaps best exemplified by Copernicus's discovery that the Earth revolved around the sun or Freud's notion of the unconscious mind. While new knowledge illuminated by these major discoveries often initially meets institutional resistance from church or state as well as individual resistance, the reproduction of knowledge in successive generations may be just as painful and traumatizing to individuals and larger social entities. The urban sustainability studies we are advocating in this book require us to confront resistance on multiple levels and to help learners (and ourselves) cope with the troubling, complex, often frightening realities of the phenomena—from devastating weather events, extreme heat, and species extinction to obscene inequality—that define our epoch.

As we drafted this introduction in the midst of the first wave of COVID 19, nine months after initiating the project, we were viscerally reminded of the racial and class inequalities embedded in urban areas. Even as COVID 19 and job loss raged through working-class communities, particularly those of color in major cities across the United States, a string of extrajudicial murders of Black people occurred, exacerbating already existing, persistent racial and economic disparities and shining a spotlight on the scourge of American racism. During this period of immense suffering and pain, with skyrocketing rates of infection, hospitalization, deaths, and unemployment, anti-Black racism, in particular, became known as the "other epidemic" as tens of millions of people worldwide poured into the streets in protests sparked by the murder of George Floyd in Minneapolis—yet another name in a tragically long list of Black victims of police and vigilante violence. These questions of social inequality and injustice in relation to environmental issues of air pollution, toxic dumps, vulnerability to severe weather events, food insecurity, and many other hazardous conditions can be considered in view of the slogan and generative theories and practices of "the right to the city," a phrase coined by French sociologist Henri Lefebvre (1996) in 1968.

Our intent in this book is to contribute to such an effort more broadly by drawing clear connections between ecological and social elements of urban sustainability across the curriculum. The troublesome aspects of the underlying processes that produce unsustainable urban ecosystems is compounded by racialized class oppression, antiscience rhetoric, fake news, and a bewildering welter of facts, predictions, and uncertainties. As a microcosm of world development with diverse multitudes and viewpoints, the city holds both the promise of solving the ecological crises that beset the planet and the worst-case scenarios for what will happen if we do not. As such, cities are sites of economic opportunity but also of widening economic inequality; cities are engines for development yet routinely destroy the very streets and relations that make them livable; cities are often the centers and origins of social justice and green movements yet continue to suffer from oppressive social conditions and environmental despoliation; city dwellers have lower per capita carbon/ecological footprints yet suffer greatly from climate change due to their location and the features of the urban landscape, such as the flooding of coastal cities from sea level rise and the "heat island" effect that causes significantly warmer temperatures in metropolitan locations. It is dualities such as these that exhibit cities' unique sustainability solutions and challenges as compared to nonurban areas. To create the sustainable city—and planet—of the future, rooted in ecological resilience, social justice, and true democracy, will require imagination and expert and community-based, cross-disciplinary, cross-border knowledge, critique, skills, and cooperation.

This book provides perspectives on these dualities and considers how we can best maintain the benefits of cities and minimize their disadvantages to move toward urban sustainability as a crucial step in repairing the Earth's ecosystem.

★ ★ ★

In the pages that follow, we begin to address these questions and to imagine the sustainable city of the future by conjoining concrete details of specific urban sustainability studies with broad concepts, practices, and implications relevant to cities in every part of the world. The book is divided into three parts: (1) The City as Ecosystem, (2) The Right to the City, and (3) The City as Classroom. Descriptions of chapters can be found at the beginning of each of these sections. We invite readers to explore all the sections of this volume as there are conceptual connections among the essays in each section. As much as possible, we sought to foster a dialogue across disciplines on urban sustainability in light of ecological, biological, sociological, and humanistic research, teaching, policy, and community-engaged practices. Unwieldy in some respects, this transdisciplinary, collaborative process gives us hope that traditional disciplinary and institutional doors are opening to engage everyone in educational and ecological practices that reimagine and enact a more egalitarian, sustainable future for us all.

Notes

1 Though still subject to debate, the Anthropocene is widely viewed as the current geological epoch in which humans are the primary cause of permanent planetary change.

2 Despite the continued popularity of his Malthusian essay, Hardin has been criticized for his nationalist, eugenicist, racist, and Islamaphobic views in both his writing and his personal life (see, e.g., Mildenberger 2019; Longo and Clark 2015).
3 We invited Loren Byrne, also a contributor to *The City is an Ecosystem*, to write the afterword and are grateful to him for summing up the book's unifying principles and ethos.
4 See, in particular, the International Panel on Climate Change (2018, 2019) reports, the U.S. Fourth National Climate Assessment in 2018 (U.S. Global Change Research Program 2018), and the 2020 Oxfam report on global inequality (Lawson, et al.).

References

Austin, Mary Hunter. (1903) 2020. *The Land of Little Rain*. New York: Warbler Classics.

Brundtland, Gro Harlem. 1987. "Report of the World Commission on Environment and Development: Our Common Future." United Nations General Assembly, document A/42/427.

Butler, Octavia. (1993) 2017. *The Parable of the Sower*. New York: Seven Stories Press.

Carson, Rachel. (1962) 2002. *Silent Spring*. New York: Houghton Mifflin.

Chavez, Cesar. 1989. "On the Perils of the Pesticides to Pacific Lutheran University." www.chavezfoundation.org/_cms.php?mode=view&b_code=001008000000000&b_no=17&page=1&field=&key=&n=8.

Clark, Timothy. 2015. *Ecocriticism on the Edge: The Anthropocene as a Threshold Concept*. New York: Bloomsbury.

Ellerbeck, Alexandra. 2021. "COP 26, Global Emissions Gaps, and Extreme Weather Disasters: These Are the Top Climate Stories of 2021." *Washington Post*, 14 December 2021. https://www.washingtonpost.com/politics/2021/12/14/cop26-global-emissions-gaps-extreme-weather-disasters-these-are-top-climate-change-stories-2021/.

Gibbs, Lois Marie. 1982. *Love Canal: My Story*. New York: Grove Press.

Hardin, Garrett. 1968. "The Tragedy of the Commons. *Science* 162 (3859): 1243–48. http://science.sciencemag.org/content/162/3859/1243.

International Panel on Climate Change (IPCC). 2018: Global warming of 1.5°C. An IPCC Special Report on the impacts of global warming of 1.5°C above pre-industrial levels and related global greenhouse gas emission pathways, in the context of strengthening the global response to the threat of climate change, sustainable development, and efforts to eradicate poverty. (2018). [V. Masson-Delmotte, P. Zhai, H. O. Pörtner, D. Roberts, J. Skea, P. R. Shukla, A. Pirani, W. Moufouma-Okia, C. Péan, R. Pidcock, S. Connors, J. B. R. Matthews, Y. Chen, X. Zhou, M. I. Gomis, E. Lonnoy, T. Maycock, M. Tignor, T. Waterfield (eds.)]. www.ipcc.ch/sr15/.

———. 2019: Climate change and land: An IPCC special report on climate change, desertification, land degradation, sustainable land management, food security, and greenhouse gas fluxes in terrestrial ecosystems. (2019). [P. R. Shukla, J. Skea, E. Calvo Buendia, V. Masson-Delmotte, H.-O. Pörtner, D. C. Roberts, P. Zhai, R. Slade, S. Connors, R. van Diemen, M. Ferrat, E. Haughey, S. Luz, S. Neogi, M. Pathak, J. Petzold, J. Portugal Pereira, P. Vyas, E. Huntley, K. Kissick, M. Belkacemi, J. Malley, (eds.)]. www.ipcc.ch/srccl/.

Krueger, Robert, Tim Freytag, and Samuel Mössner, eds. 2019. *Adventures in Sustainable Urbanism*. Albany, NY: SUNY Press.

Lawson, Max, Anam Parvez Butt, Rowan Harvey, Diana Sarosi, Clare Coffey, Kim Piaget, and Julie Thekkudan. 2020. *Time to Care: Unpaid and Underpaid Care Work and the Global Inequality Crisis*. Oxfam, January 20, 2020. www.oxfam.org/en/research/time-care.

Lefebvre, Henri. 1996. *Writing on Cities*. Edited and translated by Eleonore Kofman and Elizabeth Lebas. Hoboken, NJ: Blackwell Publishers.

Liodakis, George. 2010. "Political Economy, Capitalism and Sustainable Development." *Sustainability*, August 18, 2010. www.mdpi.com/2071-1050/2/8/2601.

Longo, Stefano B., and Brett Clark. 2015. "Ecological Crisis and the Tragedy of the Commodity." *Climate and Capitalism*, July 21, 2015. https://climateandcapitalism.com/2015/07/21/ecological-crisis-and-the-tragedy-of-the-commodity-2/.

Meyer, Jan H. F., and Ray Land, eds. 2006. *Overcoming Barriers to Student Understanding: Threshold Concepts and Troublesome Knowledge*. London: Routledge.

Mildenberger, Matto. 2019. "The Tragedy of the *Tragedy of the Commons*. Voices." *Scientific American*, April 23, 2019. https://blogs.scientificamerican.com/voices/the-tragedy-of-the-tragedy-of-the-commons/.

Neeser, Marie, Ramboll Natura, Shivani Jain, Jim Taylor, and Frans Lenglet. 2010. "Sweden's Pioneering Role in Education for Sustainable Development." In *Tomorrow Today*, edited by Michele Witthaus, Karen McCandless, and Rebecca Lambert, 89–90. Leicester, UK: Tudor Rose.

Roosevelt, Theodore. (1903) 2008. "Speech at Grand Canyon." In *American Earth: Environmental Writing since Thoreau*, edited by Bill McKibben, 132–33. New York: Library of America.

Silko, Leslie Marmon. (1977) 2006. *Ceremony*. New York: Penguin Books.

Snyder, Gary. 1969. "Smokey the Bear Sutra. Sierra Club Wilderness Conference." http://theendlessfurther.com/tag/smokey-the-bear-sutra/.

U.S. Global Change Research Program. 2018: Impacts, Risks, and Adaptation in the United States: Fourth National Climate Assessment, Volume II [Reidmiller, D. R., C. W. Avery, D. R. Easterling, K. E. Kunkel, K. L. M.]. https://nca2018.globalchange.gov.

PART I
The City as Ecosystem

Introduction

Despite being primarily defined by their numerous anthropogenic features and forces, cities still exist within the natural world. As such, the chapters in Part I of this volume—The City as Ecosystem—recognize the unique set of conditions that comprise an urban ecosystem. Using this common thread, the contributors provide historical, social, philosophical, and scientific perspectives that address key concepts at the intersection of urbanization, nature, and human well-being. This includes chapters on remote technologies and how they may accelerate or reverse our growing disconnection from nature, the social determinants of health in relation to urban climate adaptation, implications of classical Rome for contemporary climate change policy, the loss and resilience of urban biodiversity, and how urban green spaces may be developed and/or used to facilitate community engagement and promote sustainability.

In "Ecology and Technological Enframement: Cities, Networks, and the COVID-19 Pandemic," Matthew Crippen and Alice Andrea Cortés examine the effect of technology on our relationship with nature, the impact of urbanization on that relationship, and, specifically, the impact of the COVID-19 crisis and increased use of network technologies in cities on our connection with the environment. Through the lens of "technological framing," they contemplate how COVID-19 may offer "prospects for better forms of urban existence that the pandemic may trigger and . . . engender a more livable world once the crisis ends" (xx).

Anjali Sauthoff introduces the theme of social, environmental, and economic effects on public health in "Social Determinants of Health as a Framework for Addressing Urban Climate Adaptation." Focusing on urban climate adaptation and justice as well as the present threat of the COVID-19 pandemic, Sauthoff helps

DOI: 10.4324/9781003217442-2

anchor the book's arguments by defining the "sustainable city" as one that promotes well-being and equity in the face of uniquely urban health risks.

In "Climate Change Policy: Lessons From Classical Rome," Devjani Roy applies insights from Cato, Varro, and Virgil to a vision of cities that balances economic growth with environmental need to minimize the impact on climate change. She goes on to relate these classical texts with their prescient conception of urban well-being to the current coronavirus pandemic, the global public health infrastructure, and the global climate system.

Greg King and Glen Hvenegaard investigate the opportunities for sustainability in a small city in "Greenspace in Small Cities: Opportunities for Sustainability Education." They use two case studies that showcase the role of a postsecondary institution in delivering sustainability education through community-engaged research that promotes local greenspace and wildlife stewardship goals.

Next, in "The Birds and Bees of Urban Biodiversity: Patterns, Problems, and a Path Forward," Timothy Leslie and Carole Griffiths examine the role of nature in urban sustainability with a focus on biodiversity, concentrating on two wildlife groups—birds and bees—and situating the idea of the sustainable city squarely in its natural environment.

In "Action Research in a Tokyo Suburb: Building Community Ties Through Collaborative Landscaping Projects," Mariia Ermilova, Mitsunari Terada, and Isami Kinoshita report on a community-based participatory action research project located in the commuter city of Matsudo in Tokyo's urban sprawl. To create a new version of a neighborhood association, graduate students worked with local residents to promote a multigenerational collaboration in the cultivation of community gardens and designing their own curricula.

Finally, in "Community Design for Inclusive Engagement: The Old Stone House and Washington Park, Brooklyn," Claudia Joseph considers the sustainability and equity of access and ownership of public parks, school grounds, and library yards, and the diverse benefits of engaging all ages and classes in ecological education, gardening, and landscaping.

2

ECOLOGY AND TECHNOLOGICAL ENFRAMEMENT

Cities, Networks, and the COVID-19 Pandemic

Matthew Crippen and Alice Andrea Cortés

Introduction

It is impossible to summarize what makes a city tick, and we do not try. Instead, we focus on attitudes that cities and contemporary technologies foster. We further consider how the recent COVID-19 pandemic expanded the use of networked platforms and how this has shifted attitudes in ways that could promote sustainability or detract from it.

Though environmentalists see cities as key to a sustainable future because they are in many ways more efficient than rural settlements, positive sentiments about urban living have not always been widespread. Jean-Jacques Rousseau (1762) called cities "the abyss of the human species" (9). The opinion that urban centers are morally bankrupt, dirty, and environmentally destructive has additionally been echoed by American figures such as Thomas Jefferson, Edgar Allan Poe, Herman Melville, Nathaniel Hawthorne, and, more recently, the countercultural critic Ted Roszak (Pojman 2017). A good amount of the complaint may trace to discomfort with cosmopolitan values and the anonymity that helps them flourish, simultaneously enabling other things like crime. Some of the remonstration is also historical in that it was launched at a time when industrialization and its ills did not seem like the only option. While today's cities still entail heavy consumption and higher concentrations of pollution, they yield per capita benefits to the environment because of streamlined use of space and a range of corresponding advantages in transportation grids, energy consumption, and more (Angelo and Wachsmuth 2020). At the same time, cities and the technologies required to run them can fuel ecologically dubious conceptions that psychologically sever people and their built environments from nature.

Properly speaking, cities are a part of nature, and virtually nothing on Earth is outside the human sphere since the planet is ecologically bound. This is such

DOI: 10.4324/9781003217442-3

that everything from the rainforests of Brazil to the North American Great Plains were cultivated by their original inhabitants (Mann 2005), and humans continue to intentionally and unintentionally affect vast swaths. Thus, while this chapter uses the word "nature" colloquially to signify areas with trees, rivers, mountains, forests, meadows, lakes, and so on, the term should not be taken to connote pristine regions untouched by humans. That this needs stating is in fact indicative of metaphysical concepts (i.e., notions about reality) that alienate people from nature. It may further be that the divide reduces nature to an instrumentality, thereby abrogating the idea that it has inherent value. In addition to these points, we explore how technologically mediated relations in pandemic times may sharpen or diminish estrangement from nature, as well as the environmental implications of these possibilities.

Networked technologies are, of course, energy intensive and environmentally destructive to dispose of and produce (e.g., Fowler 2017; Morley, Widdicks, and Hazas 2018; Crawford 2021). Yet transatlantic flights and daily commutes are more polluting than Zoom meetings, which is why many commentators stressed the environmental benefits of working from home during the COVID-19 pandemic (Maipas, Panayiotides, and Kavantzas 2021). What has been pondered less is the extent to which electronic media may fracture our contact with the world. Thus, to give a hyperbolic example, one can imagine a future scenario in which people cease to ski on actual hills, instead pursuing the pastime in video games or virtual reality. The decrease in travel may prove ecologically advantageous, but at the cost of losing touch with unique alpine environments, along with the impacts of climate change on the winter season. An additional consequence is that love of place and protective attitudes that accompany it will almost surely decline (Yi-Fu Tuan 1974; Scruton 2012; Crippen 2021a).

On the positive side, the prominence of remote technologies during the pandemic opened certain individuals to personalized work rhythms, in turn encouraging some to take up outdoor pursuits like gardening, hiking, and so forth (Sachs 2020). The crisis also reminded urbanites what it is to dwell under clean skies and made cities less hostile to wildlife (Arora, Bhaukhandi, and Mishra 2020; Khan, Shah, and Shah 2020; Rume and Islam 2020). Such occurrences may have highlighted that neither people nor their cities are separate from nature. Thus, without diminishing the hardship and tragedy of the COVID-19 crisis, it may be that the pandemic has inched us away from harmful metaphysical divisions and toward environmentally conscious perspectives that will bring about more sustainable city living.

Technological Enframing

An entry into our analysis is Martin Heidegger's (1954) concept of "enframing"—or *Gestell* in German—which characterizes altered reasoning about reality—in other words, metaphysics—that has risen in the wake of technological advances. Heidegger asserts that "modern technology . . . puts to nature the unreasonable demand that it supplies energy that can be extracted and stored as such" (14). Thus, the Rhine is "dammed up into the power works" (16). It ceases to be a river with unique qualities and is instead leveled into an undifferentiated source of energy to be exploited, stockpiled, and held in "standing-reserve" in a hydroelectric reservoir (16–34). Likewise, "a tract of land is challenged into the putting out of coal and minerals. The earth now reveals itself as a coal extracting district, the soil as a mineral deposit" (14). Heidegger contrasts these scenarios with a windmill, observing that "its sails . . . are left entirely to the wind's blowing. But the windmill does not unlock energy from the air currents in order to store it" (14). Although Heidegger does not differentiate renewable from more destructive forms of gathering energy since he lumps hydroelectric with coal, also not recognizing that grinding grain into flour gets close to stockpiling an undifferentiated commodity; and while Heidegger seems further to suppose a clean break between older and newer technological modes instead of recognizing a gradual evolution, he correctly suggests that the windmill operates in harmony with its surroundings, whereas coal and hydroelectric energy extraction involve an almost violent slicing into the Earth.

The great American novelist John Steinbeck (1939) anticipates Heidegger in his *Grapes of Wrath* and paints a nightmare picture of an alienated, loveless, mechanical leveling of the land, giving literary form to variations of ideas that figures like Karl Marx, John Dewey, and Rousseau have expressed. Describing the industrialized farming of his day, Steinbeck writes that "the tractors came over the roads and into the fields, great crawlers moving like insects, having the incredible strength of insects" (35). Steinbeck continues: "The man sitting in the iron seat did not look like a man; gloved, goggled, rubber dust mask over nose and mouth, he was a part of the monster, a robot in the seat" (35). These machines "ignored hills and gulches, water courses, fences, houses" (35). The driver "could not see the land as it was, he could not smell the land as it smelled; his feet did not stamp the clods or feel the warmth and power of the earth" (35). What is more, "[H]e could not cheer or beat or curse or encourage the extension of his power, and because of this he could not cheer or whip or curse or encourage himself" (35). Steinbeck sums up by saying that the tractor driver:

> loved the land no more than the bank loved the land. . .. Behind the tractor rolled the shining disks, cutting the earth with blades—not plowing but surgery, pushing the cut earth to the right where the second row of disks cut it and pushed it to the left. . . . Behind the harrows, the long seeders—twelve curved iron penes erected in the foundry, orgasms set by gears, raping methodically, raping without passion. The driver sat in his iron seat and he was proud of the straight lines he did not will, proud of the tractor he did not own or love, proud of the power he could not control. . . . No man had touched the seed, or lusted for the growth. Men ate what they had not raised, had no connection with the bread. The land bore under iron, and under iron gradually died; for it was not loved or hated, it had no prayers or curses.
>
> *(36)*

As Heidegger (1954) puts it, "[A]griculture is now the mechanized food industry, so that "the cultivation of the field has come under the grip of another kind of setting-in-order, which sets upon [*stellt*] nature . . . in the sense of challenging it" (15).

Steinbeck's and Heidegger's accounts do not necessarily imply we were good stewards prior to the arrival of advanced industrialization. The point, rather, is that recent technological developments and the all-consuming economic forces driving them have brought about a leveling of nature into undifferentiated commodities, which has negative consequences, whatever the situation was before this occurrence. Accordingly, trees are conceptually and concretely reduced to woodlots; a rugged mountain similarly becomes a source of coal, measured in terms of BTUs of heat; fruit is sorted according to grade and not the highly localized conditions that produced unique flavors; soil itself is increasingly regulated by chemical and mechanical means; everything from Marvel movies to burgers is franchised, so that consumers more or less know what they are getting beforehand (Crippen 2016, 2021a). These changes bring a loss of love. A McDonald's branch is unlikely to inspire the abiding attachment that a mom-and-pop diner does, and the same goes for the latest Incredible Hulk flick in comparison to a one-of-a-kind film like *Hiroshima Mon Amour* (dir. Alain Resnais, 1959). Affectionate bonds are less likely for a woodlot that is grown to be cut than for a forest in which people spent their childhood foraging fiddleheads and wild leeks in the early spring. It is also difficult to imagine that what looked to be the permanent discontinuation of Necco candies in 2018 brought the same sense of loss as of a forest and its fiddleheads and leeks and that people yearn for the former with the same depth as the latter.

The broader point, in phenomenological parlance, is that the contemporary age may be diminishing our grasp of what it is to existentially inhabit a locale. Existential involvement is a kind of "being in" that occurs when we are in love or enmeshed in the public mood of a Christmas gathering with intimate friends (Dreyfus 1991; Crippen 2021b). Another example is that some indigenous communities have traditionally seen their relationship with the land not as one of ownership but instead as one in which they belong to it and it to them, as family members belong to one another (Whitt et al. 2001; Crippen 2021a). Existential immersion is contrasted with categorical or spatial involvement (Dreyfus 1991), whereby an individual is in surroundings as marbles are in a box (Crippen and Youssef 2022), so that intimate bonds are lacking. The decline of existential ways of experiencing forests, mountains, lakes, desert vistas, and more amounts to a diminishment of emotional involvement with these ecological niches. It thus coincides with the diminishment of what environmentally oriented scholars have called *topophilia* or *oikophilia*, meaning love of place or home (Yi-Fu Tuan 1974; Scruton 2012; Crippen 2021a). The networked age—which we now turn to—may exacerbate these losses. The risks are arguably higher for city dwellers who have less access to what is colloquially called "nature."

Digital Networks and Isolation

The COVID-19 pandemic brought about increased use of remote platforms, streaming services, and food delivery apps, with favorable but still mixed environmental outcomes. For example, packaging used in food delivery has a heavy carbon footprint (Li, Mirosa, and Bremer 2020). Streaming is energy intensive, but it may be less damaging than other forms of entertainment, such as travelling to rent or buy a DVD or see a movie (Shehabi, Walker, and Masanet 2014). Remote communication—despite also being energy intensive—reduces commuting and therewith air pollution (Maipas, Panayiotides, and Kavantzas 2021). While granting that there were net environmental benefits, the main issue we want to explore is the extent to which networked technologies are a late stage in the situation that Steinbeck and Heidegger describe. We specifically consider this question in the context of cities during the COVID-19 crisis and contemplate whether or not digital media are further cutting urbanites off from nature, thereby desensitizing them to environmental problems.

When we speak of being "cut off," we mean losing existential contact. A closely related phenomenological term is *lived experience*, which Maurice Merleau-Ponty (1945) helpfully elucidates. He does so when he advocates returning to a lived world "in relation to which every scientific schematization is an abstract and derivative sign-language, as is geography in relation to the country-side in which we have learnt beforehand what a forest, a prairie or a river is" (ix). Steinbeck's farmers, for example, have at least partly lost their lived sense of earth, seeds, plants, and land. This implies distancing, withdrawal, and hence alienation, which Steinbeck emblemizes with his otherworldly description of the tractor drivers in gloves, goggles, and masks. Their alienation is also manifested in their passionless disconnection from the land and what it produces. With no intimate relation to the fields, the machines plow through; they disregard hills and gulches, leveling the land, ignoring its unique contours in ways recalling Heidegger's account. It is not incidental, moreover, that the farming practices that Steinbeck describes contributed to environmental problems like soil erosion and thus to the disastrous Dust Bowl of the Great Depression (McLeman et al. 2013).

Let us first sketch some parallels between Steinbeck's example and the COVID-19 pandemic. Steinbeck's farmers are gloved, masked, and goggled, placing a physical barrier between them and their surroundings. The pandemic brought the same thing: people wore masks, gloves, and protective faceplates, and eerie tented test sites were pitched in cities with medical workers in hazmat suits. Remote technologies are a digital variation of masks and suchlike, for they were employed to separate and create distance to prevent viral spread. These physical and electronic barriers cause, emblemize, and are products of alienation, though one might add that remote contact is less alienating than none at all. Also, in the same way that Steinbeck's tractors level the contours of the land, digital communication strips nuances from interactions, for instance, the subtleties of a particular handshake, which is part of what makes remote platforms alienating. Food delivery apps similarly deprive us of friendly interactions with a favorite barista or server. When it comes specifically to social media, obviating uniqueness is the business model: Facebook and Twitter are designed to identify what one user has in common with one million in this category, three million in that one and so on, all in order to target ads. In Heidegger's language, social media companies generate revenues by "challenging" information out of people—mining and essentially reducing human beings to data stockpiles.

It is perhaps more than coincidental that the terms "*net*work" and "worldwide *web*" suggest entrapment and that the English translation for Heidegger's term *Gestell*—enframing—suggests being penned up. These connotations are fitting insofar as networked technologies facilitated pandemic *lock*downs. Despite the ecological advantages of the situation, environmental consciousness may wane in consequence of being confined at home in a city and spending less time outdoors and in natural settings, as seems to have sometimes occurred during the pandemic (Cindrich et al. 2021). Studies in fact suggest that the combined influence of urbanization and frequent uses of digital technologies for entertainment purposes psychologically alienates people from and degrades concern for natural environments by reducing time spent in them (Larson et al. 2018; Rosa and Collado 2019). Robert Pyle (1993) calls this growing separation from nature "the extinction of experience," adding that outdoor pursuits instilled environmental consciousness in him. This last point is amply supported by empirical evidence, which shows that outdoor play and recreation increases attachment and moral care for the environment (e.g., Wells and Lekies 2006; Evans, Otto, and Kaiser 2018; Rosa, Profice, and Collado 2018). Research also indicates that reasoning in the absence of emotional attachment is seldom enough for ecologically responsible behavior (Kals, Schumacher, and Montada 1999). All these claims are simultaneously reinforced by common sense.

Suppose, for example, that Jack lives in Toronto. Imagine further that networked ways of operating become the new norm after the pandemic, so that people increasingly stay home. Though one hopes this means Jack will spend more time outside, the reverse often happened during the COVID-19 pandemic (Cindrich et al. 2021), so it may be that he gradually loses contact with nature. He may never experience the currents of cool air that sometimes flow down hills like water, or how fragrances grow in the day, or again the way the arc of the Sun changes with the passage of seasons, to name just a few things. His knowledge may thereby become more abstract and less existential, which means an extinction of lived experience. If Jack's primary contact with wilderness is mediated and curated through Netflix documentaries, then his experience of nature is effectively reduced to standing reserve: it is on call anytime, minus the inconveniences of inclement weather, flies, and much else good and bad that goes with firsthand encounters. Therefore, it is also leveled, and it is hard to imagine Jack forming an abiding connection. Empirical research affirms something like this, indicating that video encounters with nature engender less connectivity than firsthand experiences (Mayer et al. 2009). Direct immersion also corresponds to emotional attachment to nature and heightened environmentalism (Kals, Schumacher, and Montada 1999).

To consider a contrasting illustration, imagine that Fatema dwells about 100 km northwest of Toronto in Hockley Valley, which is part of a UNESCO World Biosphere Reserve. Hockley Valley has ample forests, extensive trail systems, and numerous tributaries that form the headwaters of the Nottawasaga River, which runs along its base. From the empirical work already cited, one would expect Fatema to develop moral and emotional concern for her surroundings, though this can depend on factors such as education and upbringing (Berenguer, Corraliza, and Martín 2005; Yu 2014). Thus, while the dangers of heat waves and droughts were on the minds of many in the summer of 2021, Fatema could not help notice decreasing water levels in years previous or plants so deprived of moisture as to be wilted flat in what are normally damp forest floors. She would be very disturbed that winters are no longer cold enough to eradicate invasive insects. The emerald ash borer, an invasive beetle, has killed virtually all the ash trees. Recent summers have brought another invader—a European moth, *Lymantria dispar*—with infestations of the caterpillars so severe that their feces darken the ground.

Because Hockley Valley is her home, Fatema cannot escape seeing these problems even under lockdown conditions, which do not prevent walks in local forests, in any case. Her *topophilic* sense of place, moreover, makes the decreased water levels, dead ashes, and caterpillar infestations traumatic. We can suppose that the earlier mentioned Torontonian also sees them as problematic, but they are distant and abstract for him, so they do not have the same palpable impact. The danger, then, is that as remote technologies, streaming services, and social media platforms take on increasing roles in city life, people may become ever more desensitized to impending and current environmental disasters. Losing a sense of emotional urgency, the population may not be inclined to support drastic action to circumvent threats.

Networked Redemption?

Steinbeck (1939) describes tractor drivers as becoming "a part of the [mechanical] monster" that they rode (35). The COVID-19 crisis mirrors this insofar as it brought swaths of humankind closer to cybernetic existence by integrating them with digital technologies. In this light, COVID-19 might be seen as a techno-organic virus. Such viruses are inventions of comic book science fiction and are said to introduce machine elements into biological agents. Another sense in which the COVID-19 virus may be considered techno-organic is that it fundamentally changes urban functioning, in this sense infecting cities too (Cortés and Crippen 2021).

Such a conception supposes cities are akin to living systems, which was a common outlook in the past. An ancient Greek view was that vital life functions comprising the human soul have analogues in societal constituents that integrate into quasi-organic systems that comprise cities (cf. Plato c. 375 BC, Bks. 2–4; Aristotle c. 350 BC, Bk. 2). In some cases, for example, the human soul was seen as having rational, emotional, and appetitive components, each performing important functions but integrated in certain ways in healthy individuals. The same was said of cities, which ideally had rational rulers, emotionally committed soldiers, and wealthy producers and therewith satisfiers of appetites, who all similarly focus on their own business but in ways that integrate for the betterment of the city. The outlook was applied even more broadly, so that organic and nonorganic forces were described as ecologically uniting into a world soul (Mohr 1982), with the idea showing up in other ancient sources such as the *Tao Te Ching* (c. 400 BC). It is barely metaphorical to say that the COVID-19 virus affected the pulse and heartbeat—the soul—of urban settings. To name just a few outcomes, schools closed, restaurants went out of business, and traffic flows changed, as did social patterns and political dynamics. More positively, wildlife occasionally wandered back into urban areas when businesses were largely shuttered and road traffic light, hopefully reminding people that nature does not end at the city limits, perhaps also reducing alienation or what Pyle calls the extinction of experience.

It is obvious and hence not our central point that the business impacts were economically harmful but ecologically advantageous. Mainly, we here want to address how integration with digital networks may, somewhat ironically, disentangle us from the web of technological enframement. In this way, technology might help overcome the alienation from nature to which city dwellers are susceptible. Key to this possibility is the idea that the "immunological" reorganization of city systems is not exclusively traumatic. Among other things, remote working introduced personalized rhythms and time organization into at least white-collar labor, in this way respecting individual uniqueness. Supposing the situation does not further degrade employers' respect for private time, then the scenario may mark a movement away from what Steinbeck and Heidegger describe. This is insofar as employees are able to work around personal contours of their lives, so that they are not reduced to undifferentiated "human capital" or "resources."

An occurrence directly related to the lockdown and how remote employment allowed more personalized work rhythms for some is the fact that increased numbers of people in towns and cities took up gardening during the pandemic, to a point that there were seed shortages. Likewise, outdoor pursuits were one of the few activities that many felt they could do safely, though there were complications such as closed parks and the challenge of balancing work with childcare. Anecdotal evidence suggests this was made worse by police targeting minorities over white picnic-goers in closed parks (CBC 2021). It is worth adding that certain minority groups are inclined to use outdoor areas for social interactions, compared to whites who are often interested in solitary activity (Whiting et al. 2017), and that parks are places where people can meet safely during pandemic times if exercising some basic precautions. Introducing features that are attractive to diverse populations and not driving minorities out may be a very small way to mitigate new imbalances that are likely to arise in a postpandemic world. White-collar professionals, still stand to gain the most since they specifically have the luxury of working from home. Yet combined with less crowded roads and public transportation, along with the mass resignations of the pandemic causing labor shortages and thus more leverage for low-income earners, some modest benefits will hopefully accrue to the vulnerable and also the environment by generating greater concern for it.

So, although there is evidence that outdoor time and physical exercise actually went down during the pandemic (Xiao et al. 2020), one can imagine a future in which the reverse is so. With the flexibility of remote work, less choked public transit, outdoor areas designed for people of a variety of backgrounds, and of course the absence of pandemic obstacles, individuals would have more chances to walk, jog, or gather in city greenspaces. Some may use remote work as an opportunity to get out of the city since they can hike in the day and work in the evening or complete employment-related tasks while camping at a lake. Research suggests that the additional time spent in nature would cultivate connectedness to it (e.g., Otto and Pensini 2017), along with a stronger attachment to place and an accompanying biocentric morality (Larson, Whiting, and Green 2011; Lawrence 2012; Hahn and Garrett 2017). The fact that nature has restorative effects on humans is not of incidental importance (e.g., Mayer et al. 2009; Byrka et al. 2010; Collado and Corraliza 2013; Wyles et al. 2016). In addition to heightening well-being, health-promoting outcomes may increase caring attitudes toward nature.

By introducing flexibility into daily life, networked platforms may thereby enable people to escape from what Heidegger thinks of as technological enframing. These relatively new ways of working accordingly have the potential to subvert the environmentally costly metaphysics that divides humans and their cities from nature. If this is true, then we are perhaps moving closer to recognizing, as the psychologist James Hillman (2005) observes, that humans are always involved in broader planetary ecologies. Hillman adds that it could not be otherwise since people and the world share a common nature. These assertions are patently obvious, even though we often act contrary to them. However, Hillman's call not to divide people from nature carries a deeper imperative. In Heidegger's terms, it asks us to "let beings be" instead of mercilessly imposing our own measures, as when conceptually and concretely reducing a forest to nothing more than a woodlot. These shifts in attitude, if carried out, will heighten environmental consciousness, while reminding people that neither they nor their cities are separate from nature. Such a realization is particularly urgent in what may be the last part of the Anthropocene when half the human population already lives in urban areas; when we are therefore moving toward what is increasingly being called the "Urbanocene"; and when many are consequently blind to the world outside cities and to the ills we are inflicting on them and, by extension, on ourselves.

References

Angelo, Hillary, and David Wachsmuth. 2020. "Why Does Everyone Think Cities Can Save the Planet?" *Urban Studies* 57 (11): 2201–21. https://doi.org/10.1177/0042098020919081.

Aristotle. (c. 350) 1941. "*De Anima*, trans. J. A. Smith." In *The Basic Works of Aristotle*, edited by Richard McKeon, 533–603. New York: Random House.

Arora, Shefali, Kanchan Deoli Bhaukhandi, and Pankaj Kumar Mishra. 2020. "Coronavirus Lockdown Helped the Environment to Bounce Back." *Science of The Total Environment* 742: 140573. https://doi.org/10.1016/j.scitotenv.2020.140573.

Berenguer, Jaime, José Corraliza, and Rocío Martín. 2005. "Rural-Urban Differences in Environmental Concern, Attitudes, and Actions." *European Journal of Psychological Assessment* 21 (2): 128–38. https://doi.org/10.1027/1015-5759.21.2.128.

Byrka, Katarzyna, Terry Hartig, and Florian Kaiser. 2010. "Environmental Attitude as a Mediator of the Relationship between Psychological Restoration in Nature and Self-Reported Ecological Behavior." *Psychological Reports* 107 (3): 847–59. https://doi.org/10.2466/07.pr0.107.6.847-859.

CBC News. 2021. "How Police Choose to Intervene in Crowded Parks Raises Concerns of Racial Profiling." *CBC/Radio Canada*, April 12, 2021. www.cbc.ca/news/canada/montreal/covid-19-police-enforcement-use-of-force-racial-profiling-parks-1.5984442.

Cindrich, Sydney, Jeni Lansing, Cassandra Brower, Cillian McDowell, Matthew Herring, and Jacob D. Meyer. 2021. "Associations between Change in Outside Time Pre- and Post-Covid-19 Public Health Restrictions and Mental Health: Brief Research Report." *Frontiers in Public Health* 9. https://doi.org/10.3389/fpubh.2021.619129.

Collado, Silvia, and José Corraliza. 2013. "Children's Restorative Experiences and Self-Reported Environmental Behaviors." *Environment and Behavior* 47 (1): 38–56. https://doi.org/10.1177/0013916513492417.

Cortés, Alice, and Matthew Crippen. 2021. "Phenomenology and Ecology: Art, Cities, and Cinema in the Pandemic." *The Polish Journal of Aesthetics* 61 (2): 27–41. doi:10.19205/61.21.2.

Crawford, Kate. 2021. *The Atlas of AI: Power, Politics, and the Planetary Costs of Artificial Intelligence*. New Haven: Yale University Press.

Crippen, Matthew. 2016. "Screen Performers Playing Themselves." *The British Journal of Aesthetics* 56 (2): 163–77. https://doi.org/10.1093/aesthj/ayw044.

———. 2021a. "Africapitalism, Ubuntu and Sustainability." *Environmental Ethics* 43 (3): 235–59.

———. 2021b. "Aesthetics and Action: Situations, Emotional Perception and the Kuleshov Effect." *Synthese* 198 (S9): 2345–63. https://doi.org/10.1007/s11229-019-02110-2.

Crippen, Matthew, and Farida Youssef. 2022. "Movies as Environment." In *Aesthetic Literacy: A Book for Everyone*, edited by Valery Vinogradovs. Cambridge: Mont Publishing.

Dreyfus, Hubert. 1991. *Being-In-The-World: A Commentary on Heidegger's Being and Time, Division I*. Cambridge, MA: MIT Press.

Evans, Gary, Siegmar Otto, and Florian Kaiser. 2018. "Childhood Origins of Young Adult Environmental Behavior." *Psychological Science* 29 (5): 679–87. https://doi.org/10.1177/0956797617741894.

Fowler, Bruce. 2017. *Electronic Waste: Toxicology and Public Health Issues*. Cambridge, MA: Academic Press.

Hahn, Erin, and Marybeth Garrett. 2017. "Preschoolers' Moral Judgments of Environmental Harm and the Influence of Perspective Taking." *Journal of Environmental Psychology* 53 (November): 11–19. https://doi.org/10.1016/j.jenvp.2017.05.004.

Heidegger, Martin. 1954/1977. "The Question Concerning Technology." In *The Question Concerning Technology and Other Essays*, translated by William Lovitt, 3–35. London: Garland Publishing.

Hillman, James. 2005. *Politica della Bellezza (Policies of beauty)*, 3rd ed. Bergamo: Moretti & Vitali.

Kals, Elisabeth, Daniel Schumacher, and Leo Montada. 1999. "Emotional Affinity toward Nature as a Motivational Basis to Protect Nature." *Environment and Behavior* 31 (2): 178–202. https://doi.org/10.1177/00139169921972056.

Khan, I., D. Shah, and S. S. Shah. 2020. "COVID-19 Pandemic and Its Positive Impacts on Environment: An Updated Review." *International Journal of Environmental Science and Technology* 18 (2): 521–30. https://doi.org/10.1007/s13762-020-03021-3.

Lao Tzu (trad. cr., 400 BC). 1990. *Tao Te Ching: The Classic Book of Integrity and the Way*, translated by Victor H. Mair. New York: Bantam Books.

Larson, Lincoln, Rachel Szczytko, Edmond Bowers, Lauren Stephens, Kathryn Stevenson, and Myron F. Floyd. 2018. "Outdoor Time, Screen Time, and Connection to Nature: Troubling Trends among Rural Youth?" *Environment and Behavior* 51 (8): 966–91. https://doi.org/10.1177/0013916518806686.

Larson, Lincoln, Jason Whiting, and Gary Green. 2011. "Exploring the Influence of Outdoor Recreation Participation on Pro-Environmental Behaviour in a Demographically Diverse Population." *Local Environment* 16 (1): 67–86. https://doi.org/10.1080/13549839.2010.548373.

Lawrence, Eva. 2012. "Visitation to Natural Areas on Campus and Its Relation to Place Identity and Environmentally Responsible Behaviors." *The Journal of Environmental Education* 43 (2): 93–106. https://doi.org/10.1080/00958964.2011.604654.

Li, Charlene, Miranda Mirosa, and Phil Bremer. 2020. "Review of Online Food Delivery Platforms and Their Impacts on Sustainability." *Sustainability* 12 (14): 5528. https://doi.org/10.3390/su12145528.

Maipas, Sotirios, Ioannis Panayiotides, and Nikolaos Kavantzas. 2021. "Remote-Working Carbon-Saving Footprint: Could COVID-19 Pandemic Establish a New Working Model with Positive Environmental Health Implications?" *Environmental Health Insights* 15: 1–2. https://doi.org/10.1177/11786302211013546.

Mann, Charles. 2005. *1491: New Revelations of the Americas Before Columbus*. New York: Knopf.

Mayer, Stephan, Cynthia McPherson Frantz, Emma Bruehlman-Senecal, and Kyffin Dolliver. 2009. "Why Is Nature Beneficial?" *Environment and Behavior* 41 (5): 607–43. https://doi.org/10.1177/0013916508319745.

McLeman, Robert, Juliette Dupre, Lea Berrang Ford, James Ford, Konrad Gajewski, and Gregory Marchildon. 2013. "What We Learned from the Dust Bowl: Lessons in Science, Policy, and Adaptation." *Population and Environment* 35 (4): 417–40. https://doi.org/10.1007/s11111-013-0190-z.

Merleau-Ponty, Maurice. (1945) 1962. *Phenomenology of Perception*. Translated by Colin Smith. New York: Routledge & Kegan Paul Ltd.

Mohr, Richard. 1982. "The World-Soul in the Platonic Cosmology." *Illinois Classical Studies* 7 (1): 41–48.

Morley, Janine, Kelly Widdicks, and Mike Hazas. 2018. "Digitalisation, Energy and Data Demand: The Impact of Internet Traffic on Overall and Peak Electricity Consumption." *Energy Research & Social Science* 38: 128–37. https://doi.org/10.1016/j.erss.2018.01.018.

Otto, Siegmar, and Pamela Pensini. 2017. "Nature-Based Environmental Education of Children: Environmental Knowledge and Connectedness to Nature, Together, Are Related to Ecological Behaviour." *Global Environmental Change* 47 (November): 88–94. https://doi.org/10.1016/j.gloenvcha.2017.09.009.

Plato. (c. 375 BC) 1964. "Republic", trans. Paul Shorey. In *Plato: The Collected Dialogues*, edited by Edith Hamilton and Cairns Huntington, 575–844. Princeton: Princeton University Press.

Pojman, Louis. 2017. "The Challenge of the Future: Private Property, the City, the Globe, and a Sustainable Society." In *Environmental Ethics: Readings in Theory and Application*, edited by Louis Pojman, Paul Pojman, and Katie McShane, 7th ed. Boston: Cengage.

Pyle, Robert Michael. (1993) 2011. *The Thunder Tree: Lessons from an Urban Wildland*. Corvallis: Oregon State University Press.

Resnais, Alain, director. (1959) 2013. *Hiroshima, Mon Amour*. DVD. New York: The Criterion Collection.

Rousseau, Jean-Jacques. (1762) 1979. *Emile*. Translated by Allan Bloom. New York: Basic Books.

Rosa, Claudio, and Silvia Collado. 2019. "Experiences in Nature and Environmental Attitudes and Behaviors: Setting the Ground for Future Research." *Frontiers in Psychology* 10. https://doi.org/10.3389/fpsyg.2019.00763.

Rosa, Claudio, Christiana Cabicieri Profice, and Silvia Collado. 2018. "Nature Experiences and Adults' Self-Reported Pro-Environmental Behaviors: The Role of Connectedness to Nature and Childhood Nature Experiences." *Frontiers in Psychology* 9. https://doi.org/10.3389/fpsyg.2018.01055.

Rume, Tanjena, and S. M. Didar-Ul Islam. 2020. "Environmental Effects of COVID-19 Pandemic and Potential Strategies of Sustainability." *Heliyon* 6 (9): e04965. https://doi.org/10.1016/j.heliyon.2020.e04965.

Sachs, Naomi. 2020. "Access to Nature Has Always Been Important; With COVID-19, It Is Essential." *HERD: Health Environments Research & Design Journal* 13 (4): 242–44. https://doi.org/10.1177/1937586720949792.

Scruton, Roger. 2012. *How to Think Seriously about the Planet*. Oxford: Oxford University Press.

Shehabi, Arman, Ben Walker, and Eric Masanet. 2014. "The Energy and Greenhouse-Gas Implications of Internet Video Streaming in the United States." *Environmental Research Letters* 9 (5): 054007. https://doi.org/10.1088/1748-9326/9/5/054007.

Steinbeck, John. (1939) 2006. *The Grapes of Wrath*. New York: Penguin Books.

Tuan, Yi-Fu. 1974. *Topophilia: A study of Environmental Perception, Attitudes, and Values*. Englewood Cliffs, NJ: Prentice Hall.

Wells, Nancy, and Kristi Lekies. 2006. "Nature and the Life Course: Pathways from Childhood Nature Experiences to Adult Environmentalists." *Children, Youth and Environments* 16 (1): 41663.

Whiting, Jason W., Lincoln R. Larson, Gary T. Green, and Chuck Kralowec. 2017. "Outdoor Recreation Motivation and Site Preferences across Diverse Racial/Ethnic Groups: A Case Study of Georgia State Parks." *Journal of Outdoor Recreation and Tourism* 18 (June): 10–21.

Whitt, Laurie Anne, Mere Roberts, Waerete Norman, and Vicki Grieves. 2001. "Belonging to Land: Indigenous Knowledge Systems and the Natural World." *Oklahoma City University Law Review* 26 (2): 701–43.

Wyles, Kayleigh J., Sabine Pahl, Matthew Holland, and Richard Thompson. 2016. "Can Beach Cleans Do More than Clean-up Litter? Comparing Beach Cleans to Other Coastal Activities." *Environment and Behavior* 49 (5): 509–35. https://doi.org/10.1177/0013916516649412.

Xiao, Yijing, Burcin Becerik-Gerber, Gale Lucas, and Shawn Roll. 2020. "Impacts of Working from Home During COVID-19 Pandemic on Physical and Mental Well-Being of Office Workstation Users." *Journal of Occupational & Environmental Medicine* 63 (3): 181–90. https://doi.org/10.1097/jom.0000000000002097.

Yu, Xueying. 2014. "Is Environment 'a City Thing' in China? Rural—Urban Differences in Environmental Attitudes." *Journal of Environmental Psychology* 38 (3): 39–48. https://doi.org/10.1016/j.jenvp.2013.12.009.

3

SOCIAL DETERMINANTS OF HEALTH AS A FRAMEWORK FOR ADDRESSING URBAN CLIMATE ADAPTATION

Anjali Sauthoff

Introduction

Approximately 54% of the global population resides in urban areas today, and, following an estimated urban population growth of 2.5 billion in the next few decades, that population is projected to rise to 68% by midcentury (UN 2019). The rapid socioeconomical and agricultural changes, ecological fragmentation, and development of the built environment that accompany urbanization are influencing drivers of disease and well-being. The resulting spatial patterns of health reflect increasing social, economic, and environmental disparity and significantly impact economic potential, social mobility, and security in and around cities (WHO 2017). Multiple factors shape the adaptive changes that respond to urbanization, including the systemic and political contexts that create or influence the social and environmental context. The ongoing COVID-19 pandemic, as well as the stressors and shocks associated with climate change, amplify the need for a transformational change to mitigate the uneven impacts that increase health disparity. Equity and justice-centered frameworks that transcend communities are required for a deeper, more informed engagement and planning process to advance urban health equity and create greater resilience to future disruptions.

Social, Environmental, and Structural Determinants of Health Across Space and Time

Individual and community well-being require health equity, the state in which all people have the opportunity to attain full health potential and in which nobody is disadvantaged due to social position or any other socially defined circumstance (NASEM et al. 2017). Conversely, health *in*equities are avoidable differences in health status or in the distribution of health resources between population groups

DOI: 10.4324/9781003217442-4

(WHO 2017). The widening health disparities across urban populations cannot be explained by biological variation alone, implying that those differences can be avoided by reasonable societal-level action (WHO 2017). Though 95% of U.S. health investment is allocated to direct medical care services, McGinnis, Williams-Russo, and Knickman (2002) estimated that only 10%–15% of mortality in the United States might be prevented by better medical care. Rather, accumulating evidence has pointed to social and environmental determinants as powerful forces that act through fundamental structures to influence health outcomes (Braveman and Gottlieb 2014).

Societal factors that shape health include education, employment, income inequality, housing, crime, poverty, and social capital. Environmental contexts, including air and water quality, food access, and the built environment also drive individual and population health. Historically, research on population health largely focused on the association of a singular determinant with a specific health outcome, obscuring complex interactions, multifaceted connections between people and places, cumulative exposures, and toxic stress over different spatial and temporal scales (Coburn 2017). Such single-dimension analyses also diminish the role of institutions and processes that shape physical and social context, as well as access to health-promoting resources (Macintyre, Ellaway, and Cummins 2002). One example, a notable policy from the 1930s typically known as "redlining," has had far-reaching impacts on communities, resulting in influential patterns that persist 80 years later. The Home Owners Loan Corporation (HOLC) was formed in 1933 to help prevent home foreclosures as part of a program to lift the United States out of a recession. To assess risk, maps were developed that categorized neighborhoods into one of four categories: A (green, "Best"), B (blue, "Still Desirable"), C (yellow, "Definitely Declining"), and D (red, "Redlining," "Hazardous"). The color-coded grading was used as guidance for lenders in decisions on who should receive loans and determine which areas were safe for investments. Grading was explicitly driven by racism and classism, as demonstrated in the language that described African Americans, immigrants, and working-class whites as "undesirable" and "lower grade" populations (Nelson et al. n.d.).

Redlining and other forms of structural and cultural racism have resulted in location within a city becoming a powerful determinant of inequity. A growing body of literature has found associations between historic redlining and present-day urban disparities, including social (higher poverty rates, higher contemporary tobacco retailer density, current neighborhood lending discrimination, quality of K–12 education), environmental (higher land surface temperature, fewer trees, less open space, decreased air quality, vulnerability to natural hazards), and health disparities (later cancer diagnosis, more frequent emergency department visits for asthma, and higher rates of COVID-19) (NASEM et al. 2017; Giles-Corti, Lowe, and Arundel 2020; Mitchell and Chakraborty 2014; Schwartz et al. 2021; Lynch et al. 2021; Krieger et al. 2020; Hoffman, Shandas, and Pendleton 2020; Schell et al. 2020; Nardone et al 2020; Li and Yuan 2021). Though the United States passed fair housing laws in the 1960s and 1970s to help close the racial gap in the housing market, a recent study found that the race-associated appraisal gap effectively doubled between 1980 and 2015, indicating that persistent racial inequity is still driven in part by perpetual systemic devaluing of communities of color (Howell and Korver-Glenn 2020).

The multiple, overlapping impacts resulting from underinvestment in certain communities have led to complex interactions. Lopez and Gadsden (2016) discuss how intersecting systems of oppression including racism, classism, gender, sexual orientation, nativism, ableism, and others interact to produce major differences in the lived experiences that shape health. Furthermore, there is growing recognition of the cumulative impact that converging environmental, political, social, and economic risks have on functional ability in times of shocks and stress (Patel et al. 2020). The concept of toxic stress from multiple or persistent exposures can help highlight the challenges and approaches for understanding how urban places might influence well-being and health equity (Corburn 2017). Social and economic disadvantage, unsafe conditions, and social exclusion create chronic stressful conditions, which may increase susceptibility to illness or premature mortality (Prochaska 2014). Known toxic stressors include chronic poverty, discrimination and marginalization, physical or emotional abuse, exposure to violence, and housing instability—stressors that may begin to influence health in utero and have cumulative impacts over a lifetime (Corburn 2017). Many of the effects of traumatic stress can also be transmitted to future generations, even when individuals from these generations are not exposed to any traumatic stressor (Jawaid, Roszkowski, and Mansuy 2018).

The concept of the "syndemic" was first proposed by Merrill Singer in the mid-1990s to explain the synergistic interaction of two or more diseases and the social and environmental context within which they develop (Singer et al. 2017). As opposed to conventional methods of examining disease as a distinct entity, the syndemic approach emphasizes the ways in which social environments, particularly social inequality and injustice, contribute to the development of disease clustering. The ongoing COVID-19 pandemic is especially illustrative, as it has revealed stark racial disparities in COVID-19 morbidity and mortality among Black Americans, who constitute 13% of the US population but 30% of the cases (Poteat et al. 2020). Poteat et al. (2020) summarize explanations for the disparity, including a higher prevalence of underlying diseases like obesity, cardiovascular and respiratory disease, and diabetes; lack of health insurance; overrepresentation in essential service industries; and a greater likelihood of living in overcrowded housing, all of which have been cited as risk factors for COVID-19. Many are highlighting two categories of disease—COVID-19 and noncommunicable chronic disease—in the context of social and environmental disadvantage as an example of a syndemic, calling attention to the deeply embedded societal inequalities along with the systemic connections that perpetuate disadvantage and differences in health potential (Mendenhall 2020).

The emergence of the COVID-19 pandemic and its high social and economic costs has prompted comparisons with global climate change (Fuentes et al. 2020;

Perkins, Velazquez, and Munguia 2021). Though there are significant differences in responsibility and mitigation, time course, and nature of damage, both forces disrupt across sectors and most heavily impact already disadvantaged populations. The 2015 Paris Agreement acknowledged climate change as "an urgent and potentially irreversible threat to human societies and the planet." In 2018, the Intergovernmental Panel on Climate Change (IPCC) warned that average global temperature increase must be limited to well below 2°C, requiring transformative cross-sectoral changes in social and environmental systems that impact climate and health to mitigate and adapt to the increased frequency and severity of heat and storm events, sea level rise and associated flooding, drought, wildfires, and other impacts.

The Role of Justice in Climate and Health Equity

Notions of justice are embedded in every aspect of climate and health equity. Equitable climate adaptation will require recognition of several dimensions of justice in decision making, including distributional, procedural, and recognitional justice in order to understand how institutionalized processes modify climate and health risks across populations.

Distributional justice questions the distribution of resources, impacts, vulnerabilities, benefits, and burdens across populations. Patterns of distribution often reflect more upstream patterns of power and authority and result in differing experiences among individuals and groups. Who benefits from public parks, and where are the sources of air pollution sited? Where is there investment in green infrastructure and transportation, and where are the opportunities for fresh foods? How is investment allocated?

Procedural justice focuses on the systemic processes used to reach just decisions. Who is involved in the decision-making processes? What are the norms regarding inclusion and representation? Who is left out, and why? Procedural justice impacts the extent to which diverse voices have a role in decision-making processes.

Recognitional justice requires the fair examination and representation of the cultures, values, and considerations of all parties affected by policies or programs (Whyte 2011). It is centered on who is included in analyses of processes, benefits, and burdens. Such recognition and associated data collection and metrics are a reflection of the priorities and values embedded within decision making and the foundational knowledge upon which evidence-related decisions can be made. Who or what is being captured by the data? How are data being collected? How are metrics being defined? What and who are missing? How are spatial and temporal boundaries (i.e., future generations) considered?

Such notions of justice and equity—or lack thereof—can be recognized in historical decision making and is reflected in current structures and institutions. A recent study on "Community Based Solutions to Promote Health Equity in the United States" identified two "clusters" of root causes of health inequity in the United States: (1) intrapersonal, interpersonal, institutional, and systemic mechanisms that organize the distribution of power and resources differentially across lines of race, gender, class, and other dimensions of individual and group identity; (2) the unequal distribution of power and resources, manifesting in unequal social, economic, and environmental decisions (NASEM et al. 2017). These root causes are complex and interrelated, and they create contexts that persist across space and time.

Recognizing the importance of procedural and distributive justice in climate mitigation and adaptation planning, McDermott, Mahanty, and Schreckenberg (2013) created an equity framework to highlight the systems and processes that may initiate and propagate inequities related to climate adaptation. In the context of climate adaptation, procedural equity emphasizes representation and inclusion of affected groups in planning and decision making, while distributional equity highlights the just distribution of burdens and benefits, including differing capacities to adapt or influence the mitigation of climate change (McDermott, Mahanty, and Schreckenberg 2013). Klinsky and Mavrogianni (2020) build upon these ideas, developing a conceptual and practical context for integrating procedural, distributional, and recognitional justice into climate-related decision making on the built environment and identifying practical intervention points to address climate injustice that highlight the need to reduce gaps both in injustices and in knowledge and practices. Such frameworks that intentionally integrate notions of equity and justice are required to address structures and procedures that would otherwise perpetuate existing processes that contribute to widening health disparities across populations.

From Theory to Practice/Beyond Community Engagement: Participatory Mapping Within Communities of Practice

In "Action Pathways for Transforming Cities," Rosenzweig and Solecki (2018) suggest five pathways to the transformation necessary for urban areas to adequately address climate change impacts: integration of mitigation and adaptation; coordination of disaster risk reduction and climate change adaptation; cogeneration of risk information; a focus on disadvantaged populations; and advancement of governance, finance, and knowledge networks. Several of these pathways require new procedures that focus on data collection and knowledge sharing in vulnerable communities to ensure that stakeholders have the necessary evidence-based understanding for inclusive and collaborative planning. The importance of community involvement is widely recognized as necessary for an effective adaptation and planning process. But to be involved, communities require knowledge and understanding about the local impacts and risks associated with climate change, and, for policies to be the most effective, communities need opportunities to share their knowledge with policy makers.

Traditional approaches to decision making incorporate little to no stakeholder guidance, leading to policies that are not adequately informed by the people for whom they are intended to support—or worse, ignore them altogether. Furthermore, such approaches are not always grounded in current scientific understanding, leading to policies and actions that may stop short of full potential. Place-based participatory approaches to decision making incorporate stakeholders and have been advocated by community activists, supported in academic journals, and endorsed in official government guidance, yet in practice they lag behind more traditional top-down approaches. The dissemination of regionally appropriate knowledge about climate change risk, impacts, potential mitigation, and adaptation strategies is a priority to aid communities and government in crafting effective policy (IPCC 2018). Processes that recognize and integrate diverse representation for local knowledge, history, and the lived experience; government representation to understand the decision-making context; guidance from academia to provide scientific grounding; and facilitation by trusted community groups—all recognize expertise from each of these sectors, which can be leveraged to cocreate plans to more fully meet local needs, particularly of geographically and socially vulnerable populations.

The Community of Practice (CoP), a mechanism to disseminate new information and regularly convene individuals with different backgrounds, has been shown to be an effective model to address the "knowledge exchange gap" recognized in climate and health research and practice (El Amiri 2020). The CoP includes the essential elements of *community* to enable interaction, a shared *domain* of interest, and a shared *practice* of ways of addressing recurring problems (Wenger, McDermott, and Snyder 2002). The focus on knowledge and knowledge sharing are critical to the concept, as is attention to process and an investment in social context (Pyrko, Dörfler, and Eden 2017). A particular strength of the CoP is the inclusion and collaboration of diverse participants across disciplines and sectors, a necessity for climate adaptation planning given the transdisciplinary nature of climate and health. Bringing together research and practice enables critical multidirectional knowledge sharing, transferring scientific knowledge to inform planning, and identifying needs and experiences to shape more effective and "usable" research, tools, policy, and programs. Dialogue and understanding improve as relevant information that recognizes cultural, institutional, and personal differences between communities is exchanged. Governments, universities, and community stakeholders all have critical roles to play in developing effective CoPs, as each has an expert perspective that can be leveraged to create more informed planning.

Participatory mapping, the creation of maps by local communities, often with the involvement of supporting organizations, creates a natural opportunity for assimilation into CoPs. In addition to providing understanding of geographical area, maps can generate a rich understanding of people, history, cultural practices, social networks, and interactions over time. Originally designed as a tool to incorporate the perspective of those traditionally excluded from governance, participatory mapping has evolved conceptually and logistically throughout the 20th century (Guldi 2017). Douglas et al. (2020) describe participatory mapping approaches that range from the more rudimentary hands-on mapping that is independent of literacy and language, whereby participants draw maps of areas of concern, to reference mapping that reflects accurate scale and features of the built and social environments that shape public health.

Today, with the ubiquity of the Internet, emergence of big data, crowdsourcing capabilities, and tools and mapping platforms, traditional forms of mapping have acquired new capabilities. Geographic information system (GIS) tools that link data to place can sort information into local, regional, and global patterns and "show the community back to itself" (Guldi 2017). More sophisticated mixed methods like participatory geographic information systems (PGIS) use a framework for residents and trained specialists to synthesize local knowledge with scientifically grounded technical knowledge to coproduce more relevant mutual understanding (Douglas et al. 2020). For example, mapping the data that captures the factors driving disparity allows for greater recognition of current and historical determinants contributing to geographic patterns of inequity, creating a foundation for change. Participatory mapping can also integrate diverse voices by recognizing the expertise, preferences, and dignity of traditionally marginalized communities, enabling nontraditional participants to develop the sense of agency and empowerment, which are increasingly recognized as essential to the deliberative process necessary to address historic structural inequities (NASEM et al. 2017).

Participatory mapping within a CoP enables an iterative process for stakeholders to understand context over space and time, assimilate relevant information, cope with changing circumstances, and support response. It can reveal the multidimensional impacts on communities that create the conditions leading to toxic stress, syndemics, and overall health disparity. By highlighting the geographic areas where action is most urgent, opportunities are created to reduce health disparities, promote health justice, and craft policies that can be most effective in the short timeframe available to minimize climate impacts.

The City of Mount Vernon, New York, a historically disadvantaged community located in the southern portion of Westchester County is the focus of a developing CoP that brings together community organizations, local citizens, academics, and city and county government to identify and address community health and justice issues that will be exacerbated by climate change impacts. The city is among the most densely populated and environmentally challenged urban communities in Westchester County. Following an Environmental Justice conference, cohosted by the Elisabeth Haub School of Law at Pace University and the Federated Conservationists of Westchester County (FCWC), an umbrella environmental nonprofit, FCWC was invited to help empower Mount Vernon stakeholders to advance community-identified environmental and health priorities, including the city's lack of open space, crumbling infrastructure, its status as an urban heat island, distrust of government, and political isolation. Driven by local citizens, facilitated by FCWC, and supported by scientifically grounded data and guidance from academia and the county government, the CoP has grown to include stakeholders that comprise the fabric of the city, including schools, neighborhood associations, the Mount Vernon City Council, community organizations, and churches. Responding to community requests, GIS maps are being created to educate stakeholders about connections between the environment and health and to anchor and guide the community and local government.

These maps display some of the many factors that lead to differential exposures, sensitivities, and abilities to prepare, adapt, or recover from climate shocks or other threats. Areas particularly vulnerable to climate change are mapped, including flood and coastal zones, higher levels of health risks (chronic disease prevalence, point sources of air pollution), limited access to healthcare services and other resources (poverty status, locations of healthcare), limited ability to access information or resources in a native language, higher age-related sensitivities (population under age five and above age 65), and temperature variation across the city. In particular, these maps identify areas with multiple simultaneous exposures and vulnerabilities. For example, pockets of poverty with limited access to home air conditioning, higher prevalence of cardiovascular disease, and no tree canopies that may require interventions, such as cooling centers or tree plantings, to mitigate risk to heat stress can be readily identified.

Such maps initiate conversations that can integrate local knowledge and experiences and begin the process of rebuilding trust across community and government. The CoP structure has also enabled support for community-guided Neighborhood Health Assessment walks for citizens to better engage with their environment. Community presentations about environmental and health connections spurred a series of intergenerational walks, allowing elders to share history with school-aged children and all the more keenly observe surroundings and generate ideas for improving community health. Establishment of the Mount Vernon CoP has multiple, recognized cobenefits, including the building of social capital, knowledge sharing, deeper civic engagement, and improved governance—all necessary for the transformation required for equitable climate adaptation.

Article 25 of the Universal Declaration of Human Rights states: "Everyone has the right to a standard of living adequate for the health and well-being of himself and of his family, including food, clothing, housing and medical care and necessary social services, and the right to security in the event of unemployment, sickness, disability, widowhood, old age or other lack of livelihood in circumstances beyond his control." Prioritizing action to address social and structural determinants of health in urban areas will be essential for protecting against future shocks, whether climatic in nature or otherwise. Moreover, the threat of multiple simultaneous shocks, as exemplified by the COVID-19 pandemic, record-breaking fire and hurricane seasons, and unprecedented flooding in Europe, North America, and Asia, all during 2020–2021, emphasize the dire need to better understand the vulnerabilities anchored in inequitable structures. Creating Communities of Practice that integrate important facets of justice and utilize participatory mapping is a way to help communities and governments better recognize impacted communities and complex interdependencies, break down traditional communication and decision-making silos, and support citizen empowerment for improved governance—all essential elements to better address the historical structures and mechanisms that have led to present-day urban health inequity.

References

Braveman, Paula, and Laura Gottlieb. 2014. "The Social Determinants of Health: It's Time to Consider the Causes of the Causes." *Public Health Reports* 129 (1) (Suppl 2): 19–31. https://doi.org/10.1177/00333549141291s206.

Corburn, Jason. 2017. "Urban Place and Health Equity: Critical Issues and Practices." *International Journal of Environmental Research and Public Health* 14 (2): 117. https://doi.org/10.3390/ijerph14020117.

Douglas, Jason A., Andrew M. Subica, Laresha Franks, Gilbert Johnson, Carlos Leon, Sandra Villanueva, and Cheryl T. Grills. 2020. "Using Participatory Mapping to Diagnose Upstream Determinants of Health and Prescribe Downstream Policy-Based Interventions." *Preventing Chronic Disease* 17. https://doi.org/10.5888/pcd17.200123.

El Amiri, N., P. Abernethy, N. Spence, D. Zakus, T. A. Kara, and C. Schuster-Wallace. 2020. "Community of Practice: An Effective Mechanism to Strengthen Capacity in Climate Change and Health." *Canadian Journal of Public Health = Revue canadienne de sante publique* 111 (6): 862–68. https://doi.org/10.17269/s41997-020-00400-8.

Fuentes, Rolando, Marzio Galeotti, Alessandro Lanza, and Baltasar Manzano. 2020. "COVID-19 and Climate Change: A Tale of Two Global Problems." *Sustainability* 12 (20): 8560. https://doi.org/10.3390/su12208560.

Giles-Corti, Billie, Melanie Lowe, and Jonathan Arundel. 2020. "Achieving the SDGs: Evaluating Indicators to Be Used to Benchmark and Monitor Progress towards Creating Healthy and Sustainable Cities." *Health Policy* 124 (6): 581–90. https://doi.org/10.1016/j.healthpol.2019.03.001.

Guldi, Jo. 2017. "A History of the Participatory Map." *History Faculty Publications* 10. Accessed September 29, 2021. https://scholar.smu.edu/hum_sci_history_research/10.

Hoffman, Jeremy S., Vivek Shandas, and Nicholas Pendleton. 2020. "The Effects of Historical Housing Policies on Resident Exposure to Intra-Urban Heat: A Study of 108 Us Urban Areas." *Climate* 8 (1): 12. https://doi.org/10.3390/cli8010012.

Howell, Junia, and Elizabeth Korver-Glenn. 2020. "The Increasing Effect of Neighborhood Racial Composition on Housing Values, 1980–2015." *Social Problems*. https://doi.org/10.1093/socpro/spaa033.

IPCC. 2018: Global warming of 1.5°C. An IPCC Special Report on the impacts of global warming of 1.5°C above pre-industrial levels and related global greenhouse gas emission pathways, in the context of strengthening the global response to the threat of climate change, sustainable development, and efforts to eradicate poverty [V. Masson-Delmotte, P. Zhai, H. O. Pörtner, D. Roberts, J. Skea, P. R. Shukla, A. Pirani, W. Moufouma-Okia, C. Péan, R. Pidcock, S. Connors, J. B. R. Matthews, Y. Chen, X. Zhou, M. I. Gomis, E. Lonnoy, T. Maycock, M. Tignor, T. Waterfield (eds.)].

Jawaid, Ali, Martin Roszkowski, and Isabelle M. Mansuy. 2018. "Transgenerational Epigenetics of Traumatic Stress." *Progress in Molecular Biology and Translational Science*: 273–98. https://doi.org/10.1016/bs.pmbts.2018.03.003.

Klinsky, Sonja, and Mavrogianni, Anna. 2020. "Climate Justice and the Built Environment." *Buildings and Cities* 1 (1): 412–28. http://doi.org/10.5334/bc.65.

Krieger, Nancy, Emily Wright, Jarvis T. Chen, Pamela D. Waterman, Eric R. Huntley, and Mariana Arcaya. 2020. "Cancer Stage at Diagnosis, Historical Redlining, and Current Neighborhood Characteristics: Breast, Cervical, Lung, and Colorectal CANCERS, Massachusetts, 2001–2015." *American Journal of Epidemiology* 189 (10): 1065–75. https://doi.org/10.1093/aje/kwaa045.

Li, Min, and Faxi Yuan. 2021. "Historical Redlining and Resident Exposure to Covid-19: A Study of New York City." *Race and Social Problems*. https://doi.org/10.1007/s12552-021-09338-z.

Lopez, Nancy, and Vivian L. Gadsden. 2016. "Health Inequities, Social Determinants, and Intersectionality." *NAM Perspectives*. 6 (12). https://doi.org/10.31478/201612a.

Lynch, Emily E., Lorraine Halinka Malcoe, Sarah E. Laurent, Jason Richardson, Bruce C. Mitchell, and Helen C. S. Meier. 2021. "The Legacy of Structural Racism: Associations between Historic Redlining, Current Mortgage Lending, and Health." *SSM—Population Health* 14: 100793. https://doi.org/10.1016/j.ssmph.2021.100793.

Macintyre, Sally, Anne Ellaway, and Steven Cummins. 2002. "Place Effects on Health: How Can We Conceptualise, Operationalise and Measure Them?" *Social Science & Medicine* 55 (1): 125–39. https://doi.org/10.1016/s0277-9536(01)00214-3.

McDermott, Melanie, Sango Mahanty, and Kate Schreckenberg. 2013. "Examining Equity: A Multidimensional Framework for Assessing Equity in Payments for Ecosystem Services." *Environmental Science & Policy* 33: 416–27. https://doi.org/10.1016/j.envsci.2012.10.006.

McGinnis, J. Michael, Pamela Williams-Russo, and James R. Knickman. 2002. "The Case for More Active Policy Attention to Health Promotion." *Health Affairs* 21 (2): 78–93. https://doi.org/10.1377/hlthaff.21.2.78.

Mendenhall, Emily. 2020. "The COVID-19 Syndemic Is Not Global: Context Matters." *The Lancet* 396 (10264): 1731. https://doi.org/10.1016/s0140-6736(20)32218-2.

Mitchell, Bruce Coffyn, and Jayajit Chakraborty. 2014. "Urban Heat and Climate Justice: A Landscape of Thermal Inequity in Pinellas County, Florida." *Geographical Review* 104 (4): 459–80. https://doi.org/10.1111/j.1931-0846.2014.12039.x.

Nardone, Anthony, Joan A. Casey, Rachel Morello-Frosch, Mahasin Mujahid, John R. Balmes, and Neeta Thakur. 2020. "Associations between Historical Residential Redlining and Current Age-Adjusted Rates of Emergency Department Visits Due to Asthma Across Eight Cities in California: An Ecological Study." *The Lancet Planetary Health* 4 (1). https://doi.org/10.1016/s2542-5196(19)30241-4.

National Academies of Sciences, Engineering, and Medicine (NASEM), et al. 2017. *Communities in Action: Pathways to Health Equity*. Edited by Alina Baciu, et al. National Academies Press. doi:10.17226/24624.

Nelson, Robert K., LaDale Winling, Richard Marciano, Nathan Connolly, et al. n.d. "Mapping Inequality." *American Panorama*. Accessed July 3, 2021. https://dsl.richmond.edu/panorama/.

Patel, Ronak, David Sanderson, Pamela Sitko, and John De Boer. 2020. "Investigating Urban Vulnerability and Resilience: A Call for Applied Integrated Research to Reshape the Political Economy of Decision-Making." *Environment and Urbanization* 32 (2): 589–98. https://doi.org/10.1177/0956247820909275.

Perkins, Krystal M., Luis Velazquez, and Nora Munguia. 2021. "Reflections on Sustainable Consumption in the Context of Covid-19." *Frontiers in Sustainability* 2. https://doi.org/10.3389/frsus.2021.647542.

Poteat, Tonia, Gregorio A. Millett, LaRon E. Nelson, and Chris Beyrer. 2020. "Understanding COVID-19 Risks and Vulnerabilities among Black Communities in America: The Lethal Force OF SYNDEMICS." *Annals of Epidemiology* 47: 1–3. https://doi.org/10.1016/j.annepidem.2020.05.004.

Prochaska, John D., Alexandra B. Nolen, Hilton Kelley, Ken Sexton, Stephen H. Linder, and John Sullivan. 2014. "Social Determinants of Health in Environmental Justice Communities: Examining Cumulative Risk in Terms of Environmental Exposures and Social

Determinants of Health." *Human and Ecological Risk Assessment: An International Journal* 20 (4): 980–94. https://doi.org/10.1080/10807039.2013.805957.
Pyrko, Igor, Viktor Dörfler, and Colin Eden. 2017. "Thinking Together: What Makes Communities of Practice Work?" *Human Relations* 70 (4): 389–409. https://doi.org/10.1177/0018726716661040.
Rosenzweig, Cynthia, and William Solecki. 2018. "Action Pathways for Transforming Cities." *Nature Climate Change* 8 (9): 756–59. https://doi.org/10.1038/s41558-018-0267-x.
Schell, Christopher J., Karen Dyson, Tracy L. Fuentes, Simone Des Roches, Nyeema C. Harris, Danica Sterud Miller, Cleo A. Woelfle-Erskine, and Max R. Lambert. 2020. "The Ecological and Evolutionary Consequences of Systemic Racism in Urban Environments." *Science* 369 (6510). https://doi.org/10.1126/science.aay4497.
Schwartz, Elli, Nathaniel Onnen, Peter F. Craigmile, and Megan E. Roberts. 2021. "The Legacy of Redlining: Associations between Historical Neighborhood Mapping and Contemporary Tobacco Retailer Density in Ohio." *Health & Place* 68: 102529. https://doi.org/10.1016/j.healthplace.2021.102529.
Singer, Merrill, Nicola Bulled, Bayla Ostrach, and Emily Mendenhall. 2017. "Syndemics and the Biosocial Conception of Health." *The Lancet* 389 (10072): 941–50. https://doi.org/10.1016/s0140-6736(17)30003-x.
United Nations, Department of Economic and Social Affairs, Population Division. 2019. *World Urbanization Prospects: The 2018 Revision (ST/ESA/SER.A/420)*. New York: United Nations.
Wenger, Etienne, Richard McDermott, and William M. Snyder. 2002. *Cultivating Communities of Practice*. Boston, MA: Harvard Business School Press.
Whyte, Kyle Powys. 2011. "The Recognition Dimensions of Environmental Justice in Indian Country." *Environmental Justice* 4 (4): 199–205. https://doi.org/10.1089/env.2011.0036.
World Health Organization. 2017. *World Health Statistics 2017: Monitoring Health for the SDGs, Sustainable Development Goals*. World Health Organization. https://apps.who.int/iris/handle/10665/255336.

4

CLIMATE CHANGE POLICY

Lessons From Classical Rome

Devjani Roy

Introduction

On November 4, 2019, the United States formally left the Paris Agreement, the first and only nation among the Agreement's 75 signatories to do so, five years after the Paris Climate Agreement was signed in 2015 (Dennis, Eilperin, and Grandoni 2020). As of this writing, the United States rejoined the Paris Agreement on February 19, 2021, after President Joe Biden took office. Even so, membership in the Paris Agreement is a recognition of stark scientific reality: our planet has warmed by 1°C (1.8°F) since the start of the Industrial Revolution in the late 1800s (NASA Earth Observatory n.d.).

This essay draws upon Roman agrarian philosophy, a genre of writing more than 2,000 years old that makes a compelling argument for protecting delicate environmental ecosystems. Originally intended to be didactic in nature, agrarian philosophy has a long tradition in both Classical Greece—most notably, the works of Xenophon, Aristotle, and Hesiod—and in Classical Rome. In this essay, I focus on Roman writers, more specifically, Cato's *De Agri Cultura* (On Agriculture) written around 160 BCE, Varro's *Res Rusticae* (On Agriculture) written around 37 BCE, and Virgil's *Georgics*, written around 29 BCE.[1] Taken together, these texts cover the historical periods known as the late Roman Republic and the early Augustan Age, when the Roman state turned away from its Republican experiment to become an empire under Octavian, the first Roman emperor who is also known as Augustus Caesar. Taking an interdisciplinary and intertextual approach that draws from philosophical perspectives as well as contemporary environmental action policy, I argue that changing policy to accommodate climate-change reality is a political, social, and moral imperative that has historical precedent among the Roman philosophers.

But back to the present time. Two important developments took place in the field of environmental policy in late 2018 and early 2019. In November 2018,

DOI: 10.4324/9781003217442-5

the landmark Fourth National Climate Assessment was released (USGCRP 2018), projecting that if fossil fuel emissions are allowed to continue at their current rate unchecked, the Earth's atmosphere will get warmer by 8°F by 2090, leading to rising sea levels and widespread crop failures. Then in January 2019, 3,589 economists signed "the largest public statement of economists in history," arguing that "[g]lobal climate change is a serious problem calling for immediate national action" (Economists' Statement on Carbon Dividends 2019).

These are not isolated alarmist pieces on the likelihood of an uninhabitable Earth. In 2019 and early 2020 alone, wildfires burned over 12.35 million acres in Australia, wildfires in Brazil's Amazonian rainforests released billions of tons of carbon into the atmosphere, and climatologists warned Greenland's ice sheet and the Arctic permafrost are melting at an accelerated pace that is irreversible (Kwong 2020; Taylor 2022). We are at a tipping point in terms of a shared destiny.

Roman Agrarian Philosophy

Rome's Republican and Augustan periods collectively span roughly 509 BCE–14 CE. The Roman Republic held sway during the period 509 BCE–27 BCE and was an ancient state governed administratively from the city of Rome. During this period, elected magistrates replaced the monarchy—a form of government that continued until 27 BCE when the Roman Empire was established (Boatwright et al. 2012). The Augustan Age is named after Rome's first emperor, Augustus Caesar, also Julius Caesar's nephew, who ruled the Roman Empire from 27 BCE–14 CE, a period known for its political stability, peace, and flowering of the cultural arts after years of civil war. During the Augustan Age, Augustus, occasionally called Octavian, combined elements of republican government with those of a monarchy (Flower 2014).

Multiple agrarian philosophical tracts exist from the Classical periods in both Greece and Rome—Columella's *De Re Rustica* (On Rural Affairs) and Xenophon's *Oeconomicus*, for instance—but in this essay, I focus on texts from Classical Rome, more specifically, the late Roman Republic (509 BCE–27 BCE) and the early years of the Roman Empire (27 BCE–476 CE): those by Cato (*De Agri Cultura*), Varro (*Res Rusticae*), and Virgil (*Georgics*). All three texts share common threads. First, as agrarian advice manuals, they are comprehensive and substantial in scope: they offer timeless instructions on farming the land, herding and raising cattle and horses, viticulture, beekeeping, and the financial aspects of farming (Kronenberg 2009).

This last item is significant. These Roman writers do not ignore the concept of profit; on the contrary, they posit that natural resources cannot be used for exploitation and temporary gain precisely because doing so will deplete these limited resources. Second, they recognize the importance of climatology, or climate science, such as it existed in the pre-Industrial Age, offering specific instructions tying crop cultivation to weather patterns. Finally, and perhaps most critically, each writer recognizes that we are stewards of the Earth, charged with the task of conserving and passing on to the next generation, instead of thoughtlessly exploiting natural resources in the present.

Varro's *Res Rusticae* (1935) consists of three books—on farming, animal husbandry, and farmsteading practices—that he wrote as an advice manual for his wife, Fundania, to be followed after his death (indeed, he was past 80 when he wrote *Res Rusticae*). He writes, "Therefore I shall write for you three handbooks to which you may turn whenever you wish to know, in a given case, how you ought to proceed in farming" (161). Cato's (1935) *De Agri Cultura* (On Agriculture) is a loosely connected farmer's handbook of instructions on all manner of farm-related activities: olive cultivation, keeping vineyards, grain cultivation, growing vegetables, and rearing cattle. Virgil's (1916) *Georgics*, the last in terms of time of composition, is made up of four separate books: on farming, animal husbandry, beekeeping, and viticulture. Each author makes the case for the specialization of labor, recommends allowing the land sufficient time to regenerate itself, and suggests that being profitable in each enterprise should be the ultimate goal. In other words, sustainable environmental practices need not ignore the profit motive, or, as Varro (1935) reminds his wife Fundania, "[S]ince you have bought an estate and wish to make it profitable by good cultivation" (161). While rarely read these days outside of Classics degree programs, all three authors make the same point that environmental policy makers today have been trying to make, with varying success: that the environment's resources are limited and hence need to be sustained, protected, and used judiciously in order that the next generation may use these resources too.

All three philosopher-poets are comprehensive in scope covering nearly all aspects of agriculture, and Cato and Varro write from personal farming experience. Varro, for instance, writes, "My remarks will be derived from three sources: what I have myself observed by practice on my own land, what I have read, and what I have heard from experts" (162). And, in our contemporary age in which environmental policy has become strongly partisan and politically inflected, Varro, Cato, and Virgil offer a long-term view on patriotism: that we prove our love for our country and the depth of our patriotism by preserving the environment. Cato (1935) reminds us that "it is from the farming class that the bravest men and the sturdiest soldiers come, their calling is most highly respected" (3). Similarly, Varro (1935) notes, "What useful product is there which not only does not grow in Italy, but even grow to perfection? . . . Is not Italy so covered with trees that the whole land seems to be an orchard?" (171).

Defining the Sustainable City

It may be strange to talk about the concept of a sustainable city when discussing Roman agrarian philosophers who wrote 2,000 years ago. But indeed, there is a city in existence when Cato, Varro, and Virgil write: Rome. Rome's history is nearly 3,000 years old, and it has been one of Europe's and the Western world's oldest, most populated, and continuously inhabited cities. For most of its history, Rome has been an impeccably well-designed metropolis in terms of architecture, transportation systems, and urban design, as well as being the seat of government. One might argue Rome survived through the millennia because it was designed with survival in mind. This was not incidental. The reason was war.

While the Roman Republic endured many wars starting 509 BCE, the philosopher-poets I write about composed under the long shadows of the Punic Wars (264–146 BCE) and the Macedonian Wars (the third and second centuries BCE). These wars found farms decimated and agricultural workers conscripted to fight as soldiers. Poets from the Roman Republic, in particular the Late Roman Republic, extol the virtues of a return to farming as a return to simplicity, from the city to the country and away from the destruction of war.

Cato (1935) is quick to mention this, stating that "it is from the farming class that the bravest men and the sturdiest soldiers come, their calling is most highly respected, their livelihood is most assured and is looked on with the least hostility, and those who are engaged in that pursuit are least inclined to be disaffected" (3). But the focus remains always on farming with an eye toward maintaining profitability for the farmer, another point of concordance with our contemporary concept of sustainable agriculture.

I define the *sustainable city* as a habitat for both human beings and animals within an urban ecosystem. The sustainable city creates economic prosperity for its human population and sustained health for all its inhabitants, while also honoring and preserving natural resources. In other words, a sustainable city recognizes that prosperity is not merely economic in nature; true prosperity should accommodate the needs of the environment too.

The United Nations 2030 Agenda for Sustainable Development (United Nations Department of Economic and Social Affairs 2021), with its 17 Sustainable Development Goals and 169 Targets, devotes an entire goal to "Make cities and human settlements inclusive, safe, resilient, and sustainable" by 2030 (Goal 11). The 2030 Agenda's sub-goals establish many of the characteristics of the sustainable city: "adequate, safe, and affordable housing" (11.1); "access to safe, affordable, accessible, and sustainable transport systems" (11.2); and "participatory, integrated, and sustainable human settlement planning" (11.3), among others.

Many of the features in these goals draw upon ancient precepts that originate in agrarian philosophy. For instance, Roman philosopher-poets argue that sustainable agriculture is both inclusive and participatory in nature—i.e., it is democratic in character—benefiting all who live on the human settlement. Thus, in Virgil's *Georgics*, cattle and livestock, bees and birds are subjects of entire books. Finally, the sustainable city is also sustainable in terms of the economic destiny of its inhabitants: the outlandish wealth of the few is not more important than the livelihood of the many. Varro (1935) is quick to emphasize that farming and agrarian communities are more important than urbanization, stating that "farmers antedate city people by an enormous number of years. And no marvel, since it was divine nature which gave us the country, and man's skill that built the cities" (424).

Food Security and Food Sovereignty

Roman agrarian philosophers were ahead of their time in talking about food security—or the measure of ready and easy accessibility to food—and food sovereignty—namely, the ownership of the means of food production and distribution by those who consume the food—as integral building blocks of the sustainable city (National Family Farm Coalition 2021). Cato, Varro, and Virgil write about food in terms of sustaining the human communities who depend on the land that grows this food. In a preindustrial age, unreliable weather patterns or sickness among livestock could destroy the local communities who depended on a reliable source of food, leading to an ever-present threat of starvation. Virgil writes, "[U]nless your hoe is ever ready to assail the weeds, your voice to terrify the birds. . . you will gaze on your neighbor's large store of grain, and you will be shaking oaks in the woods to assuage your hunger" (109).

Today, food security and food sovereignty are intrinsically tied to global trade and international politics: for instance, in mid-December 2020, as the final Brexit deal between the United Kingdom and the European Union was still undergoing negotiations, freight trucks carrying food from France into the UK were held up in 62-mile-long queues, for days at a time, at the southeastern port of Dover (Reuters 2020). Ultimately, food security and food sovereignty both reflect a socially just society, one that takes into account the needs of its poorest and most vulnerable (International Food Policy Research Institute 2021). One solution is pursuing food autonomy through alternative food networks (AFNs): community gardens, organic food schemes, fair trade practices, small family-owned farms, and Slow Food that is locally sourced and produced (as opposed to industrial- and mass-produced Fast Food that is divorced from direct ties to the earth).

Global Temperature Rise

If there is one common overarching theme among all Roman agrarian philosophers, it is paying heed to weather patterns. Two thousand years ago, Cato, Varro, and Virgil each recognized that in the predictability of weather patterns lay the safety of human and animal life. Their caution was prescient. Starting with the Industrial Revolution in the late 18th century and accelerating in the 20th century, the rapid buildup of greenhouse-gas emissions (methane, nitrous oxide, carbon dioxide, and chlorofluorocarbons, among others) and the burning of fossil fuels have increased, and, with it, the Earth's capacity to trap heat. The resulting so-called greenhouse effect has progressively increased temperatures worldwide each year, even in places where such temperature rise and unpredictable fluctuations would have once been unthinkable (IPCC 2007).

The International Panel on Climate Change (IPCC), established in 1988, is the leading United Nations body for periodic scientific evaluations of human-induced climate change. An IPCC Special Report warns, "Human activities are estimated to have caused approximately 1.0°C of global warming above pre-industrial levels, with a likely range of 0.8°C to 1.2°C. Global warming is likely to reach 1.5°C between 2030 and 2052 if it continues to increase at the current rate" (IPCC 2018a). The report expresses "high confidence" for this 1.5°C increase. Severe ecological consequences will follow a greater than 2°C temperature rise, particularly in regions near the equator; for instance, nearly all the Earth's coral reefs will be gone.

Pandemics and Emergent Infectious Diseases

Animals and human beings have always lived in close proximity to each other, with each species dependent on the other. In the *Georgics*, Virgil writes, "No scruples ever forbade us. . . to set snares for birds . . . or to plunge bleating flocks into the health-giving stream. Oft, too, the driver loads his slow donkey's sides with oil or cheap fruits" (Virgil 1916, 119). The one fact that has remained unchanged since the days of the Roman agrarian philosophers is that human beings continue to depend on animals for economic and dietary needs, no matter the rapid spread of technology or the development of new means of transportation. In many parts of the world, horses, bullocks, and oxen are still used by the poorest of the poor. Hence Virgil's reminder to treat these animals well: "[D]o not fodder your unbroken steers only on grass or willow leaves and sedge from the swamp but also add kernels of grain picked by hand" (189).

But, as recent years have shown, pandemics and other emergent infectious diseases (EIDs) such as COVID-19, the Swine flu, the Ebola virus, and Zika spread when living conditions between human beings and animals are so close as to permit the easy spread of human–animal infection. As I write this in 2021, the current COVID-19 pandemic has had ripple effects across the world for nearly two years, showing vulnerabilities in the interconnectedness of global economies and the inadequacy of public healthcare systems.

The close habitat of humans and animals within densely packed urban and suburban areas makes human populations increasingly vulnerable to new strains of EIDs. Pandemics demonstrate also that the socioeconomic impact of extreme events is unequal, skewed by race, ethnicity, location, and economic vulnerability (Keating, Cha, and Florit 2020). Regions and populations that are already poor and disenfranchised tend to descend even further into poverty when faced with public health crises.

Roman Agrarian Philosophy and the City as an Ecosystem

Why do so many poets and philosophers from the late Roman Republic and the early years of the Roman Empire write didactic tracts on farming? In particular, they are at pains to contrast the wealth but also the anxiety of the city with the simplicity, if relative poverty, of the country. While Cato the Elder, Varro, and Virgil are my focus in this essay, these poet-philosophers are not our only extant sources. Lucius Columella who writes the 12-volume *De re Rustica* (On Agriculture) and also the volume titled *De arboribus* (On Trees) is one among notable others. Another famous source is Horace, whose Second Epode and also his Sixth Satire about the country mouse and the city mouse emphasize the country-versus-city dialectic wherein the countryside is praised as the locus of innocence and authenticity that contrasts with the wheeling and dealing of influence in Rome. Much of this dialectic would be subsumed, centuries later, into Renaissance pastoral poetry and then into the Romantic movement of the 18th and early 19th centuries.

Roman agrarian philosophy rejects the city—Rome, in this case—as the political center of Roman life in favor of the life of the working farmer. The central underpinning of this dialectic is work. Successive centuries of war—the Macedonian Wars and the Punic Wars, to name only two—had taught Romans that true sustenance was to be found, not in power and conquest, which served only to destroy, but in restoring the earth through farming; or, as Virgil observes in the First Georgic, "so many wars overrun the world. . . [that] respect for the plough is gone; our lands, robbed of the tillers, lie waste, and curved pruning hooks are forged into straight blades [swords]" (135). Virgil is even more explicit in his Second Georgic, observing that "happy, too, is he who knows the rural gods. . .. [U]ntroubled is he by the Dacian incursion swooping from a Danube leagued in war, untroubled by Rome's policies spelling doom to kingdoms" (171). Virgilian scholar L.P. Wilkinson (1978) reminds us, "The effect of the civil wars on agriculture could not fail to be [appreciated]. It was not so much crops and property that were damaged, as in modern warfare; rather that weeds and decay got the upper hand." (52).

While some of the imagery associated with agrarian life is utopian and contains the idealized tinge of the pastoral, the city, with its soldiers and politicians, are nearly always represented as inferior to farming communities, brimming with the quest for power and the grandiosity of conquest but inferior to the life of the farmer. Virgil expresses this sentiment toward the conclusion of the Second Georgic: "One wreaks ruin on a city and its wretched homes. . .. Meanwhile the husbandman has been cleaving the soil . . . hence comes sustenance for his country and his little grandsons" (173). The connection between Roman agrarian philosophy 2,000 years old and environmental policy today is this same sense of decay and havoc. The city today must be rejuvenated using restorative practices from farming.

Policy Recommendations

In the years between 133 BCE and 31 CE, we see a literary profusion of agrarian philosophy—the writings of Vegetius, Columella, and Pliny—in addition to those of Cato, Varro, and Virgil. The very presence of these agrarian advice manuals in a preurban and preindustrial world tells us the advice was considered important enough to record and pass down to future generations, teaching that it was critical to preserve what was limited. This overarching impulse, to preserve and protect and nurture what nature provides, has been lost or, perhaps more accurately, left behind on civilization's upward march. One reason is that the Industrial Revolution in the 18th century created the possibility that machines could create what nature could not and replicate better than nature itself. Human beings, one might argue, became reckless with this possibility.

The writings of Roman agrarian philosophers often represent farmers' notebooks but are replete with instructions important enough to write down and bequeath to the next generation. It might be worth distilling the essence of their advice for environmental policy makers today.

Unitary intervention systems: Roman agrarian philosophers describe nature as a unified system. The constituent parts of the system—the land, trees, water, birds and bees, fish and fowl, horses, cattle and livestock—form one whole natural system and not separate parts. Varro (1935) makes this point explicitly:

> Bees are not of a solitary nature. . . but are like human beings . . .. [I]t is from these [bees] that men learn to toil, to build, to store up food. [The bees] have three tasks: food, dwelling, toil; and the food is not the same as the wax, nor the honey, nor the dwelling.
>
> *(501)*

This is a useful approach in rejuvenating our cities. For instance, urban greenspaces such as trees, parks, and community gardens, developed in high-carbon-emission areas, offer health benefits, both physical and cognitive, in addition to conferring environmental benefits (Wolf et al. 2020, 8). Parks, gardens, and human-made lakes confer social benefits too: they become spaces for congregation and physical exercise. Thus future policy interventions must include nature into public-health planning.

Adaptation and adaptation tracking: Adaptation efforts are advanced interventions to potential climate-based changes. It is "the process of adjustment to actual or expected climate and its effects" (IPCC 2018a). Roman agrarian philosophers stress the importance of adjustment to anticipated changes in nature and that nature's cyclicality may be used for human benefit if we work with and not against this cyclicality. This, in short, is the essence of adaptation (Araos et al. 2016).

Resilience: This is the

> capacity of social, economic, and environmental systems to cope with a hazardous event or trend or disturbance, responding or reorganizing in ways that maintain their essential function, identity and structure, while also maintaining the capacity for adaptation, learning, and transformation.
>
> *(IPCC 2018b)*

Tyler and Moench (2012) describe three characteristics of urban climate resilience:

1 Strengthening systems to reduce their fragility in the face of climate impacts and to reduce the risk of cascading failures
2 Building the capacities of social agents to anticipate and develop adaptive responses, to access and maintain supportive urban systems
3 Addressing the institutional factors that constrain effective responses to system fragility or undermine the ability of agents to take action (319)

As cities become complex systems, developing resilience to system-wide shocks, whether human or natural, becomes increasingly important.

Vulnerability: This refers to the likelihood or predisposition of a complex system to come to harm. Roman agrarian philosophers recommend building invulnerability, or the ability to adapt and cope quickly, on an ongoing basis. Indeed, their advice, written for a preindustrial society, is based on the understanding that human beings are dependent upon nature and weather patterns and that unpredictable changes in these patterns render us vulnerable. Reducing vulnerability increases resilience (characteristic 3), but doing so requires the constant evaluation of exposure through stress testing—or, systematic and periodic tests designed to assess in advance whether a system will survive when faced with unusual pressures—at both the local and the global levels.

Planning: While they may be unlikely sources of advice today, Roman agrarian philosophical texts each recommend sound planning practices. They tell us a good harvest or successful animal husbandry is not the result of random chance but of advance strategic planning. Today, as urban populations grow, more cities than ever need formal planning practices. Such plans need the collaboration of local groups and institutions that know their vulnerable populations best. In other words, interdisciplinary stakeholders need to create evidence-based interventions at the local level.

Creating green urban spaces and other nature-based solutions: In contrast with the traditional gray, steel-and-concrete structures within urban jungles, greenspaces offer health benefits in addition to being visually pleasing (Barron et al. 2019, 11). Finding nature-based solutions at the intersection of environmental policy and public health is a rich future area of research. However, these solutions require concerted, coordinated efforts at local and municipal levels, as well as multiple layers of cooperation, such as between greenspace managers, urban planners, designers and architects, and local health authorities, all of which is currently without precedent.

With family farms being acquired by large corporations and rural land being encroached by commercial properties, modern-day policy makers would do well to heed Roman agrarian philosophers' argument for the value of farming. Virgil scholar Janet Lembke (2005) correctly points out the sharp psychological portrait in the *Georgics* when she writes, "In his nostalgia for the lost golden age, Virgil shows an intuitive grasp of the havoc wrought in human life by urbanization and warfare, which appeared—and still appears—to be unending" (xix). The Second Georgic's concept of pruning—when to hold back and when to cut—is a lesson for contemporary environmental policy makers.

Acknowledgment: I am grateful to John Morillo for encouraging an interest in Roman agrarian philosophy and for his insightful comments on an earlier version of this project.

Note

1 All references in this chapter to Cato (1935), Varro (1935), and Virgil (1916) are to the translations noted in the list of references.

References

Araos, Malcolm, Lea Berrang-Ford, James D. Ford, Stephanie E. Austin, Robbert Biesbroek, and Alexandra Lesnikowski. 2016. "Climate Change Adaptation Planning in Large Cities: A Systematic Global Assessment." *Environmental Science & Policy* 66: 375–82. https://doi.org/10.1016/j.envsci.2016.06.009.

Barron, Sara, Sophie Nitoslawski, Kathleen L. Wolf, Angie Woo, Erin Desautels, and Stephen R. J. Sheppard. 2019. "Greening Blocks: A Conceptual Typology of Practical Design Interventions to Integrate Health and Climate Resilience Co-Benefits." *International Journal of Environmental Research and Public Health* 16 (21, 4241): 1–21. https://doi.org/10.3390/ijerph16214241.

Boatwright Mary T., Daniel J. Gargola, Noel Lenski, and Richard J. A. Talbert. 2012. *The Romans: From Village to Empire*. New York: Oxford University Press.

Cato, Marcus Porcius. 1935. *Marcus Porcius Cato, On Agriculture*. Translated by William Davis Hooper. Revised by Harrison Boyd Ash. Loeb Classical Library 283. Cambridge: Harvard University Press.

Dennis, Brady, Juliet Eilperin, and Dino Grandoni. 2020. "As U.S. Leaves Paris Accord, Climate Policy Hangs on Election Outcome." *The Washington Post*, November 4, 2020. www.washingtonpost.com/climate-environment/2020/11/04/climate-change-election-paris-agreement/.

Economists' Statement on Carbon Dividends. 2019. "Climate Leadership Council." Accessed January 13, 2021. https://clcouncil.org/economists-statement/.

Flower, Harriet I. 2014. *The Cambridge Companion to the Roman Republic*. Cambridge: Cambridge University Press.

International Food Policy Research Institute. 2021. *Food Security*. International Food Policy Research Institute. Accessed February 20, 2021. www.ifpri.org/topic/food-security.

IPCC. 2007. "FAQ 1.3 What Is the Greenhouse Effect?" *IPCC Fourth Assessment Report Climate Change 2007*. Accessed February 20, 2021. https://archive.ipcc.ch/publications_and_data/ar4/wg1/en/faq-1-3.html.

———. 2018a. "Annex I: Glossary" [Matthews, J. B. R. (ed.)]. In *Global Warming of 1.5°C. An IPCC Special Report on the Impacts of Global Warming of 1.5°C Above Pre-Industrial Levels and Related Global Greenhouse Gas Emission Pathways, in the Context of Strengthening the Global Response to the Threat of Climate Change, Sustainable Development, and Efforts to Eradicate Poverty*, edited by V. Masson-Delmotte, P. Zhai, H.-O. Pörtner, D. Roberts, J. Skea, P. R. Shukla, A. Pirani, W. Moufouma-Okia, C. Péan, R. Pidcock, S. Connors, J. B. R. Matthews, Y. Chen, X. Zhou, M. I. Gomis, E. Lonnoy, T. Maycock, M. Tignor, and T. Waterfield. Geneva, Switzerland: World Meteorological Organization.

———. 2018b. "Global Warming of 1.5°C." Accessed February 2019. www.ipcc.ch/site/assets/uploads/sites/2/2019/06/SR15_Headline-statements.pdf.

Keating, Dan, Ariana Eunjung Cha, and Gabriel Florit. 2020. " 'I Just Pray God Will Help Me': Racial, Ethnic Minorities Reel from Higher Covid-19 Death Rates." *The Washington Post*, November 20, 2020. www.washingtonpost.com/graphics/2020/health/covid-race-mortality-rate/.

Kronenberg, Leah. 2009. *Allegories of Farming from Greece and Rome*. Cambridge, UK: Cambridge University Press.

Kwong, Emily. 2020. "Australia Deploys Military Reservists to Combat Wildfire, As Thousands Evacuate." *NPR*, January 5, 2020. www.npr.org/2020/01/05/793754335/australia-deploys-military-reservists-to-combat-wildfire-as-thousands-evacuate/.

Lembke, Janet. 2005. *Virgil's Georgics: A New Verse Translation*. New Haven, CT: Yale University Press.

NASA Earth Observatory. n.d. "World of Change: Global Temperatures." Accessed January 12, 2021. https://earthobservatory.nasa.gov/world-of-change/global-temperatures.

National Family Farm Coalition. *What is Food Sovereignty?* National Family Farm Coalition. Accessed February 20, 2021. https://nffc.net/what-we-do/food-sovereignty/.

Reuters. 2020. "As Brexit Cliff Edge Looms, Miles of Trucks Stack Up Near Southern English Port." *Reuters*. Accessed December 10, 2020. www.reuters.com/article/

uk-britain-eu-trucks/as-brexit-cliff-edge-looms-miles-of-trucks-stack-up-near-southern-english-port-idUSKBN28K2PW.

Taylor, Chloe. 2022. "Why Deforestation in Brazil's Amazon Has Soared to its Highest Level in 15 Years." *CNBC.com*, January 4, 2022. www.cnbc.com/2021/12/30/why-deforestation-in-brazils-amazon-has-soared-to-its-highest-level-in-15-years.html/.

Tyler, Stephen, and Marcus Moench. 2012. "A Framework for Urban Climate Resilience." *Climate and Development* 4 (4): 311–26. https://doi.org/10.1080/17565529.2012.745389.

United Nations Department of Economic and Social Affairs. 2021. *The 2030 Agenda for Sustainable Development*. United Nations. Accessed February 14, 2021. https://sdgs.un.org/2030agenda.

USGCRP. 2018. *Impacts, Risks, and Adaptation in the United States: Fourth National Climate Assessment*, Vol. II. Washington: U.S. Global Change Research Program. https://doi.org/10.7930/NCA4.2018.

Varro, Marcus Terentius. 1935. *Marcus Terentius Varro, on Agriculture*. Translated by William Davis Hooper. Revised by Harrison Boyd Ash. Loeb Classical Library 283. Cambridge: Harvard University Press.

Virgil. 1916. *Eclogues. Georgics. Aeneid: Books 1–6*. Translated by H. Rushton Fairclough. Revised by G. P. Goold. Loeb Classical Library 63. Cambridge: Harvard University Press.

Wilkinson, L. P. 1978. *The Georgics of Virgil: A Critical Survey*. London: Cambridge University Press.

Wolf, Kathleen L., Sharon T. Lam, Jennifer K. McKeen, Gregory R. A. Richardson, Matilda van den Bosch, and Adrina C. Bardekjian. 2020. "Urban Trees and Human Health: A Scoping Review." *International Journal of Environmental Research Public Health* 17 (12, 4371) (2020): 1–30. https://doi:10.3390/ijerph17124371.

5

GREENSPACE IN SMALL CITIES

Opportunities for Sustainability Education

Greg King and Glen Hvenegaard

Introduction

This chapter presents two case studies in a small Canadian city that showcase sustainability education and the role of a postsecondary institution in community-engaged research to promote local greenspace and wildlife stewardship goals. Both of these ongoing projects highlight the importance of interdisciplinary perspectives when considering pathways toward urban sustainability.

Environmental and sustainability education have many definitions (Ontario Institute for Studies in Education 2015). Common themes include learning that explores the interdependencies of ecological, societal, and economic systems; developing attitudes, skills, and knowledge to empower both teachers and learners; and actively participating in creating a sustainable future. Ideally, sustainability education should also emphasize the transfer and application of environmental knowledge across all sectors of society. Civic ecology theory helps us study the role of sustainability practices in integrating "social and environmental values within a social-ecological systems framework" (Krasny and Tidball 2009, 466). This theory focuses on active engagement with nature, using notions such as biophilia (Wilson 1986) and nature deficit disorder (Louv 2008). At the individual scale, people actively engaging in nature restoration, for example, contribute to their individual well-being (Miles, Sullivan, and Kuo 1998). At the community scale, individual actions merge into civic engagement. Sustainability education can be flexible and creative while supporting critical reflection and opportunities to take concrete action (Aguayo and Eames 2017). Community partnerships and civic engagement directly support the concept of sustainability contributing to changed participants (e.g., raised consciousness, altered behavior) and changed politics (e.g., processes of involvement, communication with local leaders) (Portney 2005).

DOI: 10.4324/9781003217442-6

Partnerships with local communities are a core part of the mission of the Augustana Campus of the University of Alberta. Augustana is an undergraduate, liberal arts and science campus located in Camrose, a small city (population 18,520) in east-central Alberta, Canada (Statistics Canada 2017a). Small cities have unique limitations and opportunities related to sustainability. Although they face limited funds and, crucially, limited capacity or expertise, small cities offer opportunities for nimbleness and community connectivity via local social networks. Cities that leverage and support this citizen engagement adopt many more sustainable policies (Homsy and Warner 2015).

Camrose is a scenic community with many greenspaces that contribute toward an exceptional quality of life. Public greenspaces include parks, schoolyards, a golf course, and other municipal lands (e.g., public utility lines, cemeteries, etc.) totaling 840 hectares, or 20% of city land (Dillon Consulting Limited 2014). Considering only parks, Camrose has 16 hectares of greenspace per 1,000 residents, one of the highest ratios in Canada. However, the area has experienced high population growth (~16%) over the past 15 years, which has led to continued development and expansion of the city, posing a potential threat to the greenspaces, natural features, and wildlife populations of the area (Hvenegaard and Barr 2016). Ecologically, Camrose is located within the aspen parkland natural region that is undergoing rapid change (Michaelian et al. 2011). All of these characteristics offer a unique opportunity to undertake sustainability education programs linking Augustana's faculty and students with the broader community. Our case studies reveal how residents view and connect with Camrose's greenspaces and how those connections might foster wider urban sustainability perspectives.

Case Study 1: A Neighborhood Urban Forest Inventory

Over 80% of Canada's population resides in urban areas (Statistics Canada 2017b), directly benefiting from ecosystem services provided by urban forests, including urban heat island mitigation, energy use reduction, stormwater interception, wildlife and pollinator habitat provision, air pollution removal, and carbon sequestration (Roy, Byrne, and Pickering 2012; Jansson 2013; Endreny 2018; Nowak and Greenfield 2020). The urban forest is also the most accessible avenue for exposure to the natural environment for most Canadians, providing additional aesthetic, recreational, and psychological benefits (Peckham, Duinker, and Ordóñez 2013).

Several neighborhoods and greenspaces in Camrose feature extensive canopies of urban forest, highly valued by residents as enhancing their quality of life (City of Camrose 1999). However, this appreciation is often limited to the aesthetics of mature tree-lined streets and lacks awareness of the other extensive urban forest ecosystem benefits. Camrose's urban forest is under a range of pressures and threats, including city growth and pests and pathogens (e.g., poplar borer and black knot), some of which are invasive (e.g., Dutch elm disease, emerald ash borer) (McKenney et al. 2012). As such, there is a need to measure and communicate the value of these wider benefits. Without building awareness and greater civic engagement on the topic, any municipal plan regarding the preservation, maintenance, or expansion of the urban forest will lack broad support and likely be unsuccessful (Clark and Matheny 1998). Thus, urban forests represent an opportunity for collaborative community-based research and sustainability education to increase awareness and individual sustainability actions.

The Augustana neighborhood, one of the oldest in the city, is home to many students and university employees, increasing potential engagement between the university and the wider Camrose community (Mosier and Ruxton 2018). The Augustana Neighborhood Association (ANA) has worked actively on development plans with the city and recognizes the mature urban forest canopy as a major asset. Similar to most North American cities, the neighborhood trees include both publicly planted and maintained trees as well as private trees managed by homeowners or institutions. Most urban forest research focuses solely on publicly owned trees, often neglecting those on privately owned property, despite the importance of decision-making at the individual household level (Shakeel and Conway 2014).

In the summer of 2019, a collaborative research project was initiated with the ANA and the City of Camrose to inventory all public and private trees in the neighborhood. Partner-informed goals for the project included knowing more about the composition of the urban forest and its current state; piloting an approach to collect data on public trees and their benefits; and comparing the ecosystem services provided by public and private trees. Working for and with the community was a core consideration to foster local connection to trees and greenspaces as well as to provide new perspectives and knowledge on associated benefits. To support this connection, we presented the project idea at the ANA annual general meeting in fall 2018, adjusting the project based on feedback from residents. We used ANA's social networks to communicate and increase awareness about the project in the spring prior to sampling. Following training in species identification and measurement protocols, a team of three undergraduate summer research assistants started sampling block by block. We delivered letters to each home in the week prior to sampling, seeking permission to measure private trees. While conducting sampling, we knocked on doors to seek permission and invite participation from those residences who had not responded. All of our communications provided an overview of the project including definitions of the urban forest, an explanation of project goals, and an option for residents to receive results about their individual property.

The research team located, identified, and measured over 2,700 public and private trees across the neighborhood (covering approximately 36 hectares and 275 properties; Figure 5.1). We used iTree Eco software (USDA Forest Service 2019) to model the structural and functional ecosystem benefits of neighborhood trees. Structural benefits have accumulated over the lifetime of the tree and can be interpreted as the approximate replacement value of a tree for a given size and age. Functional benefits reflect tree ecosystem services (e.g., carbon sequestration, reduction of stormwater runoff, removal of air pollution, and energy savings) and are presented at annual values. Importantly, the program is a lower-end estimate as

FIGURE 5.1 Map showing the distribution of trees sampled in the Augustana neighborhood urban forest inventory from 2019 (image oriented north).

Source: Image produced by G. King.

it does not provide an economic value for benefits such as mental health improvement that have been observed with urban forests (Collins et al. 2020).

The inventory revealed that 79% of trees are on private property, while the remaining 21% were public trees managed by the city, with more than 60 species identified across the neighborhood. The most common species on private property was white spruce (*Picea glauca*—representing 16% of total trees), while the most common public tree was green ash (*Fraxinus pennsylvanica*—representing 50% of all trees). Across the entire neighborhood, we found a total structural value of more than $3.9 million CAD. The total functional benefits for the neighborhood were $63,900 per year, the majority in the form of energy savings through summer shading and shelter from winter winds. In terms of specific benefits, neighborhood trees reduced total surface water runoff by 1,190 m^3/year, sequestered 13,370 MT/year of carbon, and removed 341 kg of air pollution per year. An average property in the neighborhood had 10.5 trees with a structural value of approximately $15,000 and a functional value of $245 each year. The analysis showed the immense value that the urban forest brought to both the neighborhood and the wider city.

We shared a full neighborhood report with the ANA, the City of Camrose Parks Department staff, members of the Camrose City Council, and other stakeholders. This report included infographics that addressed some of the key questions asked at the start of the project, such as the number of trees and the most common species found in the neighborhood. We also prepared over 50 reports for individual homeowners that described their trees and provided a breakdown of tree benefits with comparisons to their block and the full neighborhood. The goal of these reports was to broaden local residents' knowledge and spark greater interest in their own trees and the wider urban forest. This is important as there are many motivations for removing private trees, but the major ones include a lack of information on trees and on the benefits they provide (Conway 2016). Our study found that private trees provided approximately 70% of the total benefits in the neighborhood. We are currently developing an interactive and publicly accessible neighborhood map of all of the trees and their respective benefits. We have incorporated this research into public tree tours of the neighborhood in which unique species and examples of tree benefits are presented to participants, many who come from other neighborhoods across Camrose.

Conversations on doorsteps were some of the most insightful and valuable aspects of the project for everyone involved. Our research team often heard unique stories about trees in yards or around the neighborhood, and many individuals were interested in our work, with some residents accompanying us to complete measurements to learn more about their trees and the research project (Figure 5.2).

FIGURE 5.2 Neighborhood residents helping measure tree canopy during our urban forest inventory in the Augustana neighborhood.

Source: Photograph by G. King.

This engagement offered one-on-one opportunities to communicate about the important services that these trees provided and also hear more about the perspectives of residents, answer questions about trees, and refer them to certified arborists for further consultation. It was also an opportunity to inform residents that people are an important part of the urban forest, shaping it through integrating with the built environment of our homes and other infrastructure, and that these practices influence the forest's health (Ordóñez and Duinker 2014). In the face of a climate crisis, well designed and managed urban forests will be key to supporting healthy, sustainable, and resilient communities. One of the best ways to support these efforts are to directly involve residents in sustainability education opportunities focused on urban forests.

Case Study 2: Camrose Wildlife Stewardship Society

We have also been asking questions about how to engage students and the public in wildlife stewardship. Many answers to these questions arose out of activities of the Camrose Wildlife Stewardship Society (CWSS), formed in 2002 to promote wildlife and greenspace conservation based on education, policy development, stewardship, and research (Hvenegaard and Barr 2016). The CWSS benefits from a diverse board (representing the city, local environmental nongovernmental organizations, government agencies, Augustana campus, local service organizations, and local outdoor organizations) to promote a wildlife and greenspace network that enhances community values and quality of life for Camrose and area residents (Hvenegaard and Barr 2016). The CWSS hosts community-based stewardship activities, facilitates an environmental educational program, conducts research, supports greenspace planning and management, and develops wildlife and greenspace policy. Based on its work, the CWSS was awarded a 2015 Alberta Emerald Award for environmental excellence in the community category.

As the focus of our wildlife stewardship efforts, we chose a flagship species, the Purple Martin (*Progne subis*), because it attracts public attention, generates sympathy for nature, promotes fundraising, encourages tourism, and can provide focus for conservation programs (Rempel and Hvenegaard 2013). The Purple Martin, North America's largest swallow species, nests in cavities and has a long relationship with humans in North America. Prior to colonization, Indigenous people hung gourds around their camps in which the Martins would nest and scare scavengers away (Doughty and Fergus 2002). Over time, this relationship developed into a dependency in which today the eastern population of Purple Martins (east of the Rocky Mountains) relies almost fully on human-made nesting structures and supportive stewards whom we call "landlords" (Raleigh et al. 2019). Each nest box can hold multiple nesting pairs. Eastern Purple Martin populations are in decline across Canada but are increasing in Alberta (Sauer et al. 2017). The key threats are limited nest houses, nest site competition with European Starlings (*Sturnus vulgaris*) and House Sparrows

(*Passer domesticus*; Raleigh et al. 2019), and changes in the quantity and timing of available food (primarily flying insects) due to pesticide use and climate change (Nebel et al. 2010).

Camrose has had a long connection to Purple Martins. It was the first community in Alberta to successfully use nest boxes to attract Purple Martins in 1918 (Finlay 1975). Farley (1932) describes early efforts to raise the profile of Martins, educate others, and build nesting sites. To conserve Purple Martin populations, it is helpful to understand how to use sustainability education to engage the public and Martin landlords to support Martin nesting sites. At the community level, a group of Purple Martin enthusiasts in Camrose led and encouraged wider participation by interested residents. Here, we describe six strategies used by Camrose landlords and residents in stewardship activities and sustainability education to support Purple Martin conservation.

First, an annual census of Purple Martins in Camrose helped organizers and landlords understand long-term trends and the impact of stewardship efforts (Figure 5.3). Some trends reflected continent-scale conservation concerns, such as climate change and food abundance, while other trends reflected local efforts to provide nest houses for these birds.

Second, we facilitated the landlord program through recruitment and mentoring. Individual community members were engaged as volunteer Martin landlords, responsible for maintaining nest boxes, monitoring Martin activity, and collecting standardized data. Of the 86 nest houses in Camrose, 70 were on private land managed by landowners, and 16 were on public land managed by nearby landlords. All of these landlords benefited from information and support to increase occupancy by Martins. For example, education about where to locate houses is critical to increased use by Martins and their reproductive success

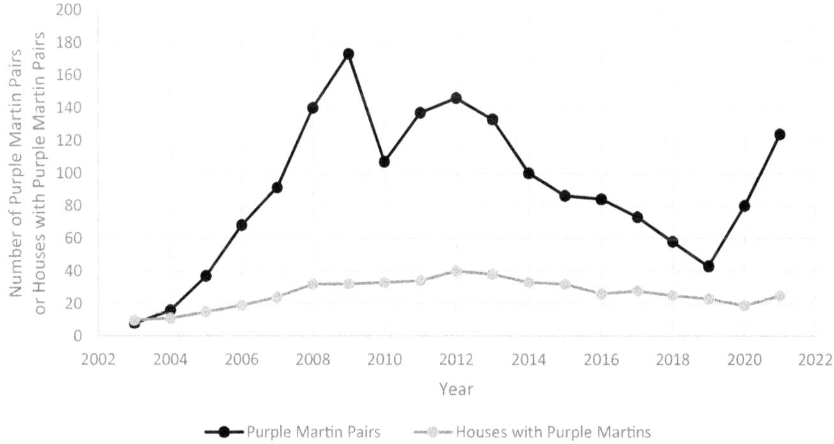

FIGURE 5.3 Purple Martin population trends for Camrose (2002–2021).

Source: Image produced by R. Henault.

(Anderson and Hvenegaard 2021). Education on various stewardship practices (e.g., cleaning nests, removing blowfly larvae, removing competitor nests, and adding nest material) also helps improve Martin nesting success (Kelly and Hvenegaard 2022).

Third, stewardship benefited from healthy interactions among students, landlords, professors, visiting scientists, and citizen scientists. Students conducted a variety of research projects focused on the social and natural science aspects of Martins that can promote long-term conservation. For example, an early student-led study on landlord motivations provided clear direction about what landlords needed in order to stay engaged in Martin stewardship (Tremblay and Hvenegaard 2008). This research provided opportunities for local landlords and nature enthusiasts to help gather data through citizen science (Hvenegaard and Fraser 2014). The initial motivations of volunteers focused on helping a species in need and learning about Purple Martins, but meeting like-minded people became a key benefit. This demonstrated that we can design engagement programs that attract citizen scientists with promises of helping wildlife and learning about the target species while ensuring positive social interactions.

Fourth, local concern and public interest in Martin conservation helped develop many valuable and durable partnerships (Hvenegaard and Barr 2016). The CWSS developed collaborative activities to promote community stewardship among project partners. Local organizations donated money, in-kind support, and expertise for Martin conservation activities. Service clubs and local volunteers supported research and an annual Purple Martin festival (Hvenegaard and Kaiser 2015). The Augustana campus provided office space and volunteer students. The city's parks department provided funding for a greenspace stewardship coordinator who supported Martin conservation projects.

Fifth, in the summer, the CWSS engaged the public in sustainability education through weekly events that helped to generate a broad constituency of support for stewardship activities. Attendance at the events ranged widely, but a committed core of people attended regularly. In the following years, outcomes from these educational efforts included a new Greenspace Management Plan, a City Green Action committee, and advocacy for urban sustainability (Hvenegaard and Barr 2016).

Lastly, starting in 2010, the CWSS has offered an annual Purple Martin Festival to deliver social, learning, mentoring, demonstration, and public support benefits. The purpose of the festival is to "provide a high profile, community-based nature tourism event to showcase the vision and work of the Camrose Wildlife and Stewardship Society" (Hvenegaard and Kaiser 2015, 4). Through presentations, workshops, field trips, and social interactions, the festival raises awareness about Purple Martins and other wildlife species, promotes networking among nature enthusiasts, and supports a Martin landlord and nest box program (Hvenegaard 2012). The festival generates education for participants, tourism revenue for the city, and stewardship volunteers for Purple Martins (Figure 5.4).

FIGURE 5.4 Leaders at the Camrose Purple Martin Festival demonstrate techniques to manage Martin nest boxes.

Source: Photograph by G. Hvenegaard.

Conclusions

Common threads from our two case studies reveal key requirements and opportunities for urban sustainability that small cities can address and embrace when engaging with local communities. First, partnerships with diverse organizations and individuals are critical for stewardship and education projects as they provide valuable resources and sustain interest over time. Second, quantifying the benefits of sustainability and educational projects (e.g., benefits of shade trees, Martin population trends) helps tell engaging stories of why sustainability-centered actions are important. Third, meaningful engagement with the community can be either formal or informal. For example, we have provided many talks for the CWSS educational series but have also provided informal doorstep tree discussions and Purple Martin landlord advice. Fourth, an interdisciplinary approach to sustainability education that builds on the strengths of and the synergies among the natural sciences (e.g., forestry, ornithology, and ecology) and social sciences (e.g., economics, social psychology, marketing) improves outcomes. Fifth, multiple perspectives are tremendously valuable when planning, executing, and evaluating sustainability education projects. Last, it is important to tell stories in an experiential way. Facilitating and telling stories about children who recently held a Purple Martin in their hands or how a resident helped measure trees allowed researchers and community members to create a lasting and deep emotional connection with nature and with their community. This kind of connection will engage volunteers, donors, and learners in working for long-term urban sustainability and help residents envision cities as places for nature (Beatley 2020).

References

Aguayo, Claudio, and Chris Eames. 2017. "Community Partnerships in Sustainability Education Research." In *Realising Innovative Partnerships in Educational Research*, edited by Rachel McNae and Bronwen Cowie, 235–44. Rotterdam: Sense Publishers. https://doi.org/10.1007/978-94-6351-062-2_24.

Anderson, Britt C., and Glen T. Hvenegaard. 2021. "Purple Martin Nest House Selection in Central Alberta, Canada." *Wildlife Society Bulletin* 45 (1): 55–61. https://doi.org/10.1002/wsb.1161.

Beatley, Timothy. 2020. *The Bird-Friendly City*. Washington, DC: Island Press.

City of Camrose. 1999. "Municipal Tree Care Policy. TREE-1." https://camrose.civicweb.net/document/46300#.

Clark, James R., and Nelda P. Matheny. 1998. "A Model of Urban Forest Sustainability: Application to Cities in the United States." *Journal of Arboriculture* 24 (2): 112–20.

Collins, Rebecca M., Rebecca Spake, Kerry A. Brown, Booker O. Ogutu, Dianna Smith, and Felix Eigenbrod. 2020. "A Systematic Map of Research Exploring the Effect of Greenspace on Mental Health." *Landscape and Urban Planning* 201: 103823. https://doi.org/10.1016/j.landurbplan.2020.103823.

Conway, Tenley M. 2016. "Tending Their Urban Forest: Residents Motivations for Tree Planting and Removal." *Urban Forestry & Urban Greening* 17 (June): 23–32. https://doi.org/10.1016/j.ufug.2016.03.008.

Dillon Consulting Limited. 2014. "City of Camrose Green Space Master Plan." *Camrose, AB: City of Camrose*. www.camrose.ca/en/your-government/resources/Documents/Green-Space-Master-Plan.pdf.

Doughty, Robin, and Rob Fergus. 2002. *The Purple Martin*. Austin: University of Texas Press.

Endreny, Theodore A. 2018. "Strategically Growing the Urban Forest Will Improve Our World." *Nature Communications* 9 (1): 1160. https://doi.org/10.1038/s41467-018-03622-0.

Farley, Frank L. 1932. *Birds of the Battle River Region (with Notes on Their Present Status, Migrations, Food Habits and Economic Value*. Edmonton: Institute of Applied Arts.

Finlay, J. Campbell. 1975. "Nesting of Purple Martins in Natural Cavities and in Man-Made Structures in Alberta." *Canadian Field-Naturalist* 89 (3): 454–55.

Homsy, George, and Mildred Warner. 2015. "Defying the Odds: Sustainability in Small and Rural Places." Briefing Paper. ICMA Centre for Sustainable Communities. https://icma.org/articles/article/sustainability-small-places.

Hvenegaard, Glen T. 2012. "Purple Martin Festival Promotes Partnerships for Conservation." *Purple Martin Update* 21: 26–27.

Hvenegaard, Glen T., and Michael T. Barr. 2016. "Engaging the Public in Wildlife and Greenspace Stewardship in Camrose, Alberta: An Analysis of Outcomes, Drivers, and Lessons Learned." In *Taking the Next Steps in Rural Sustainability in Canada*, edited by Lars Hallstrom, M. Beckie, Glen T. Hvenegaard, and K. Mundel. Edmonton: University of Alberta Press.

Hvenegaard, Glen T., and Laura Fraser. 2014. "Motivations and Benefits of Citizen Scientists Engaged in Purple Martin Migration Research." *Human Dimensions of Wildlife* 19 (6): 561–63. https://doi.org/10.1080/10871209.2014.940562.

Hvenegaard, Glen T., and Alanna N. Kaiser. 2015. "Ecotourism, Conservation, and Sustainability: A Case Study of the Camrose Purple Martin Festival." In *Innovative Leisure Practices: Cases as Conduits between Theory and Practice*, Vol. 1, 1–10. Nanaimo, BC: World Leisure Center of Excellence in Sustainability and Innovation, University of Vancouver Island.

Jansson, Åsa. 2013. "Reaching for a Sustainable, Resilient Urban Future Using the Lens of Ecosystem Services." *Ecological Economics* 86: 285–91. https://doi.org/10.1016/j.ecolecon.2012.06.013.

Kelly, Braeden D., and Glen T. Hvenegaard. 2022. "Impacts of Purple Martin Landlord Stewardship Activities on Nest Box Occupancy." *Wildlife Society Bulletin*. https://doi.org/10.1002/wsb.1247.

Krasny, Marianne E., and Keith G. Tidball. 2009. "Applying a Resilience Systems Framework to Urban Environmental Education." *Environmental Education Research* 15 (4): 465–82. https://doi.org/10.1080/13504620903003290.

Louv, Richard. 2008. *Last Child in the Woods: Saving Our Children from Nature-Deficit Disorder*. Updated and Expanded edition. Chapel Hill: Algonquin Books.

McKenney, Daniel W., John H. Pedlar, Denys Yemshanov, D. Barry Lyons, Kathy L. Campbell, and Kevin Lawrence. 2012. "Estimates of the Potential Cost of Emerald Ash Borer (*Agrilus planipennis* Fairmaire) in Canadian Municipalities." *Arboriculture & Urban Forestry* 38 (3): 81–91.

Michaelian, Michael, Edward H. Hogg, Ronald J. Hall, and Eric Arsenault. 2011. "Massive Mortality of Aspen Following Severe Drought along the Southern Edge of the Canadian Boreal Forest." *Global Change Biology* 17 (6): 2084–94. https://doi.org/10.1111/j.1365-2486.2010.02357.x.

Miles, Irene, William C. Sullivan, and Frances E. Kuo. 1998. "Ecological Restoration Volunteers: The Benefits of Participation." *Urban Ecosystems* 2 (1): 27–41. https://doi.org/10.1023/A:1009501515335.

Mosier, Samantha, and Megan Ruxton. 2018. "Sustainability University—Community Partnerships: Lessons for Practitioners and Scholars from Highly Sustainable Communities." *Environment and Planning C: Politics and Space* 36 (3): 479–95. https://doi.org/10.1177/2399654417749593.

Nebel, Silke, Alex Mills, Jon D. McCracken, and Philip D. Taylor. 2010. "Declines of Aerial Insectivores in North America Follow a Geographic Gradient." *Avian Conservation & Ecology* 5 (2): 1–14. https://doi.org/10.5751/ACE-00391-050201.

Nowak, David J., and Eric J. Greenfield. 2020. "The Increase of Impervious Cover and Decrease of Tree Cover within Urban Areas Globally (2012–2017)." *Urban Forestry & Urban Greening* 49 (March): 126638. https://doi.org/10.1016/j.ufug.2020.126638.

Ontario Institute for Studies in Education. 2015. *Definitions of Environmental and Sustainability Education Traditions*. ESE: Environmental and Sustainability Education. www.oise.utoronto.ca/ese/About/Definitions.html.

Ordóñez, Camilo, and Peter N. Duinker. 2014. "Assessing the Vulnerability of Urban Forests to Climate Change." *Environmental Reviews* 22 (3): 311–21. https://doi.org/10.1139/er-2013-0078.

Peckham, Shawna C., Peter N. Duinker, and Camilo Ordóñez. 2013. "Urban Forest Values in Canada: Views of Citizens in Calgary and Halifax." *Urban Forestry & Urban Greening* 12 (2): 154–62. https://doi.org/10.1016/j.ufug.2013.01.001.

Portney, Kent. 2005. "Civic Engagement and Sustainable Cities in the United States." *Public Administration Review* 65 (5): 579–91. https://doi.org/10.1111/j.1540-6210.2005.00485.x.

Raleigh, Daniel, James D. Ray, Blake A. Grisham, Joe Siegrist, and Daniel U. Greene. 2019. "Nest Survival Data Confirm Managed Housing Is an Important Component to the Conservation of the Eastern Purple Martin." *Wildlife Society Bulletin* 43: 93–101. https://doi.org/10.1002/wsb.941.

Rempel, Stacey, and Glen T. Hvenegaard. 2013. "Can Purple Martins Be Used as Flagship Species for Wildlife Conservation?" *Purple Martin Update* 22 (3): 18–21.

Roy, Sudipto, Jason Byrne, and Catherine Pickering. 2012. "A Systematic Quantitative Review of Urban Tree Benefits, Costs, and Assessment Methods across Cities in Different Climatic Zones." *Urban Forestry & Urban Greening* 11 (4): 351–63. https://doi.org/10.1016/j.ufug.2012.06.006.

Sauer, J. R., D. K. Niven, J. E. Hines, K. L. Pardieck, J. E. Fallon, W. A. Link, and D. J. Jr. Ziolkowsk. 2017. *The North American Breeding Bird Survey, Results and Analysis 1966–2015*. Laurel: USGS Patuxent Wildlife Research Center. https://pubs.er.usgs.gov/publication/70180972.

Shakeel, Tooba, and Tenley M. Conway. 2014. "Individual Households and Their Trees: Fine-Scale Characteristics Shaping Urban Forests." *Urban Forestry & Urban Greening* 13 (1): 136–44. https://doi.org/10.1016/j.ufug.2013.11.004.

Statistics Canada. 2017a. "Camrose [Population Centre], Alberta and Alberta [Province] (Table). Census Profile. 2016 Census." *Statistics Canada*. https://www12.statcan.gc.ca/census-recensement/2016/dp-pd/prof/details/page.cfm?Lang=E&Geo1=POPC&Code1 = 0123&Geo2=PR&Code2 = 48&SearchText=Camrose&SearchType=Begins&SearchPR=01&B1=All&GeoLevel=PR&GeoCode=0123&TABID=1&type=0.

———. 2017b. "Population Size and Growth in Canada: Key Results from the 2016 Census." *The Daily*, February 8, 2017. https://www150.statcan.gc.ca/n1/daily-quotidien/170208/dq170208a-eng.htm.

Tremblay, Candice A., and Glen T. Hvenegaard. 2008. "For Me or the Martins? Motivations for Being a Purple Martin Landlord." *Purple Martin Update* 17 (3): 10–13.

USDA Forest Service. iTree Eco. V. 6.0.16. USDA Forest Service. PC. 2019.

Wilson, Edward O. 1986. *Biophilia*. Revised ed. Cambridge: Harvard University Press.

6

THE BIRDS AND BEES OF URBAN BIODIVERSITY

Patterns, Problems, and a Path Forward

Timothy Leslie and Carole Griffiths

Loss of Biodiversity and Ecosystem Services

Biodiversity—the variety of all living organisms—is declining worldwide, a fact well documented for nearly all forms of plant and animal life (Butchart et al. 2010). Even when using conservative estimates, the current rates of biodiversity loss far exceed expected background rates, leading some scientists to claim that we may be witnessing the Earth's sixth mass extinction event (Ceballos et al. 2015). The causes of the current biodiversity crisis are myriad and varied, yet they contain one unifying feature: they are human-induced (IPBES 2019). Anthropogenic drivers of biodiversity loss include habitat loss and fragmentation through urban development and agriculture, overexploitation of species, climate change, invasive species introductions, and pollution (Pereira, Navarro, and Martins 2012). These drivers may be amplified in cities where the amount of development results in extreme habitat loss and changes in microclimate (e.g., heat island effect), where population density causes concentration of pollutants and where international commerce facilitates the introduction of non-native species (Fenoglio et al. 2021).

In addition to species declines, we are experiencing reductions in the ecosystem services that biodiversity provides and upon which human well-being relies (Cardinale et al. 2012). Ecosystem services, or the natural processes and products that directly or indirectly benefit humanity, are categorized as provisioning, regulating, cultural, or supporting (Millennium Ecosystem Assessment 2005). While efforts to conserve ecosystem services have historically focused more on wilderness or rural areas (van der Plas 2019), there is increasing recognition that maintaining and enhancing urban ecosystem services makes cities more resilient and livable and can be economically advantageous (Elmqvist et al. 2015). Urban examples include ensuring reliable urban food production (*provisioning*); promoting natural means of air and water purification and flood control (*regulating*); introducing more

DOI: 10.4324/9781003217442-7

greenspaces to promote community cohesion, connect urban residents with nature, and improve mental health and physical well-being (*cultural*); and, finally, provisioning natural habitats and maintaining biogeochemical cycles for all other ecosystem services (*supporting*) (Gomez-Baggethun et al. 2013). These services and others are foundational to urban sustainability and undoubtedly linked to biodiversity (Sirakaya, Cliquet, and Harris 2018). However, understanding how urbanization influences biodiversity patterns—and the mechanistic links between these patterns and ecosystem services—is a relatively new area of research, and further exploration is needed (Behm 2020).

Within the last decade, humans have officially become an urban species with greater than 50% of humans living in urban areas. Furthermore, the United Nations (2019) projects that 68% of people will live in urban areas by 2050 and that the global footprint of these areas may triple by as early as 2030 (Seto, Güneralp, and Hutyra 2012). As more of the world's population resides in cities, the functioning and stability of urban ecosystems becomes even more important, and cities will need to take a leading role in the conservation of biodiversity (Oke et al. 2021). In addition, if conservation-based policy and efforts are to be supported and prioritized, then connecting urban dwellers with nature is imperative (Dunn et al. 2006). We hope this chapter will help facilitate such a connection and draw attention to the importance of biodiversity in cities.

The remainder of this chapter will examine general biodiversity patterns in relation to urbanization—illustrated with investigations of two faunistic groups (birds and bees)—and conclude with ideas on how to support biodiversity in urban areas.

Urban Biodiversity Patterns

Understanding how species respond to urbanization is needed to inform conservation efforts in cities. The tolerance of species to urban environments may be revealed in patterns related to abundance, distribution, species richness, and community composition, as well as how these patterns differ from rural and/or wilderness areas. The effect of urbanization on biodiversity can be conceptualized as an environmental filtering process (e.g., Williams et al. 2009), in which one imagines an urban landscape—and all its contingent abiotic conditions—imposed upon the natural landscape and its regional pool of biodiversity. Only a subset of the species, based upon their ecological requirements, will be able to survive and possibly thrive. Add to this the introductions of human-associated species (e.g., non-native plants found in urban parks and gardens) and a complex mix of interspecific interactions, and the outcome is a unique urban community assemblage. Using New York City (where our university is located) as an example, this filtering process can be envisioned when reading and viewing the images from Eric Sanderson's (2009) *Manahatta: A Natural History of New York City*, in which he uses historical ecology to recreate the New York City bioregion from 400 years in the past. These modeling efforts identified 53 unique ecosystems within Manhattan alone and thousands of plant, mammal, bird, and aquatic species that may have flourished in this area.

While vestiges of these habitats and some of these species remain, the consequences of extreme urbanization on the native flora and fauna are readily apparent.

While no single unifying principle explains the effect of urbanization on biodiversity, to identify patterns one needs to consider the scale of observation, the extent of urbanization, and the type of organisms being studied. At broad scales, urbanization tends toward biotic homogenization, characterized by a preponderance of generalist species, losses of native species, and increases in the number of non-native species (McKinney 2006). However, many urban areas, especially those featuring moderate levels of urbanization, can exhibit high levels of spatial and temporal heterogeneity and thus host a diverse array of species with functions related to the particular habitat (Fortel et al. 2014). In areas of extreme urbanization, where impervious surface is the dominant land use type, the persistence of species may be limited by high levels of habitat loss and fragmentation (Rochat et al. 2017). At the genetic level, isolation and barriers to dispersal can reduce the genetic diversity and potential resilience of some urban populations (Delaney, Riley, and Fisher 2010). Constraints of extreme urbanization on species diversity have also been revealed in gradient studies, which examine changes in biodiversity across geographically proximate rural, suburban, and core urban areas. Indeed, in urban cores, the number of species for a wide range of taxa—including plants, mammals, birds and insects—may be less than half of what is found in less developed areas along the gradient (McKinney 2002).

The following sections examine two diverse groups of organisms—birds and bees—and how they are affected by urbanization. For each group we provide (1) information on their importance and natural history, (2) examples of how urbanization influences their diversity and community structure, and (3) ideas on how to support these groups within cities.

Birds

A Snowy Owl appeared last winter in Central Park for the first time in 130 years, and it attracted hundreds of people. Several years ago, a Mandarin duck resided in Central Park for a month or so, and hundreds stopped by to see this beautiful animal. For many city people who never look at birds, these rarities were an introduction to wildlife and to nature. Parks and other greenspaces in the city offer this possibility—giving urban residents a chance to connect to nature—either as an initial introduction or to maintain a long-standing relationship (Spotswood et al. 2021).

Nationally, interest in birds has increased. As of 2016, there were 45 million birdwatchers in the United States (U.S. Department of the Interior 2018), with numbers surging during the pandemic (Fortin 2020). This growing interest makes birding a valuable tool for gathering support for conservation, an effort that is increasingly important due to the alarming decline in bird species in North America (Rosenberg et al. 2019). Understanding the causes of these declines, including the impact of urbanization, is needed to develop conservation practices (Lepczyk et al. 2017). Accordingly, research in urban birds has been increasing over the last two decades; for example, the Science Direct database lists 435 articles published in 2000 and 3,095 in 2021.

This research highlights the essential role that birds play in maintaining ecosystems, serving as predators, scavengers, pollinators, nutrient cyclers, and seed dispersers (Whelan, Wenny, and Marquis 2008). Research also illuminates factors that influence urban bird distribution, reproduction, and survival. These include the amount of greenspace, proportion and type of land cover, habitat heterogeneity, and specific factors including noise, light, tall buildings and windows, rodenticides, and outdoor cats (La Sorte et al. 2020; Walters et al. 2019).

Greenspace and area size have the greatest impact on diversity—as urbanization increases, natural land cover decreases, impervious surface increases, and species richness and biodiversity decline. In addition to general vegetation, urban species richness is correlated with tree cover (La Sorte et al. 2020) and water cover (Sol et al. 2020; La Sorte et al. 2020; Aronson et al. 2014).

Functional diversity is correlated with land cover (Sol et al. 2020), and cities with structurally complex habitats may be more functionally diverse than natural areas with limited heterogeneity (e.g., grasslands; Sultana, Corlatti, Storch 2021). Within cities, species composition reflects the degree of habitat urbanization, and species have been classified based on their habitat's urbanization. Thus, urban dependents and urban-tolerant species occur in highly urbanized areas (Rock Pigeons: sociable, widely distributed with broad diets; Croci, Butet, and Clergeau 2008), urban exploiters occur in moderately urbanized areas (e.g., House Finches), and the least urbanized areas attract urban-avoider species (Riley and Gehrt's classification in Rodewald and Gehrt 2014).

There is also regional and seasonal variation in species composition. For cities in the temperate regions of the Northern Hemisphere, species tend to be omnivores and granivores, with species richness lowest during breeding season and highest during spring and autumn migration (La Sorte and Horton 2021). The majority of species that conduct seasonal migration within this region are nocturnal migrants (e.g., warblers), and urban greenspace provides critical stopover habitat during daylight hours (Spotswood et al. 2021).

The variation in seasonal species richness and the importance of urban greenspaces is highlighted in data derived from eBird (2021), a citizen science web application from Cornell University. Birders provide checklists each time they bird, with the location time, date, species, and number of individuals observed. Checklists over the last ten years for four iconic birding locations in Manhattan recorded two to over three times as many species during migration compared to breeding season. Bryant Park, the smallest (9.6 acres) had 25 species in July and February and 93 in May and 94 in September—prime migration months. Central Park, the largest (843 acres), had spring and fall migration totals of 210 species and 191, respectively, and 105 in breeding season in July. Similar differences occur in two Brooklyn hotspots: Prospect Park (May: 234, July: 118) and Green-Wood Cemetery (May: 184, July: 80).

The novel urban environment can provide distinct advantages and disadvantages for birds. Cities can positively impact survival by buffering extreme environmental conditions, such as drought buffered with artificial sources of freestanding water. Anthropogenic foods, supplemental bird feeding, and landscaped yards can counteract environmental swings in natural food availability. In winter, the heat island effect can reduce the impact of winter weather (e.g., Rodewald and Gehrt 2014).

However, urbanization increasingly exposes birds to multiple sources of anthropogenic mortality. Collisions with buildings, with windows in particular, are a top bird mortality source, causing 365–988 million fatalities annually in the United States (Riding, O'Connell, and Loss 2020). Collisions vary in their impact, depending on time of day. Nocturnal migrants are impacted by artificial night light interfering with the ability of migrants to navigate effectively (La Sorte and Horton 2021; Zhao et al. 2020). In New York City, during the September 11 Memorial & Museum's "Tribute in Light" illumination, birds aggregated in high densities and were disoriented with decreased flight speeds, circular flight paths, and frequent vocalization. These disruptions disappeared when the lights were extinguished (Van Doren et al. 2017). Mortality from daylight strikes tends to be correlated with the amount of reflected vegetation in windows. Parkins, Elbin, and Barnes (2015) estimated 243,000 window collision deaths occur each year in the buildings surrounding New York City's Bryant Park, most during the morning. These were correlated with interior building light, in turn correlated with the amount of window glass. Gelby and Delacretax (2006) found similar results for a low (six-story) building and hypothesized that the reflected vegetation is perceived by birds as a continuation of actual vegetation.

Additional causes of mortality for urban birds include road traffic, feral cats, and use of rodenticides to kill rats (impacting birds of prey, especially Red-Tailed Hawks; Okoniewski et al. 2021). Other urban impacts are more ambiguous in their effect on birds. Urban noise interferes with bird vocalizations, their major method of communication, and birds adjust frequency, composition, and timing to adapt (e.g., Walters et al. 2019). However, the impact on birds may depend on scale or be species specific.

For breeding birds, the impacts of urbanization are broader and affect all aspects of social behavior from establishing territories to finding mates, breeding, and raising young. Urban Northern Cardinals form smaller groups during winter, sing earlier in the spring, and occupy territories at higher densities than their rural counterparts (Kucera et al. 2019). Urban female cardinals are more similar in their reproductive success than rural breeders that are more varied in their pattern of winners and losers. Rodewald and Arcese (2017) suggested that these behaviors result from greater anthropogenic food availability and the urban heat island effect providing earlier nest sites.

Urban species may breed earlier than their rural counterparts, correlated with artificial night lighting in European Blackbirds (Russ, Lučeničová, and Klenke 2017) and with urban social and environmental conditions advancing hormone secretion and gonadal development in Dark-Eyed Juncos (Fudickar et al. 2017). Delays in breeding onset are less common. Shustack and Rodewald (2010) reported delays in Acadian Flycatchers (long-distance migrants), possibly resulting from competition forcing inexperienced birds into less desirable breeding territories. Urban Great and Blue Tits may produce fewer young, reflecting poorer food quality (Solonen 2001) or more polluted environments (Chatelain et al. 2021). Alternatively, for Northern Mockingbirds, brood size as well as number of young fledged may be greater in urban than in rural sites (Stracey and Robinson 2012).

Physiology and genetics research illustrates the effect of city living on avian health. Urban American Crows have elevated cholesterol levels resulting from the abundance of high-fat anthropogenic foods (Townsend, Staab, and Barker 2019). Giraudeau et al. (2020) found that urban birds living in a carcinogenic environment did not upregulate gene expression for cancer defense, possibly predisposing them to higher cancer risk. Finally, urban birds respond rapidly to the environment by adjusting genes epigenetically through DNA methylation (Watson et al. 2021).

City planners and city residents can take several major actions to encourage diverse bird populations. Maximizing green area and tree cover, planting a diversity of native trees and shrubs, maintaining dead trees (natural nesting sites), and providing artificial nesting structures will help maintain biodiversity (Cooper, Shultz, and Blumstein 2020). Specific urban features that impact bird survival can be mitigated. Window collisions can be reduced by minimizing glass on new buildings and by reducing glare and reflectiveness through etching windows or putting nonreflective film on them to significantly reduce the size of the reflective surface. However, additional research is needed to assess the seasonal and species-specific effect of window collisions (Riding, O'Connell, and Loss 2020). Some cities are reducing artificial light from skyscrapers at night during migration seasons, and this should be adopted more widely. Additionally, the color of light is also important in mitigating the impact of night light. Zhao et al. (2020) found that birds were rarely attracted to long-wavelength red light, whereas blue light attracted birds, particularly during nights with fog and headwinds. Switching to longer-wavelength lights is a convenient and economically effective way to reduce bird collisions.

Bees

In terms of ecosystem services, the importance of pollination services provided by insects and other animals is immense. Almost 90% of all plants require animal-mediated pollination to reproduce (Ollerton, Winfree, and Tarrant 2011). This includes many types of fruit, vegetable, and nut crops, representing about one-third of the human diet (Klein et al. 2007). In the United States alone, insect-mediated pollination services in the agricultural sector are valued at $34 billion (Jordan et al. 2021). Although many different types of insects (e.g., flies) and other animals (e.g., bats) contribute to pollination services, bees are generally considered the most important and effective pollinators. Bees are taxonomically and functionally diverse with more than 17,500 described species representing a wide range of morphological and behavioral characteristics (Michener 2007). They vary greatly in size and coloration. Some are pollen-feeding specialists in that they only visit specific types of flowers (i.e., oligolectic), whereas others are generalist feeders (i.e., polylectic). Bees may nest in self-made hives, existing cavities, hollowed-out stems, or—as is most common—in excavated tunnels in the soil. And while some species may exhibit social behavior in the form of complex colonies with divisions of labor, the vast majority of bees are actually solitary in nature.

Beginning in 2006, commercial beekeepers in the United States reported mysterious large-scale losses of honeybees at rates more than three times that under normal conditions (vanEnglesdorp et al. 2009). Although overall bee diversity has likely been in decline for over 50 years (Goulson et al. 2015), the magnitude and rapid onset of honeybee losses in the United States—and the fact that honeybees are of such economic importance to agricultural production—served as a catalyst to address threats to bee health and raised public awareness of the importance of pollinators. The Obama White House even established a Pollinator Health Task Force in 2014 to coordinate research, education, and implementation of practices that promote the health of honeybees and other pollinators. As a result, interest and action in bee conservation efforts have grown substantially in recent years, even if many people are still not aware of the full extent of the diversity of bees and their pollination services (Hall and Martins 2020).

As with other species-rich groups of organisms, it is difficult to fully ascertain the breadth, persistence, and status of bee populations. Barriers include lack of data (Goulson et al. 2015), lack of taxonomic expertise (Kim and Byrne 2006), and reductions in collections-based funding (Dalton 2003). Nonetheless, there is still strong evidence of bee declines. Studies of larger-bodied bees, such as bumblebees, are more common and have shown bumblebee declines and even species extinctions around the world (Goulson et al. 2015). In instances where historical records can be compared to current records, wild bee diversity may be greatly reduced (Burkle, Marlin, and Knight 2013). Modeling efforts also predict where bees may be experiencing population declines—even if these areas haven't been adequately monitored—based on known threats to bee survival (Koh et al. 2016). Similar to the causes of biodiversity loss in general, some of the major drivers of bee declines include habitat loss, climate change, invasive species, pesticide use, and pests and pathogens (Goulson et al. 2015).

Urban bee ecology—the study of how environmental factors unique to urban areas influence the abundance, diversity, and assemblages of bees—is a relatively new area of study. In fact, outside of research on pest insects, insect ecology in urban environments has largely been ignored until relatively recently (McIntyre 2000). However, substantial progress has been made toward identifying patterns and processes that define urban bee assemblages. The way in which bees are influenced by the urban landscape exemplifies the environmental filtering process, as studies have revealed noticeable changes in both bee diversity and community structure linked to urbanization (Ayers and Rehan 2021).

As is common in ecological research, results across studies are variable, but recent review papers have been able to identify trends related to urbanization and bee diversity. Hernandez, Frankie, and Thorp (2009) note three general patterns for bee communities in urban areas (relative to nonurban areas): (1) reductions in species richness, (2) fewer soil-nesters and more cavity nesters, and (3) fewer floral specialists. Reductions in richness, soil-nesters, and feeding specialists are likely linked to extreme habitat loss, increases in impervious surface, and reductions in native floral diversity in cities, whereas increases in cavity nesters may be linked to a greater number of nesting sites in structurally complex urban landscapes. A more recent review of urban pollinators also notes that the relationship between urbanization and pollinator diversity is not linear: pollinator diversity is often highest in areas of moderate urbanization and lower in high-density areas with little greenspace at the landscape scale (Wenzel et al. 2020). Clearly, then, in terms of urbanization and bee diversity, conservation efforts should focus on core urban areas of cities, where bee diversity may be most drastically affected.

New York City is the largest metropolitan area in the United States, and studies on bee ecology within the city have mirrored the aforementioned trends. Some of these studies were conducted in community gardens as they represent a substantial amount of greenspace within the city. In addition, these gardens often devote a large amount of their area to food production and more than 90% of the crops typically grown in urban gardens require or benefit from bee pollination services (Matteson and Langellotto 2009). Many urban gardeners also rely on the produce grown in their gardens for a substantial portion of their household produce needs, and this reliance is higher among low-income and food-insecure households (Gregory, Leslie, and Drinkwater 2016). Thus, supporting bee diversity and pollination services in cities is a concern not only strictly of biodiversity conservation but also one of urban sustainability and food security.

Multiyear studies on bee diversity in New York City community gardens found 50 to 54 species of bees (Matteson, Ascher, and Langellotto 2008; Alamoudi 2017). By comparison, the statewide pool of bee species is at least 416 species (Pollinator Network @ Cornell 2021). When compared to bee surveys in nearby suburban and natural areas, the urban bee community exhibited lower species richness, fewer oligolectic species, and a greater proportion of non-native species and cavity-nesting species. High levels of variation in bee diversity were also found among the gardens themselves, which could primarily be explained by factors such as sunlight availability, floral abundance/diversity, and greenspace in the surrounding landscape (Matteson and Langellotto 2010; Alamoudi 2017).

These trends can be used as a guide when considering a path forward for supporting bee diversity in cities. Overall richness noticeably declines in areas of extreme urbanization, and thus it is imperative to increase the amount of greenspace and connectivity in core urban areas. Since it may not be realistic to move or eliminate many of the features of the built environment, rooftops provide opportunities to introduce novel greenspace in cities that may serve as habitat for highly mobile flying organisms such as bees. When green roofs incorporate pollinator-friendly plants and nesting resources, bee diversity found in these locations can be substantial (Kratschmer, Kriechbaum, and Pachinger 2018). In addition, when vacant lots are present within cities, community groups can lead efforts to maintain these areas as bee-friendly greenspaces.

Although floral plantings are often a focus of conservation initiatives, it is important to remember that nesting sites are also a limiting factor for bee survival. One of the greatest challenges for wild bees in urban settings is the lack of nesting sites for ground-nesting bees. In cities where greenspace is often lacking, there may be a temptation to "manage" every part of an urban garden or yard with mulch or various plantings. However, leaving some areas relatively untended (beyond clearing vegetation) can be important for ground nesting bees. Areas that are south-facing and have well drained soil are especially desirable.

Lastly, since urbanization may limit specialist feeders, supporting these species should be another area of focus. Purposeful introduction of a diverse array of plants known to support various floral specialists should be considered when pollinator habitats are introduced. For example, in the United States, thousands of bee species exhibit some form of feeding specialization, and their host plant associations have been compiled in publicly available databases (e.g., Fowler and Droege 2020).

Conclusion

Despite the drastic impacts that urbanization can have on the environment, studies in urban ecology have revealed that cities are not, in fact, biological deserts but instead are areas of important ecological value (Hall et al. 2016). However, an increasingly urbanized world still poses a serious threat to biodiversity that will need to be addressed through three intersecting areas: (1) education, (2) research, and (3) action.

Educating the public, especially urban dwellers, on the importance of biodiversity and ecosystem services will be essential for supporting conservation practices and policies. From our perspective as educators, this should start in the classroom through coursework that allows students to interact with nature and study biodiversity concepts. As biology faculty members at an urban university, we have seen firsthand how field-based classes, research mentoring, and use of conservation-themed lab exercises (e.g., see the journal *Lessons in Conservation*) can raise awareness among students on the link between biodiversity and urban sustainability. City dwellers may also benefit from easy access to educational and conservation-minded institutions commonly found in metropolitan areas, such as zoos, natural history museums, and botanical gardens.

Despite growth in the fields of urban ecology and sustainability, there are still gaps in knowledge related to the mechanistic underpinnings of how urbanization affects biodiversity and ecosystem services. Increased funding for research in urban ecology with an interdisciplinary focus and explicit links to urban planning are necessary. In addition, urban universities can invest in programs in sustainability and environmental science with an emphasis on cross-disciplinary collaboration and public outreach. Research should include projects that involve participatory action research (PAR) or citizen science to better engage local citizens in areas related to conservation biology.

Lastly, prioritizing biological conservation in urban planning, with a focus on measurable targets for both biodiversity and ecosystem services (as well as human well-being), will be essential for meeting global sustainability goals as urban areas expand (Nilon et al. 2017). As this chapter reveals, after looking at just two groups—birds and bees—different communities may exhibit divergent and even contrasting responses to urban environments. As such, conservation practices in any given urban area need to be tailored to specific environmental and socioeconomic needs and priorities.

References

Alamoudi, Alaa. 2017. "Environmental Filtering of Wild Bee Assemblages in New York City Parks and Community Gardens." Master's thesis, Long Island University.

Aronson, Myla F. J., Frank A. La Sorte, Charles H. Nilon, Madhusudan Katti, Mark A. Goddard, Cristopher A. Lepczyk, Paige S. Warren, Nicholas S. G. Williams, Sarel Cilliers, Bruce Clarkson et al. 2014. "A Global Analysis of The Impacts of Urbanization on Bird and Plant Diversity Reveals Key Anthropogenic Drivers." *Proceedings of the. Royal Society. B* 281: 20133330.

Ayers, Anthony C., and Sandra M. Rehan. 2021. "Supporting Bees in Cities: How Bees are Influenced by Local and Landscape Features." *Insects* 12 (2): 128.

Behm, Jocelyn E. 2020. "Is Biodiversity Needed for Sustainability? A Spotlight on Urban Landscapes." *American Journal of Botany* 107 (5): 703–6. doi:10.1002/ajb2.1465.

Burkle, Laura A., John C. Marlin, and Tiffany Knight. 2013. "Plant-Pollinator Interactions Over 120 Years: Loss of Species, Co-Occurrence, and Function." *Science* 339 (6127): 1611–15.

Butchart, Stuart H. M., Matt Walpole, Ben Collen, Arco van Strien, Jörn P. W. Scharlemann, Rosamunde E. A. Almond, Jonathan E. M. Baillie, et al. 2010. "Global Biodiversity: Indicators of Recent Declines." *Science* 328 (5982): 1164–68. doi:10.1126/science.1187512.

Cardinale, Bradley J., J. Emmett Duffy, Andrew Gonzalez, David U. Hooper, Charles Perrings, Patrick Venail, Anita Narwani, et al. 2012. "Biodiversity Loss and Its Impact on Humanity." *Nature* 486: 59–67. doi:10.1038/nature11148.

Ceballos, Gerardo, Paul R. Ehrlich, Anthony D. Barnosky, Andres Garcia, Robert M. Pringle, and Todd M. Palmer. 2015. "Accelerated Modern Human-Induced Species Losses: Entering the Sixth Mass Extinction." *Science Advances* 1 (5): e1400253. doi:10.1126/sciadv.1400253.

Chatelain, Marion, Sylvie Massemin, Sandrine Zahn, Eliza Kurek, Ewa Bulska, and Marta Szulkin. 2021. "Urban Metal Pollution Explains Variation in Reproductive Outputs in Great Tits and Blue Tits." *Science of the Total Environment* 776: 1459663. doi:10.1016/j.scitotenv.2021.145966.

Cooper, Daniel S., Allison J. Shultz, and Daniel T. Blumstein. 2020. "Temporally Separated Data Sets Reveal Similar Traits of Birds Persisting in a United States Megacity." *Frontiers in Ecology and Evolution* 8 (July): 251. doi:10.3389/fevo.2020.00251.

Croci, Solène, Alain Butet, and Philippe Clergeau. 2008. "Does Urbanization Filter Birds on the Basis of Their Biological Traits." *Condor* 110 (2): 223–40. doi:10.1525/cond.2008.8409.

Dalton, R. 2003. "Natural History Collections in Crisis as Funding is Slashed." *Nature* 423: 575. doi:10.1038/423575a.

Delaney, Kathleen S., Seth P. D. Riley, and Robert N. Fisher. 2010. "A Rapid, Strong, and Convergent Genetic Response to Urban Habitat Fragmentation in Four Divergent and Widespread Vertebrates." *PLoS One* 5 (9): e12767. doi:10.1371/journal.pone.0012767.

Dunn, Robert R., Michael C. Gavin, Monica C. Sanchez, and Jennifer N. Solomon. 2006. "The Pigeon Paradox: Dependence of Global Conservation on Urban Nature." *Conservation Biology* 20 (6): 1814–16. doi:10.1111/j.1523-1739.2006.00533.x.

eBird. 2021. *eBird: An Online Database of Bird Distribution and Abundance [Web Application]. eBird, Cornell Lab of Ornithology*. Ithaca, NY. Accessed: June 25, 2021. www.ebird.org.

Elmqvist, T., H. Setälä, S. N. Handel, S. van der Ploeg, J. Aronson, J. N. Blignaut, E. Gómez-Baggethun, D. J. Nowak, J. Kronenberg, and R. de Groot. 2015. "Benefits of Restoring Ecosystem Services in Urban Areas." *Current Opinion in Environmental Sustainability* 14: 101–8.

Fenoglio, Maria S., Ana Calvino, Ezequiel Gonzalez, Adriana Salvo, and Martin Videla. 2021. "Urbanisation Drivers and Underlying Mechanisms of Terrestrial Insect Diversity Loss in Cities." *Ecological Entomology* 46 (4): 757–71. doi:10.1111/een.13041.

Fortel, Laura, Mickaël Henry, Laurent Guilbaud, Ann Laure Guirao, Michael Kuhlmann, Hugues Mouret, Orianne Rollin, and Bernard E. Vaissière. 2014. "Decreasing Abundance, Increasing Diversity and Changing Structure of the Wild Bee Community (Hymentopera: Anthophila) Along and Urban Gradient." *PLoS One* 9 (8): e104679. doi:10.1371/journal.pone.0104679.

Fortin, Jacey. 2020. "The Birds Are Not on Lockdown, and More People Are Watching Them." *New York Times*, May 29, 2020.

Fowler, Jarrod, and Sam Droege. 2020. "Pollen Specialist Bees of the Eastern United States." Accessed July 2021. https://jarrodfowler.com/specialist_bees.html.

Fudickar, Adam M., Timothy J. Greives, Mikus Abolins-Abols, Jonathan W. Atwell, Simone L. Meddle, Guillermo Friis, Craig A. Stricker, and Ellen D. Ketterson. 2017. "Mechanisms Associated with an Advance in the Timing of Seasonal Reproduction in an Urban Songbird." *Frontiers in Ecology and Evolution* 5: 1–13. doi:10.3389/fevo.2017.00085.

Gelby, Yigal, and Nicole Delacretaz. 2006. "Avian Window Strike Mortality at an Urban Office Building." *The Kingbird* 56 (3): 190–98.

Giraudeau, Mathieu, Hannah Watson, Daniel Powell, Orsolya Vincze, Frederic Thomas, Tuul Sepp, Beata Ujvari, Guillaume Le Loc'h, and Caroline Isaksson. 2020. "Will Urbanisation Affect the Expression Level of Genes Related to Cancer of Wild Great Tits?" *Science of the Total Environment* 714 (135793): 1–6. doi:10.1016/j.scitotenv.2019.135793./.

Gómez-Baggethun, Erik, Åsa Gren, David N. Barton, Johannes Langemeyer, Timon McPhearson, Patrick O'Farrell, Erik Andersson, Zoé Hamstead, and Peleg Kremer. 2013. "Urban Ecosystem Services." In *Urbanization, Biodiversity and Ecosystem Services: Challenges and Opportunities*, edited by T. Elqvist et al., 175–251. Dordrecht: Springer. doi:10.1007/978-94-007-7088-1_11.

Goulson, Dave, Elizabeth Nicholls, Cristina Botias, and Ellen L. Rotheray. 2015. "Bee Declines Driven by Combined Stress from Parasites, Pesticides, and Lack of Flowers." *Science* 347 (6229): 1255957. doi:10.1126/science.1255957.

Gregory, Megan M., Timothy Leslie, and Laurie Drinkwater. 2016. "Agroecological and Social Characteristics of New York City Community Gardens: Contributions to Urban Food Security, Ecosystems Services, and Environmental Education." *Urban Ecosystems* 19 (2): 763–94. doi:10.1007/ s11252-015-0505-1.

Hall, Damon M., Gerardo R. Camilo, Rebecca K. Tonietto, Jeff Ollerton, Karin Ahrne, Mike Arduser, John Ascher, et al. 2016. "The City as a Refuge for Insect Pollinators." *Conservation Biology* 31 (1): 24–29.

Hall, Damon M., and Dino J. Martins. 2020. "Human Dimensions of Insect Pollinator Conservation." *Current Opinion in Insect Science* 38: 107–14. doi:10.1016/j.cois.2020.04.001.

Hernandez, J. L., G. W. Frankie, and R. W. Thorp. 2009. "Ecology of Urban Bees: A Review of Current Knowledge and Directions for Future Study." *Cities and the Environment* 2 (1): Article 3.

IPBES. 2019. *Global Assessment Report on Biodiversity and Ecosystem Services of the Intergovernmental Science-Policy Platform on Biodiversity and Ecosystem Services*. Edited by Eduardo S. Brondizio, Sandra Díaz, and Josef Settele, 1148. Bonn, Germany: IPBES Secretariat. doi:10.5281/zenodo.3831673.

Jordan, Alex, Harland M. Patch, Christina M. Grozinger, and Vikas Khanna. 2021. "Economic Dependence and Vulnerability of United States Agricultural Section on Insect-Mediated Pollination Service." *Environmental Science and Technology* 55 (4): 2243–53.

Kim, Ke Chung, and Loren B. Byrne. 2006. "Biodiversity Loss and the Taxonomic Bottleneck: Emerging Biodiversity Science." *Ecological Research* 21: 794. doi:10.1007/s11284-006-0035-7.

Klein, Alexandra-Maria, Bernard E. Vaissiere, James H. Cane, Infolf Steffan-Dewenter, Saul A. Cunningham, Claire Kremen, and Teja Tscharntke. 2007. "Importance of Pollinators in Changing Landscapes for World Crops." *Proceedings of the Royal Society B* 274 (1608): 303–13.

Koh, I., E. V. Lonsford, N. M. Williams, C. Brittain, R. Isaacs, J. Gibbs, and T. H. Ricketts. 2016. "Modeling the Status, Trends, and Impacts of Wild Bee Abundance in the United States." *Proceeding of the Natural Academy of Sciences* 113 (1): 140–45.

Kratschmer, Sophie, Monkia Kriechbaum, and Bärbel Pachinger. 2018. "Buzzing on Top: Linking Wild Bee Diversity, Abundance, and Traits with Green Roof Qualities." *Urban Ecosystems* 21: 429–46.

Kucera, Abigail J., Megan C. Smith, Gigi S. Wagnon, Caleb Cantwell, Andrew Eaton-Clark, Britney Elliott, Lara Ghanem, Eric Ko, Melissa Waggener, and Charles R. Brown. 2019. "Group Size and Singing Phenology of Northern Cardinals (*Cardinalis cardinalis*) Varies Between Urban and Rural Habitats." *Southwestern Naturalist* 64 (1): 56–60.

La Sorte, Frank A., Myla F. J. Aronsonb, Christopher A. Lepczyk, and Kyle G. Horton. 2020. "Area Is the Primary Correlate of Annual and Seasonal Patterns of Avian Species Richness in Urban Green Spaces." *Landscape and Urban Planning* 203: 1–10. doi:10.1016/j.landurbplan.2020.103892.

La Sorte Frank A., and Kyle G. Horton. 2021. "Seasonal Variation in the Effects of Artificial Light at Night on the Occurrence of Nocturnally Migrating Birds in Urban Areas." *Environmental Pollution* 270: 116085. doi:10.1016/j.envpol.2020.116085.

Lepczyk, Christopher A., Frank A. La Sorte, Myla F. J. Aronson, Mark A. Goddard, Ian MacGregor-Fors, Charles H. Nilon, and Paige S. Warren. 2017. "Global Patterns and Drivers of Urban Bird Diversity." In *Ecology and Conservation of Birds in Urban Environments*, edited by Enrique Murgui and Marcus Hedblom, 13–33. Boston: Springer. doi:10.1007/978-3-319-43314-1_2.

Matteson, Kevin C., John S. Ascher, and Gail A. Langellotto. 2008. "Bee Richness and Abundance in New York City Urban Gardens." *Annals of the Entomological Society of America* 101 (1): 140–50.

Matteson, Kevin C., and Gail A. Langellotto. 2009. "Bumblebee Abundance in New York City Community Gardens: Implications for Urban Agriculture." *Cities and the Environment* 2 (1): Article 5. doi:10.15365/CATE.2152009.

———. 2010. "Determinates of Inner City Butterfly and Bee Species Richness." *Urban Ecosystems* 13: 333–47. doi:10.1007/s11252-010-0122-y.

McIntyre, Nancy. 2000. "Ecology of Urban Arthropods: A Review and a Call to Action." *Annals of the Entomological Society of America* 93 (4): 825–35.

McKinney, Michael L. 2002. "Urbanization, Biodiversity and Conservation: The Impacts of Urbanization on Native Species Are Poorly Studied but Educating a Highly Urbanized Human Population about These Impacts Can Improve Species Conservation in All Ecosystems." *BioScience* 52 (10): 883–90.

———. 2006. "Urbanization as a Major Cause of Biotic Homogenization." *Biological Conservation* 127 (3): 247–60. doi:10.1016/j.biocon.2005.09.005.

Michener, C. 2007. *The Bees of the World*. Baltimore, MD: The John Hopkins University Press.

Millennium Ecosystem Assessment. 2005. *Ecosystems and Human Well Being: Synthesis*. Washington, DC: Island Press.

Nilon, Charles H., Myla F. J. Aronson, Sarel S. Cilliers, Cynnamon Dobbs, Lauren J. Frazee, Mark A. Goddard, Karen M. O'Neill, et al. 2017. "Planning for the Future of Urban Biodiversity: A Global Review of City-Scale Initiatives." *BioScience* 67 (4): 332–42.

Oke, Cathy, Sarah A. Bekessy, Niki Frantzeskaki, Judy Bush, James A. Fitzsimons, Georgia E. Garrard, Maree Grenfell, et al. 2021. "Cities Should Respond to the Biodiversity Extinction Crisis." *npj Urban Sustainability* 1: 11. doi:10.1038/s42949-020-00010-w.

Okoniewski; Joseph C., Christine VanPatten; Ashley E. Ableman; Kevin P. Hynes; Angela L. Martin, and Peter Furdyna.2021. "Anticoagulant Rodenticides in Red-Tailed Hawks (*Buteo jamaicensis)* from New York City, New York, USA, 2012–18." *Journal of Wildlife Diseases* 57 (1): 162–67. doi:10.7589/JWD-D-19–00003.

Ollerton, Jeff, Racheal Winfree, and Sam Tarrant. 2011. "How Many Flowering Plants are Pollinated by Animals?" *Oikos* 120 (3): 321–26. doi:10.1111/j.1600-0706.2010.18644.x.

Parkins, Kaitlyn L, Susan B. Elbin, and Elle Barnes. 2015. "Light, Glass, and Bird—Building Collisions in an Urban Park." *Northeastern Naturalist* 22 (1): 84–94.

Pereira, Henrique M., Laetitia M. Navarro, and Ines S. Martins. 2012. "Global Biodiversity Change: The Bad, the Good, and the Unknown." *Annual Review of Environment and Resources* 37: 25–50. doi:10.1146/annurev-environ-042911-093511.

Pollinator Network @ Cornell. 2021. "Species List—Bees of New York." Accessed July 2021. https://pollinator.cals.cornell.edu/wild-bees-new-york/species-list-bees-new-york/.

Riding, Corey S., Timothy O'Connell, and Scott R. Loss. 2020. "Building Façade-Level Correlates of Bird-Window Collisions in a Small Urban Area." *The Condor: Ornithological Applications* 122 (1): 1–14. doi:10.1093/condor/duz065.

Rochat, E., S. Manel, M. Deschamps-Cottin, I. Widmer, and S. Joost. 2017. "Persistence of Butterfly Population in Fragmented Habitats Along Urban Density Gradients: Motility Helps." *Heredity* 119: 328–38. doi:10.1038/hdy.2017.40.

Rodewald, Amanda D., and Peter Arcese. 2017. "Reproductive Contributions of Cardinals Are Consistent with a Hypothesis of Relaxed Selection in Urban Landscapes." *Frontiers in Ecology and Evolution* 5 (77): 1–7. doi:10.3389/fevo.2017.00077.

Rodewald Amanda D., and Stanley D. Gehrt. 2014. "Wildlife Population Dynamics in Urban Landscapes." In *Urban Wildlife Conservation: Theory and Practice*, edited by Robert A. McCleery, Christopher E. Moorman, and M. Nils Peterson, 117–47. Boston: Springer. doi:10.1007/978-1-4899-7500-3_8.

Rosenberg, Kenneth V., Adriaan M. Dokter, Peter J. Blancher, John R. Sauer, Adam C. Smith, Paul A. Smith, Jessica C. Stanton, Arvind Panjabi, Laura Helft, Michael Parr, and Peter P. Marra. 2019. "Decline of the North American Avifauna." *Science* 366 no. 6461: 120–24. doi:10.1126/science.aaw1313.

Russ, Anja, Terézia Lučeničová, and Reinhard Klenke. 2017. "Altered Breeding Biology of The European Blackbird Under Artificial Light at Night." *Journal of Avian Biology* 48: 1114–25. doi:10.1111/jav.01210.

Sanderson, Eric. 2009. *Manahatta: A Natural History of New York City*. New York: Abrams.

Seto, Karen C., Burak Güneralp, and Lucy R. Hutyra. 2012. "Global Forecasts of Urban Expansion to 2030." *Proceedings of the National Academy of Sciences* 109 (40) (October): 16083–88. doi:10.1073/pnas.1211658109.

Shustack, Daniel P., and Amanda D. Rodewald. 2010. "Attenuated Nesting Season of The Acadian Flycatcher (*Empidonax Virescens*) in Urban Forests." *The Auk* 127 (2): 421–29.

Sirakaya, Aysegül, An Cliquet, and Jim Harris. 2018. "Ecosystem Services in Cities: Toward the International Legal Protection of Ecosystem Services in Urban Environments." *Ecosystem Services, Part B* 29: 205–12.

Sol, Daniel, Christopher Trisos, Cesc Múrria, Alienor Jeliazkov, Cesar González-Lagos, Alex L. Pigot, Carlo Ricotta, Christopher M. Swan, Joseph A. Tobias, and Sandrine Pavoine. 2020. "The Worldwide Impact of Urbanisation on Avian Functional Diversity." *Ecology Letters* 23 (6): 962–72. doi:10.1111/ele.13495.

Solonen, Tapio. 2001. "Breeding of the Great Tit and Blue Tit in Urban and Rural Habitats in Southern Finland." *Ornis Fennica* 78: 49–60.

Spotswood, Erica N., Erin E. Beller, Robin Grossinger, J. Letitia Grenier, Nicole E. Heller, and Myla F. J. Aronson. 2021. "The Biological Deserts Fallacy: Cities in Their Landscapes Contribute More Than We Think to Regional Biodiversity." *Bioscience* 71 (2): 148–60. https://academic.oup.com/bioscience/article/71/2/148/6102678.

Stracey, Christine M., and Scott K. Robinson. 2012. "Are Urban Habitats Ecological Traps for a Native Songbird? Season-Long Productivity, Apparent Survival, and Site Fidelity in Urban and Rural Habitats." *Journal of Avian Biology* 43 (1): 50–60. doi:10.1111/j.1600-048X.2011.05520.x.

Sultana, Marufa, Luca Corlatti, and Ilse Storch. 2021. "The Interaction of Imperviousness and Habitat Heterogeneity Drives Bird Richness Patterns in South Asian Cities." *Urban Ecosystems* 24 (3): 35–344. doi:10.1007/S11252-020-01037-8.

Townsend, Andrea K., Hannah A. Staab, and Christopher M. Barker. 2019. "Urbanization and Elevated Cholesterol in American Crows." *Condor* 121: 1–10 doi:10.1093/condor/duz040.

United Nations, Department of Economic and Social Affairs, Population Division. 2019. *World Urbanization Prospects: The 2018 Revision, Highlights*. (ST/ESA/SER.A/421). New York: United Nations.

U.S. Department of the Interior, U.S. Fish and Wildlife Service, and U.S. Department of Commerce, U.S. Census Bureau. 2018. "2016 National Survey of Fishing, Hunting, and Wildlife-Associated Recreation." www.census.gov/content/dam/Census/library/publications/2018/demo/fhw16-nat.pdf.

van der Plas, Fons. 2019. "Biodiversity and Ecosystem Functioning in Naturally Assembled Communities." *Biological Reviews* 94 (4): 1220–45. doi:10.1111/brv.12499.

van Doren, Benjamin M., Kyle G. Horton, Adriaan M. Doktera, Holger Klincke, Susan B. Elbin, and Andrew Farnsworth. 2017. "High-Intensity Urban Light Installation Dramatically Alters Nocturnal Bird Migration." *Proceedings of the National Academy of Sciences* 114 (42): 11175–80.

van Englesdorp, Dennis, Jay D. Evans, Claude Saegerman, Chris Mullin, Eric Haubruge, Bach Kim Nguyen, Maryann Frazier, et al. 2009. "Colony Collapse Disorder: A Descriptive Study." *PLoS One* 4 (8): e6481. doi:10.1371/journal.pone.0006481.

Walters, Mitchell J., Robert P. Guralnick, Nathan J. Kleist, and Scott K. Robinson, 2019. "Urban Background Noise Affects Breeding Song Frequency and Syllable-Type Composition in the Northern Mockingbird." *Condor* 121: 1–13. doi:10.1093/condor/duz002.

Watson, Hannah, Daniel Powell, Pablo Salmón, Arne Jacobs, and Caroline Isaksson. 2021. "Urbanization Is Associated with Modifications in DNA Methylation in a Small Passerine Bird." *Evolutionary Applications* 14: 85–98. doi:10.1111/eva.13160.

Wenzel, A., Ingo Grass, Vasuki V, Belavadi, and Teja Tscharntke. 2020. "How Urbanization is Driving Pollinator Diversity and Pollination—A Systematic Review." *Biological Conservation* 241: 108321. doi:10.1016/j.biocon.2019.108321.

Whelan, Christopher J., Daniel G. Wenny, and Robert J. Marquis. 2008. "Ecosystem Services Provided by Birds." *Annals of the New York Academy of Sciences* 1134: 25–60. doi:10.1196/annals.1439.003.

Williams, Nicholas, S. G., Mark W. Schwartz, Peter A. Vesk, Michael A. McCarthy, Amy K. Hahs, Steven E. Clemants, Richard T. Corlett, et al. 2009. "A Conceptual Framework for Predicting the Effects of Urban Environments on Floras." *Journal of Ecology* 97 (1): 4–9. doi:10.1111/j.1365-2745.2008.01460.x.

Zhao, Xuebing, Min Zhang, Xianli Che, and Fasheng Zou. 2020. "Blue Light Attracts Nocturnally Migrating Birds." *Condor* 122: 1–12. doi:10.1093/condor/duaa002.

7

ACTION RESEARCH IN A TOKYO SUBURB

Building Community Ties Through Collaborative Landscaping Projects

Mariia Ermilova, Mitsunari Terada, and Isami Kinoshita

Introduction

In this chapter, we present a case study from the greater Tokyo area that shows how growing plants can facilitate communication between generations and cultures within an urban neighborhood. The neighborhood, Iwase, is located in a suburban commuter city of Matsudo, which was developed in several stages after World War II, from paddy fields to detached housing in a residential area. Our study contains two interrelated narratives. The first concerns neighborhood redevelopment and destruction of green commons and how this can upset the fragile balance of a community. The second narrative is about the alternative ways of participating in a local community and, in particular, empowering women and children to interact in ways different from conventional male-oriented models. In addition, this chapter discusses how "outsiders," such as university students and foreign residents, can facilitate these democratic and ecologically beneficial processes.

Digging Hidden Stories

Our project consisted of university–community collaboration. As two graduate students from the Chiba University Graduate School of Horticulture who were interested in urban planning, we embedded ourselves in the community by living in a building owned by the local neighborhood association, called the Iwase Neighborhood Association (INA), consisting of 630 households. We worked with citizens to understand local issues and then, using our background in sociology, ecology, and horticulture, tried to improve conditions through collaborative, community-led greening efforts.

DOI: 10.4324/9781003217442-8

First, we observed the community life for approximately a year. Conversations with citizens happened while meeting on official occasions like regular cultural and management events held by the neighborhood association or casually when crossing paths on the street. From observations and preliminary interviews, we found that the INA was quite active but mainly consisted of elderly men participating in and managing events. We observed low participation levels among people who live close to the INA building and pondered the reason for this. Among the other local issues that emerged in these discussions were heavy car traffic on the street, disunity of residents in terms of generations and family structure, and the fear of strangers and passersby as the community is located close to the train station and is used as a transit way. We observed a lack of identity uniting tradition or place, which people would care about or would be proud of, as Iwase had withdrawn from the traditional Japanese territorial Shrine celebrations decades ago and performed only secular seasonal celebrations.

To approach these issues and discuss possibilities for collaboration, we held participatory mapping workshops with residents and individual interviews with them about the area's past, present, and future. Our project focused on the ordinary life functions of a community and about maintaining a healthy environment at the microlevel, encouraging residents to engage in self-help activities.

We held the first workshop utilizing "Gulliver's Map" (Figure 7.1), introduced by Nakamura (1989, 511). For this workshop, a large map of the area was printed and assembled on the floor. Participants walking on the paper streets of their city felt "like Gulliver." Using small color-coded sticky notes, every participant ($N = 30$) marked their house, the places they liked the most, the places they had a memory of, the dangerous/unpleasant spots, and natural/green areas that they recognized as valuable. We mapped 52 positive comments, 24 negative comments, and 64 greenspaces of various sizes in the area. Participants were then encouraged to discuss each of the places they marked, their perception of community in terms of past versus present, and their hopes for the future of the area. Comments were recorded to allow us to evaluate how perceptions of the neighborhood differ geographically and may have changed over time.

The workshop participants were elderly members of the community (from 65 and older) who lived there over a long time period. Only 33% of the positive comments related to the current period, while 80% of the negative comments were about current conditions in the area. Similarly, the present moment layer was full of dangerous and unpleasant spots. At the time, we realized that such observations might vary for different age groups, which could strongly influence the agenda. This difference also explains why one should actively involve younger generations in the planning process (Chawla 2002, 13). Therefore, we organized additional workshops, which included children. After inputting the data from the handwritten notes on the wide paper "Gulliver Map" into GIS, we realized that the slope area on the south side of the valley collected the most positive comments (Figure 7.2). However, these positive memories belonged to the past. Such results

FIGURE 7.1 Process of the Gulliver Map workshop, in which we listened to the citizens about the place where they live.

Source: Photograph by the author.

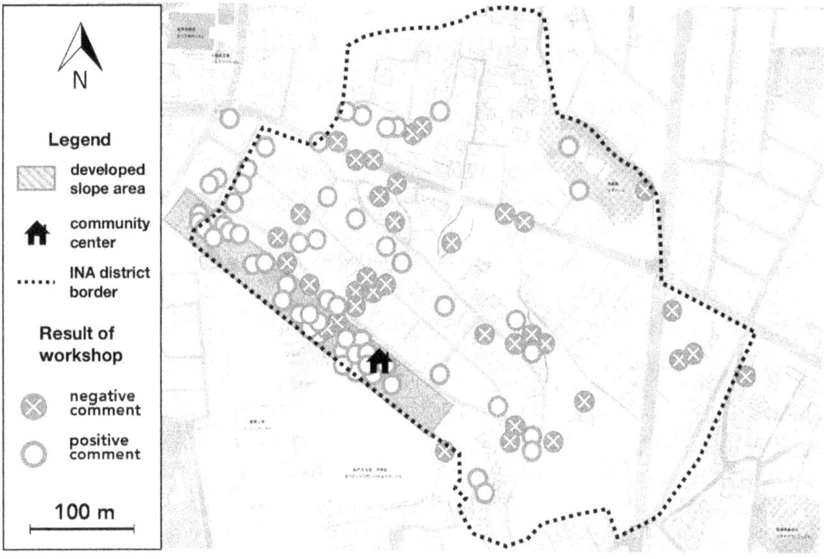

FIGURE 7.2 Results of Gulliver Map workshop, pointing out the significance of the valley's green slopes for the local people.

Source: Image created by the author.

prompted us to explore local history through in-depth, face-to-face conversations with the locals, using an ethnographic method. We organized individual hearings several times and invited people to come and share the past, recording them on a voice recorder. Overall, we had 18 in-depth interviews with the people of the area, which were transcribed. However, having lived in the place for five years, we had countless encounters and conversations with people on the street, in the park, at the garbage can (managed by a group of houses, where neighbors unwittingly meet and engage in small talk), on the road, and during the maintenance of the gardens. At some point, it was difficult to separate life from the research process.

Iwase was a village in the *yatsu* valley (a narrow landform with rice paddies) for over 300 years. From local residents, we learned that the valley slopes originally had numerous water springs. Water from the springs was first used for rice paddies, and after the removal of the rice paddies, water in the area was used for washing vegetables, drinking, and recreation. Covered by forests and grass, the slopes also provided plants, soil resources, and a place for annual community events and daily events. However, currently, the land on the slopes mostly belongs to a private university and is therefore gated, not available for use. Local people who grew up in this area (30% of the total population of Iwase) felt the loss of their natural commons and talked about it with nostalgia.

We gathered further details about the area through conversations with the residents living close to the slope. In the 2000s, the mountain slope was gradually redeemed and rebuilt; this was done in stages, first building the Private University, then the Elderly House, each time with considerable negative impacts, both social and ecological. When land development first became a topic of discussion, INA members stated that the community supported the idea of retaining the land. However, when developers offered substantial compensation in exchange for the land, many community members changed their minds, resulting in the sale of their land and migration to another area. This incident undermined the trust among people in the neighborhood. Although those living in the immediate vicinity were compensated for this disturbance, households located at a distance from the future building were not, even though they faced inconvenience resulting from the impact of construction on the landscape. Such unequal benefits to the community members resulted in alienation.

Owing to the construction, the natural flow of water to the valley was disturbed, causing subsidence of the soil and damage to the walls and foundations of the residential buildings. Furthermore, the height of the newly constructed buildings blocked the antenna signal for television, and residents in the valley had to solve this problem on their own. This caused excessive discontent among them. Although a solution was offered with the introduction of cable TV, the company providing this service required a sum of money to create the infrastructure. At the time, a group of residents came together and covered the costs. After several years, the same company offered its services without charging for infrastructure to all the remaining community members, which caused outrage too. All along, the Neighborhood Association did not provide substantial support to citizens against developers. Gradually, the feelings of "betrayal," "conceding defeat," "cannot be helped," and "I do not want to talk about it" accumulated and soured the hearts of people living in the same neighborhood.

We conceptualized this narrative as "collective trauma," which has been applied in the context of natural, human disasters and cultural studies (Perry 2012, 81–85). Usually, it is associated with the negative effect of development on a bigger scale—such as a river basin. Even though our case unfolded at a smaller level in the urban community, we still describe it as "collective trauma." The trauma resulted from the loss of trust, declining community connectedness caused by the invasive development, and inconsistent support of INA.

It took time and effort to learn this history, as people were reluctant to return to those times in their memory. We literally rebuilt the puzzle piece by piece—a picture of the events of those years. So the "disunity" in the area had a real reason, even if it was decades old. Then we thought of how green activities can help.

Plants, Aliens, and Alternative Ways of Participation

Gardens are known to facilitate intercultural communication, for example, helping to build bridges of understanding between immigrants and local people (Francis and Hester 1990, 160, 166). Community gardens are recognized as able to foster the social capital of the area.

Isami Kinoshita's lab at Chiba University Graduate School of Horticulture was actively involved with local communities in Matsudo. Professor Kinoshita advised citizens, sometimes using the city as a classroom for his students. One of the lab members, Mitsunari Terada, researching child-friendly planning, started communication with children in Iwase and was asked by NA leaders to organize activities to attract children and support multigenerational communication. The Iwase NA leader asked if Mitsunari would like to live in the empty rooms of the association's building. Mitsunari invited Mariia Ermilova to join the project as an ecologist, and they moved in March 2016.

For one of us—Mariia Ermilova—being Russian, a foreigner, this opened channels of communication with community members who were curious about different cultures. The "outsider" position thus made it easier for her to engage community members who were eager to talk to a foreigner and wanted to help her understand their language and culture.

Having collected herbs for tea with her mother from gardens and in the wild, she was able to interact meaningfully with community members, asking about Japanese plants and sharing her own herbal heritage. Thus her foreigner/outsider/exchange student status, as well as her exposure to urban agriculture, partly helped spark interest among various citizen groups in the revitalization project.

In 2016, our laboratory team started an EDIBLE WAY (Eguchi et al. 2020, 2183) project, where we grow vegetables on the street in felt planters, creating an edible landscape to facilitate communication among the locals close to Chiba University campus. An edible landscape approach was inspired by various projects abroad, including Incredible Edible Todmorden[1] and Japanese potted gardening. Together with the local women in the Iwase area, we discussed the type of plants that should be grown to gain people's interest. We aimed to involve local women in the community design. They advised that the cultivation of herbs, with distinctive aromas, like rosemary, lavender, and mint would fit best. These herbs are associated with Western cuisine, tea, aromatherapy and cosmetics, and the culture in general. For example, the woman leader in our project said she had read *Anne of the Green Gables* and imagined a European garden landscape as her ideal garden, so she was inspired by foreign literature in her own herbal garden pursuit.

Given the popularity of herbal therapy and tea in Japan, where overworking is a widespread malady, our focus on herbs contributed to creating a peaceful, relaxed atmosphere. Moreover, planting and growing various herbs stimulated the local creative production of handmade crafts and cosmetics with herbal ingredients. In addition, herbs are easy to manage, and the plant material can be rapidly increased to share with others; thus it is ideal for neighborhood social network creation.

We negotiated to spread the EDIBLE WAY planters in the Iwase area as a distinct neighborhood group. As a neighborhood research group, we assumed we needed to build strong relations in this area, and therefore the name of the district, Iwase, was printed on the planter bag. We filled felt planters with herbs, aiming for more diverse use (other than edible) of the plants, and labeled this EDIBLE WAY chapter the Iwase Herb Project. We proceeded with a series of workshops in which local citizens created their own planters by printing and listening to instructions on growing and using herbs, while socializing and drinking herbal tea. Using the Neighborhood Association's information resources and word of mouth, planters were distributed among citizens who wished to join the project. To be a part of the project, they just have to place a planter in front of their home and care for the plants. Since launching the project, about 100 planters were managed by citizens in the area (Figure 7.3). Any further involvement, like attending events, was voluntary. Even though potted gardening is popular in Japan, felt planters sharing the same color and logo stood out on the street landscape, creating an image of connectedness. Edible plants and the necessity of owners to go out on the street for management facilitated spontaneous conversations (Figure 7.5) (Eguchi et al. 2020, 2183). We hoped planters would serve not only as mere beautification but also as an indicator of "a healthy community"—a crime preventive measure.

In one year, the activity grew from planters to community gardens and reached various stakeholders. In particular, local women organized two independent groups to support two generations of the community: the elderly and children. These groups developed their own curricula based on greening activities, planters, and gardens.

The curricula of the younger women's group, ASUKEN (an abbreviation of Afterschool Research Group in Japanese), consisted of activities that used locally grown Western herbs to attract young mothers. Their community practice focused

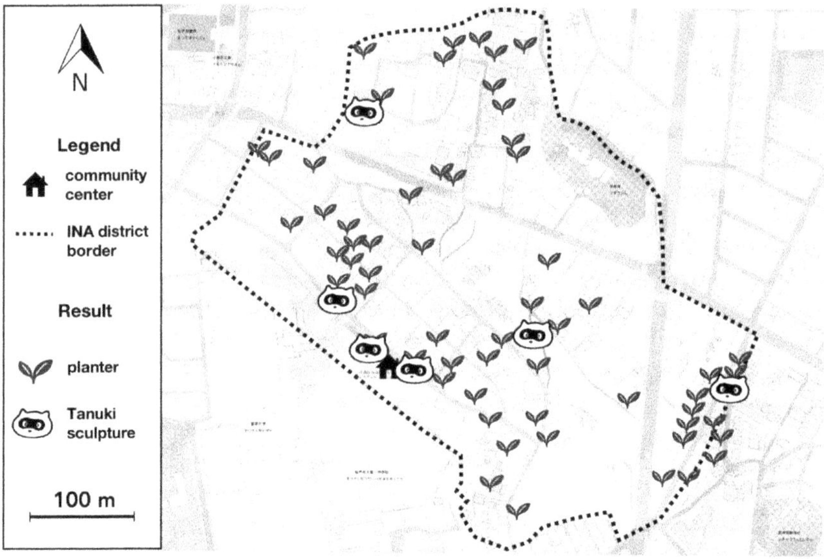

FIGURE 7.3 Distribution of planters and handmade sculptures of Tanuki mascot in the area.

Source: Image created by the author.

FIGURE 7.4 During the spring Sakura festival, women organize the Easter egg hunt, hiding eggs in Iwase Herb Project's planters and private gardens of local people, who welcome and guide kids.

Source: Photograph by the author.

FIGURE 7.5 Black felt planters with edible herbs, flowers, vegetables, and other culturally significant plants, spread around in an Iwase neighborhood as the EDIBLE WAY project chapter, facilitate conversations on the streets.

Source: Photograph by the author.

on gaining knowledge of Western herbs, making cosmetic products, aromatherapy, and gardening technology, while also supporting children's play and general community well-being. They introduced a new annual children's event, the Easter egg hunt[2] (Figure 7.4), which utilized the planters' network and citizens' gardens to hide the eggs for local children to find. They also created innovative products, such as the mix of dokudami and lavender herb essences for cosmetic use. They also held clay sculpture workshops, where Tanuki mascot sculpture lamps were created, and placed them in the gardens. Tanuki, or raccoon dogs, are seen in the area, and it was a way of celebrating the connection with the local wilderness, or "sacredness" (Hester 2010, 64).

From March 2017 to June 2018, we held meetings with the local women's group almost every week. These weekly meetings created a space for any community member to join, a kind of third space,[3] especially for children. Children could play in the territory near the community center. We also discussed the development of the Herb Project and other activities in Iwase.

After the distribution of felt planters, we were able to attract the attention of the collaborators in the community who owned the land and start work on a larger landscape scale. It was successful compared to previous Neighborhood Association performances in connecting to people in the area.

In sum, we created many options for participation, leaving people free to do what they liked best. They could grow herbs and flowers in a bag with the project logo, and it counted as a contribution. If they preferred, they could skip meetings and socializing. In a modern society where people strive to protect their privacy and independence, this degree of flexibility is a great advantage and attracts supporters.

Planting Seeds of Conversation

In the years following the summer of 2017, we implemented a weeklong workshop on practical design that met three times in August of 2017, 2018, and 2019. This workshop is called Cross-Cultural Design Collaboration (CCDC). The CCDC is a summer program hosted by Chiba University, which brings landscape architectural students together from around the world to create practical design projects in an international environment. It features design/built service learning (Porpora 1999, 121), an activity where students produce output upon community requests, thus combining societal contribution and learning (Winterbottom 2011, 189). In three years, we had 85 students from Japan, China, Switzerland, the UK, Germany, Malaysia, Thailand, the United States, and Russia, working in three teams at three locations. During the studio, local children joined the design/built process and communicated with foreigners.

In these three years, the students created a community garden with native plant cover and herbs, which became the basis for the activity of the Genki Ouen (Support of Elderly) women's group. They also created a native plant garden in front of the NA center and updated the Minnanoniwa community garden. Additionally, the impressive work was done in a private forest located on the valley slope, where students under the guidance of Associate Professor Ryosuke Shimoda made a selective "landscape elements elimination with reuse": the concrete pillars from the barbed wire fence have been moved from vertical to horizontal, becoming steps on the slope; cutting dense thickets of bamboo obtained material to make benches, chairs, a hut, lanterns, bamboo wind chimes, and bamboo wicker partitions in a traditional style. Students especially appreciated learning the bamboo crafts from the Chiba University gardener, Mr. Yamazaki, and the head of Nakada Seeds and Nursery Ltd., Mr. Nakada. As a result, dense dark forest became open and bright, upgraded with play and leisure objects.

Several gardens with culturally significant native plants were created in the area, thanks to Nakada Seeds and Nursery Ltd., a landscape company pioneering in designing native flower "mats" in Japan. Newly planted native flowers attracted local elderly citizens, especially women. Dozens of them stopped by and told me that the plants look nostalgic, like those they had seen as children in the countryside. Many of these flowers appear in Japanese folklore and classical poems, such as "Collection of Ten Thousand Leaves," "The Tale of Genji," etc. Typically educated in poetry, tea ceremony, and flower arrangement, this generation of women was drawn to the flowers in front of the community building where the gardens are located and often stopped to chat about them.

The Genki Ouen (Support of Elderly) Suzuki Herb Garden group developed their curricula, which consisted of gardening activities every weekend, socializing, visiting outside facilities based on the interest in herbs, and learning about native flowers. This group emphasized the process of increasing their knowledge of plants and gardening as a way to improve their mental and physical health and to socialize (loneliness is an acute problem in Japan). Such activities are in keeping with the Matsudo city government's active promotion of preventive medicine in a general framework of care for Japan's aging population.

Throughout the project, we played the role of coordinators (Figure 7.6), who provided meaningful interventions in the community, enabled by careful prior research and negotiation (Ermilova et al. 2018, 25–34) with various groups like ASUKEN, Genki Ouen Club, and others.

We found that, as students, we possessed the "power" of being in a "weak position" (lacking any official job or status that might be perceived as overwhelming, students are persons who need help). Local people treated us kindly and felt free to confide in us without stress or fear of being judged, after overcoming the first surprise at our being there, living together in the NA building. In this respect, bringing students to a community through a financial aid program can help bridge gaps between elderly executives, local citizens, nonassociation members, and younger generations, including children, who usually do not take part in community planning.

Conclusion

Our residence as graduate students in the community building can be seen as a form of use of vacant housing and land for community revitalization. This approach has been applied in Japan in response to depopulation trends that have resulted in a growing number of vacant houses and land in the countryside and regional cities. However, this is the first case in Japan of using the Neighborhood Association building manager's room for the researchers' residence. First, not every NA has its own building and a manager's room (*kanrininshitsu*) in it. Second, the level of activity of NA may vary upon the area, and Iwase is an active one, with the leaders being companies' retirees (*salary men*), highly educated, motivated, and open to new approaches. The fact that this is an unusual confluence of possibilities does not prevent us from extrapolating an approach with small iterations in the set of initial elements: student-researchers, housing owners, and active community members, all wishing to improve conditions.

We emphasize this residence-based action research through collaboration between students and local people as a socially transformative inquiry and design method for urban environments. This combination of simply living in the area and doing action research allowed us to apply academic knowledge and resources to a long-term community-based landscape project that engaged local people in codesigning greenspaces. As doctoral student researchers, we spent five years in Iwase, side by side with community members transforming landscapes, little by little creating visible change.

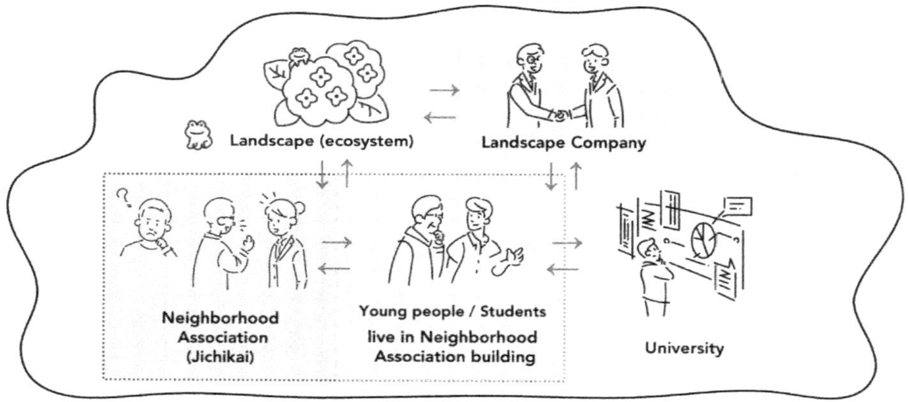

	Landscape	Neighborhood Association	Students / Young people	University	Company
Role	Platform for involvement, visible changes, facilitating communication	Local governance, actors, co-creators	Intergenerational communication facilitators, coordinators, counselors, co-creators	Education/Research	Provide material (product)
Needs/ Problems	Degrading ecosystems, need management for productive functioning	Aging society (mental and physical health, lack of manpower in area management), generation gap	Financial struggle paying university fee, housing fee. Need to gain skills, third place outside of home or school	Meaningful research and education. Practical contribution to society	Advertise an innovative landscape product, which restore ecosystems by use of native plants
Project		Students live in Neighborhood Association building	Students live in Neighborhood Association building	Cross-cultural design collaboration landscape studio	Cross-cultural design collaboration landscape studio
	↓	↓	↓	↓	↓
Output	Functioning as a habitat for all living beings, provide ecosystem services	Quality of life, health (socialization, safety, citizenship etc.)	Skills Research data Housing fee covered Social and cultural capital	Education, research, international exchange	Developing and testing product Corporate social responsibility

FIGURE 7.6 Scheme of the landscape-based collaboration in the Iwase Neighborhood Association; roles and value outputs for different stakeholders of the project.

Source: Image created by author.

The greatest challenge for implementing local grassroots landscape projects is environmental management. Aging residents have difficulty cutting tree branches in their gardens. Local communities motivated by a sense of duty are weakening and cannot gather people for weeding neighborhood park as easily as before. Greenspace management is not only a problem in urban areas, which still contain private forests and gardens, but also in rural areas in association with population decline. Researchers report a declining trend in traditional knowledge of forest management across generations. An individually oriented and busy working younger generation often eschews traditional community activities, which may feel obligatory and burdensome. For example, a community garden that existed for nearly 15 years close to Chiba University campus was closed in 2019 because all the managers were aging, and no one stepped up to replace them. Rather exceptionally, some urban citizens in Japan strive to return to their past agricultural experience in a new form (Terada, Yokohari, and Amemiya 2016, 164), producing alternative lifestyles like being half-farmer/half-professional. However, these models are still uncommon.

In our case, we tried to prevent our project from being unrealistic. For that, we made sure that there were locals who were interested in initiating and continuing the work. Importantly, we used materials for which the management effort would be minimal: native plants, which do not need extensive watering, and a stone garden with landscape fabric, which is easy to weed.

In the eyes of citizens, NA is responsible for neighborhood order. However, recently, NA has lacked popularity due to its old-style management/participatory process. Participation in any local community NA, regardless of gender, still involves attending long meetings of elected neighborhood representatives where leaders talk about directives from the government, reports on the budget, etc. On the festival days, it usually means physical work without the opportunity to lead the initiative. Our study led to the discovery of alternatives to the traditional participation of women in local communities: using the Herb Project, women created their own events and products and gained recognition for their contribution in the form of greens.

But much work remains to be done to improve conditions. In many Japanese communities with a patriarchal structure, it is still common to think that a woman's place is in the kitchen. During festivities, they eat in the kitchen, while guests and men enjoy themselves outside. These things drive the younger generation away. In the case of modern INA, located in the Matsudo city center, with a high standard of living, education, and culture, the situation differs in a good way. For that reason, it was beyond our study to address issues of the most vulnerable women in society.

Many working-age people could not join community activities due to work and family responsibilities. However, some participated in our project because it was as simple as just individually growing herbs by their front doors and sometimes attending the events. Two working-age project leaders from INA participated by planning, organizing events and daily activities, as well as reflecting together, making up a research team.

Busy "working" at school and private lessons until evening, children, too, are less involved in local communities. One advantage of our study is that university students engaged with us and assumed the role of "older friends" or "counselors" to children and involved them playfully in landscape work.

At the other end of the spectrum, an aging society, in our case, provided resources in the form of highly qualified community members. Those with time and energy achieved outstanding results in gardening for the common good; at the same time, they alleviated their loneliness. As doctoral students, we helped bridge generation gaps and sensitively mediated the process.

Landscape-based projects can provide direct benefits by producing raw materials for consumption, but considering the size of the planters we had, this was secondary to the social benefits. Participants reported an increased feeling of safety after knowing more people in the area through herbal endeavors.

In our partnership with the Iwase Neighborhood Association, local women took the opportunity to improve their well-being and organize "communities of practice" as participant researchers utilizing the vacant land and planters to cultivate plants. They addressed issues of alienation in their community and helped people participate in a variety of visible ways, including those residents who were not members of the Neighborhood Association or were excluded from the process (children). They planted seeds of conversations all along community roads. As doctoral students, we facilitated research and coordination and offered sustainable solutions like native plant selections that contribute to a healthy ecosystem and simplify greenspace management. We hope this mostly informal, noninstitutionalized student-community action research project can serve as a model for university–community partnerships that study and promote urban sustainability.

Notes

1 Incredible Edible Todmorden: www.incredible-edible-todmorden.co.uk/
2 Even though Christianity in Japan is not the most common religion, many customs were accepted by the Japanese. The Easter egg hunt was probably popularized by a British expatriate, Venetia Stanley-Smith Kajiyama, living in the Ohara village near Kyoto and streaming a program on national television for almost a decade (NHK).
3 In community design, the third place is the social surroundings separate from the two usual social environments of home ("first place") and the workplace ("second place").

References

Chawla, Louise. 2002. "Insight, Creativity and Thoughts on the Environment: Integrated Children and Youth into Human Settlement Development." *Environment and Urbanization* 14 (2): 11–22.

Eguchi, Aiko, Ermilova Mariia, Abe Kenichi, and Kinoshita Isami. 2020. "Study on Community Participatory Edible Landscaping Along Streets in Housing Area." *Journal of Architecture and Planning* 5 (776): 2183–92. https://doi.org/10.3130/aija.85.2183.

Ermilova, Mariia, Mitsunari Terada, Ryosuke Shimoda, and Isami Kinoshita. 2018. "Improving the Practice of Landscape Design Collaboration between University and Local Community: Case Study in Matsudo, Japan." 4th Association-of-Behavioural-Researchers-on-Asians (ABRA) International Conference on Quality of Life (AQoL), Environment-Behaviour Proceedings, Istanbul, Turkey, Vol. 3, No. 9, pp. 25–34.

Francis, Mark, and Randolph T. Hester Jr, eds. 1990. *The Meaning of Gardens: Idea, Place, and Action*. Cambridge, MA: MIT Press.

Hester, Randolph T. 2010. *Design For Ecological Democracy*. Cambridge, MA: MIT Press.

Nakamura, Masahiro. 1989. "The New Phase of Participation and 'Gulliver Map' as the Tool." [in Japanese]. *Journal of the City Planning Institute of Japan* 24: 511–16. https://doi.org/10.11361/journalcpij.24.511.

Perry, Simona L. 2012. "Development, Land Use, and Collective Trauma: The Marcellus Shale Gas Boom in Rural Pennsylvania." *Culture, Agriculture, Food and Environment* 34 (1): 81–92.

Porpora, Douglas V. 1999. "Action Research: The Highest Stage of Service-Learning." In *Cultivating the Sociological Imagination*, 121–33. Washington, DC: American Association for Higher Education.

Terada, Toru, Makoto Yokohari, and Mamoru Amemiya. 2016. "Urban Farming in Tokyo." In *Green Asia: Ecocultures, Sustainable Lifestyles, and Ethical Consumption*, edited by Tania Lewis, 155–68. New York: Routledge.

Winterbottom, Daniel. 2011. "Effecting Change through Humanitarian Design." In *Service-Learning in Design and Planning: Educating at the Boundaries*, edited by Tom Angotti, Cheryl Doble, and Paula Horrigan, 189–204. New York: NYU Press.

8

COMMUNITY DESIGN FOR INCLUSIVE ENGAGEMENT

The Old Stone House and Washington Park, Brooklyn

Claudia Joseph

Introduction

This chapter details how approximately an acre of public land came to be used as a community commons. Greening was accelerated through educational opportunities and public outreach that supported improvement, redesign, and renovation of the gardens at Washington Park. A spiral rather than a line best represents the project's growth, so sections within this chapter are arranged by theme, not chronology. Stormwater mitigation measures, resilient plantings, and a midscale compost system are modeled on the site. Green infrastructure was established and maintained through free public workshops, fee-based trainings, and daily care during which visitors and neighbors were encouraged to interact with the landscape. The installation of native shrubs and flowers restored habitat for songbirds and pollinators and added indigenous context to the landscape. Culinary plants that had been introduced by Dutch colonists were fundamental to the historic theme, though we favored perennial food and herbs over swaths of annual crops. Botanical medicine and craft species expanded programming possibilities and fundraising potential. Educational activities were inclusive for low-income participants. Practical constraints, like budget, were creatively met while art, music, and events made the gardens more visible and inviting.

Originally constructed in 1699 by the Vechte family on land taken from the Lenape, the Old Stone House (OSH) was a modest residence on a working farm until the mid-19th century. In August of 1776, it was the culminating site of the Battle of Brooklyn, the largest battle of the American Revolutionary War. The damaged house was moved in 1935 and reconstructed at the center of the three-acre park. The Old Stone House & Washington Park is a Historic House Trust of New York City site and is listed on the national Register of Historic Places. The small, smartly updated museum is open to the public and used for school programs.

DOI: 10.4324/9781003217442-9

Classes and events are held in the Great Room upstairs, which doubles as a gallery. Attendance grew exponentially due to public programming, increasing from 3,000 people to more than 30,000 annually between 2004 and 2008, as the gardens were initially expanding. OSH and the surrounding landscape are now a cultural center for the neighborhood. See Figures 8.1 and 8.2.

Relationship and Proximity

The site has transitioned from a sleepy weekend museum to a vibrant town square, largely due to the hard work and embrace of community by OSH Executive Director Kimberly Maier, who was hired in 2004. I volunteered my services when I heard that the historic house, just two blocks from my apartment, wanted a colonially referenced kitchen garden. One garden can jump-start another, and the Garden of Union (GOU), my community garden nearby, donated compost, plants, and the help of experienced gardeners. My proximity allowed frequent observation that informed the design of the new gardens. I soon became an independent contractor for OSH and, with the community, established, maintained, or improved many of the gardens from 2004 to 2018. Classes, workshops, courses, and events were developed in parallel with the plantings. Hundreds of volunteers and students built beds, managed compost, distributed mulch, planted, pruned, and harvested with me. The gardens evolved slowly as participant use influenced their form and function. OSH raised funds for my salary through annual appeals and events. Fee-based programs, grants, and generous donations from our community supported the creation of several gardens where visitors may smell, touch, and even taste the plants. By encouraging active engagement through cultural exchange, shared harvests, and skills training, we developed a loyal base of support.

The New York City Department of Parks and Recreation (DPR), Grow NYC, GreenThumb, the City Council, Citywide Nursery, Greenbelt Native Plant Nursery, and numerous other partners, made the expanding scale, diversity, and beauty of the gardens possible. As a pioneer of the landscape, my work seemed finished in 2018, and I retired from my position as environmental educator for OSH. A new garden director was hired in 2020, and his work began when OSH and the surrounding park were closed due to the pandemic. Irrigation and bed borders were added while the park was quiet, and a compost manager was hired recently. Staff continues to engage extensively with visitors, and dedicated volunteers still help tend the gardens.

The Right to a Public Commons

While DPR officials were hesitant to approve food cultivation in 2004 when we proposed it, they have embraced the landscape at OSH wholeheartedly over the years. The site gained a Dooryard Garden referencing the colonial period

FIGURE 8.1 The landscape was mostly grass before gardens were installed (2004).

Source: Photograph by the author.

FIGURE 8.2 Plantings provide historic context to the site (2017).

Source: Photograph by the author.

when OSH administrators agreed to oversee the gardens adjacent to the House, at the start of the project. OSH is an independent nonprofit, which has since expanded its role at the historic site to become the conservancy organization for the entirety of the surrounding park. Over the years much has changed as OSH increased exhibits, programs, and public history offerings and has received more than $9 million in capital improvements to transform substantial spaces in the landscape.

Unlike community gardens, NYC parks are recreational spaces, not community food sources or horticultural education centers. Food in public spaces poses a number of liabilities, from allergic reactions to attracting rodents. Bridging this difference was necessary at the start since in most NYC parks, visitors are forbidden to harvest anything. Administrative views can shift over the course of several years, and our model helped demonstrate the soundness of creating a variety of yields for educational purposes and public benefit. Empowering neighborhood groups to productively steward public land is an emerging trend. Beacon Food Forest in Seattle, Washington, broke ground in 2012 and at seven acres may be the largest publicly accessible food forest in the country. OSH donated plant material to SWALE, founded in 2016 in the South Bronx, a floating art installation and edible landscape that explores our right to a public commons. Land access and long-term rights are an ongoing issue for those who forage on public lands and for NYC community gardeners. In 2017, the Bronx River Foodway (BRF) in Concrete Plant Park was founded under the direction of a permaculture organization from Florida, using inroads created within DPR by SWALE and mirroring the work at OSH. See Figures 8.3 and 8.4. BRF was designed to bring an abundance of fresh food to the park for local residents to harvest.

Permaculture Theory: Applying Interdisciplinary Design

Permaculture design practice addresses our basic human needs: food, shelter, energy, water, and waste systems are obvious fundamental requirements. Purpose, customs, economy, well-being, and other influences are also considered in whole systems design. Permaculture differs from conventional landscaping, gardening, and farming in many ways due to its cross-disciplinary practice, ethics, and directive to work with nature. The ethics of permaculture—"Earth Care, People Care and Fair Share"—are necessary to the practice. Fair share means that we share excess materials, time, and knowledge whenever possible. It is customary to allow 20% or more of course participants to enroll through scholarship or work-trade opportunities or by sliding-scale fees, making a course or class affordable to all. Mentoring students and training teachers, as well as honoring contributions from every participant, is common practice. Decolonizing permaculture is a recent educational movement to credit appropriated knowledge used in permaculture to the indigenous and peasant cultures who originated the practices. This movement

FIGURE 8.3 Soil was badly damaged during renovation (2011).

Source: Photograph by the author.

FIGURE 8.4 Visitors harvest hyacinth beans in the farm garden (2016).

Source: Photograph by the author.

focuses on food sovereignty and land rights, antiracist training, and environmental justice initiatives.

Bill Mollison and David Holmgren cofounded the concept of permaculture in the mid-1970s. Mollison began modeling resilient ecological systems and teaching globally, while Holmgren developed long-term community models in Australia. Social injustice, environmental degradation, climate change, and the spread of pandemic disease are topics that have been discussed in the global permaculture community for decades. Awareness of relationships, prolonged observation, mimicking natural patterns, and the use of appropriate technologies are some of the strategies for designing positive outcomes. A "Permaculture Design Consultant" course consists of 72 hours of combined classroom and site visits, with a culminating design presentation. The OSH site was reimagined many times through creative brainstorming, practical scrutiny, mapping, and conceptual designs presented by groups at the conclusion of a course. Ten hours of on-site field training were added to the standard curriculum. Field training was also a requirement for the 40-hour "Urban Permaculture Accelerated Training" Course. Twenty or so students at a time helped shape the new gardens that emerged each year through the capital improvement process. Eventually, a fee-based permaculture field training certificate was offered for 40 hours of participation in the landscape. Students came weekly to learn, share, improve their own health, and find community: quite a few returned repeatedly, long after their training was completed.

A few events became annual traditions, supported by students and members of our permaculture guild who met at OSH regularly. The Seed Celebration & Swap, held from 2004 to 2018, offered free seeds to everyone, as well as seed education. Gardeners and students demonstrated how to clean and save seeds, created seed art, taught seed starting, and offered free garden advice. Importantly, we offered education on genetically modified seed and on the consolidation of seed ownership by international agrochemical corporations. As buyouts of traditional seed companies accelerate, historic genetic material is at risk of being patented or lost. While flavor, nutrition, and regional adaptation were selection criteria in the past, a uniform appearance, shipping life, and tolerance of chemical applications are now favored traits. Maintaining seed stocks by growing them out has become critical to preserving many heirloom varieties. Open pollinated seeds that reproduce true to form were featured at our event, and we explained the difference between these and hybrid seeds that revert to characteristics possessed by one or the other parent plant. Small East Coast companies and a few specialty seed companies from the West were invited to donate catalogs for people to take home. Everyone attending received some free seeds from our collection, other gardeners, or seed companies. Remainders were used in garden and school programs or gifted to caregivers. Seeds are abundant, and the joy they brought was palpable, year after year.

Balancing the financial bottom line while offering free access to programs pushed us to create fundraising opportunities. There is much to share from gardens, and the reciprocal donations are often generous. Specialty seed packs and plants purchased through Grow NYC sold alongside those grown at OSH at our May plant sale. The post-Halloween pumpkin smash and leaf shredding event was centered on compost education for several years. Children joined in enthusiastically, smashing, shredding, and mixing material to help fill our bins. An outgrowth of this event, the NYC Permaculture Festival and Skill Share, was held every fall from 2010 to 2018. Hands-on workshops, lectures, craft demonstrations, vendors, movies, and food attracted permaculture enthusiasts from a wide area and introduced our neighbors to permaculture principles and design techniques. Divide Plants and Unite Neighborhoods was a spring event to thin shrub material and harvest plant divisions. Perennial plants often reproduce at the root, and we invited schools and groups to dig plants for their community projects.

Compost Is Essential

The initial strategy, before OSH had compost bins, was to haul compost from the abundant GOU system. Soon, some standard city bins were available for free from a nearby garden, and I eagerly brought those to the site. Even small bins allowed me to begin compost education and produce garden amendments. Large-capacity compost bins were constructed in partnership with the Brooklyn Botanic Garden (BBG) in 2010 and with Park Slope Food Coop (PSFC) in 2014. I had longstanding relationships with each organization through teaching and membership, respectively. PSFC is a member-worker cooperative, and for many years my work shift was managing the compost system at GOU where 14 tons of waste material were moved from the produce aisle to bins, annually. OSH replicated the original bin design by Paula Hewitt Amram of Open Road. The bins have vent pipes to circulate air, can hold up to 1,000 pounds of material, and exclude rodents. They are made of recycled plastic lumber which lasts 20 years, and the lids are unbreakable plastic. A cement slab is required to prevent the bins from twisting when heavily loaded and to keep rats from burrowing underneath. Rats are an urban constant, so we took every measure to keep the site hygienic. PSFC sent haulers to process and load material. The browns team delivered carbon rich materials: giant bags of wood shavings were picked up for free from carpenters who worked with hardwoods. Additionally, wood chips that haulers had bagged at a public site were needed to improve ventilation at the bottom of the bins. The active bins were adjusted often by adding wood shavings or water or by introducing more air into the material.

A combination worm compost box and bench was added in the picnic garden in 2016. Park users were encouraged to deposit apple cores and banana peels in the worm box under the hinged seat. Lessons on soil, compost, and waste stream management occurred both spontaneously and in workshops. The box was adjusted periodically to maintain the correct moisture level for the worms. Our specific guidelines encouraged use while preventing mistakes. The worms and bugs were accessible to children, even when no gardeners were present. And because it was wide and deep, the compost bench became the most coveted seat in the garden. Fee-based membership in our Compost Club was offered to neighbors in 2017: individuals and families received hands-on tutorials by BBG master composter certificate candidates. Membership resulted in material gains and more friends for the site. A considerable number of mistakes are possible, and I made the small corrections that were needed. Processing kitchen waste became a social opportunity and reason for neighbors to visit the garden, as well as an act to benefit the environment.

City soils are degraded, and it can take years to bring them to a point of high biological function. We improved ours with straw or leaf mulch and by applying compost. To make hot compost in bins requires air, water, carbon-rich material, and nitrogen-rich material combined in the right proportions. On-site aerobic compost systems eliminate trucking pollution and fees while, compared to anaerobic landfill conditions, reducing methane emissions. Almost all of our landscape material went into the compost system and needed to be cut into small pieces. A multitude of volunteers made this possible. Most who participated, whether chopping material, turning a bin, or harvesting the finished product, connected with the process. It was one of the most beneficial activities for the site and one of the most popular. Children frequently visited the worms and bugs in the farm garden's public bins. This reduced digging in the beds and taught parents and children alike that healthy soil is full of living organisms. Nutrient levels, water holding capacity, and biological life in the finished compost will vary, according to the inputs and process used. Our nutrient-rich inputs and available labor pool resulted in a soil amendment that was better than any product we might purchase. Organisms assist the soil's performance by breaking down and digesting organic matter. Nutrients are delivered in available forms directly to plant roots. The channels created by organisms as they move through soil open airspace, improve drainage and moisture retention, and soften the soil for better root penetration. The increased plant vitality resulting from compost application makes synthetic chemical use unnecessary. (In our historic model, we would not use chemicals or bagged amendments, anyway.) Gardens flourished for hundreds of years in cooperation with nature.

Partnership Gardens: Managing Flows

The gardens expanded to the adjacent middle school, MS 51, in 2005. Whenever school is in session, students flood out of the school and into the park at lunchtime. Their energy was harnessed for setting bed edges, planting, compost management, path making, and mulching. A science teacher who gardened with students following 9/11 and a special education teacher who centered his curriculum on the environment brought students to the schoolyard first. I began visiting classes to explain the history and importance of our relationship with plants and to invite students to participate in our project. We developed programs and classes, and, as participation expanded, we planted the schoolyard with historically useful trees, shrubs, food, herbs, and flowers.

Teachers construct their lessons for standardized tests and often have little leeway in subject matter. An engaged parent cross-referenced the sixth-grade science curriculum with garden activities. Classroom lessons already acknowledged the benefits of plants, especially trees. The intake of carbon dioxide and release of oxygen, temperature buffering attributes, and the ability to filter pollution are commonly taught. Thermal dynamics led us to study the effects of slope and sun angle on soil and cement temperatures. The wheelbarrow, levers, fulcrum, rollers, and axle fit into the unit on simple machines. Meanwhile, at OSH, a narrative about the colonial residents of the House was developed and used for school visits by Director of Education Margaret Weber. A story about the Vechte family led students to the gardens on a scavenger hunt for culinary and medicinal plants. Also led by OSH staff, students in after-school science programs planted the rolling raised boxes and studied the compost system. OSH's Garden Explorers program brought younger children to the gardens. History and Community Studies students at MS 51 pondered how colonists survived without grocery stores or drugstores. The long history of plant use from wild and cultivated landscapes was discussed, alongside conceptual changes in our relationship with nature and agriculture.

Projects and classes at MS 51emphasized our place in the watershed at the upper edge of the flood evacuation zone. The Gowanus Canal neighborhood, located two blocks west of the Park, is a Superfund site and highly prone to flooding. Students learned that combined sewer overflows (CSOs) dump raw sewage into our waterways during storm events. The impacts of plastics in waterways and their blockage of storm drains in our neighborhood were noted. We were able to talk directly to students about litter: trash tends to overflow the 20 receptacles that are emptied daily. Some environmentally aware parents worked to get a filtered drinking water station installed at the school, and a local business donated canisters to every student. Reducing waste involves shifting individual consumer habits, passing legislative measures, and regulating producers.

As the landscapes of MS 51 and Washington Park were being planted to increase stormwater infiltration, the director of OSH worked with the Department of Transportation to install permeable pavers in the cul-de-sac on Fourth Street, between the two sites. A catchment basin was recently retrofitted into the park playground to prevent water from pooling in front of the House during storm events. City engineers are learning to direct water into the landscape rather than simply sending it downhill. The popularity of rain gardens is a welcome sign that this is now widely recognized by the lay public as well. Planted troughs along curb cuts are also effective in allowing stormwater time to infiltrate, and they have been installed widely in the Park Slope neighborhood. Students dug holes and measured the infiltration rates of water in tree beds. Compacted soil limits root growth and contributes to runoff and soil erosion. Caring for street tree areas and installing resilient plants in public spaces increases stormwater infiltration. By opening soils and introducing diverse plants with deep and shallow roots, water is captured and held in the landscape. We invited a street tree specialist to provide lessons since the tree areas near the park and school are mistreated by dog walkers and children. By

FIGURE 8.5 Beds placed on contour slow and absorb stormwater flowing downslope during storm events.

Source: Photograph by the author.

combining the site history with related topics in science and community studies, we were able to bring more unity to the site and encourage students to consider the greater ecological system in which they participate (See Figure 8.5).

Caregivers, Children, and Gardens

Whether a relative, domestic worker, au pair, or baby sitter, adults accompanied our youngest visitors to the park every day. Many caregivers who visited on weekdays were from other neighborhoods and native to other countries. Some were familiar with harvesting from gardens and landscapes, using traditional herbs, or growing food. They often stayed a few hours in the park, sometimes sitting in the picnic garden while the children ate and played. Several of the gardens allowed children to stretch their legs, get close to the plants, and explore freely. Attentive adults taught the children not to pull on the plants or walk in the beds. They were offered flowers, herbs, or food whenever the opportunity arose. See Figure 8.6. The children were often attracted to our work, and we invited them to move a small bucket of chips, sweep a sidewalk, or stuff leaves in a bag with the assistance of their caregivers. Handling the wheelbarrow was an exciting challenge for some children who enjoyed testing their strength and skills. See Figure 8.7.

Many plants were grown with our caregiver constituency in mind: callaloo, Cuban oregano, hot peppers, rosemary, thyme, and lavender were often requested. Callaloo is a term for leafy green vegetables, common to Jamaica. Amaranth is a popular edible plant and is consumed internationally. Visitors from Italian, Asian, and Caribbean cultures moved to it in reflexive recognition. Amaranth is grown easily and spreads through the thousands of seeds it produces. The seed, which contains protein and fiber, is used in baked goods. The leaf is not widely available, and we offered it free and fresh to its many fans. Hyacinth bean, also called lablab, is a delicate vine in the pea family that was grown by Thomas Jefferson. Indonesian and African visitors explained to me how it is prepared in their countries. A routine visitor indicated which volunteer plants, ones we might call weeds, are useful as medicine in Asian culture. Continual exchange took place over the years, enriching cultural understanding and informing our plans.

Adding Structure

Rolling raised beds were constructed in 2010 through a green jobs training program. During the renovation of the Park in 2011, many of the gardens were destroyed and redesigned. The rolling boxes allowed us to continue growing food. Easily accessible, they are planted by classes and soften the starkness of the expansive stone plaza on the west side of the House. A multifunctional trellis hut, also on the west side, was constructed in 2011 at the center of the South Dutch Garden. Originally built to protect a Three Sisters planting program, it was converted to a field office, with a table for cleaning seeds and writing notes. Children were attracted to the leaf-covered hut that sheltered us. The frame supported hops, gourds, winter squash,

FIGURE 8.6 Diverse yields bring children and their caregivers to the garden.

Source: Photograph by the author.

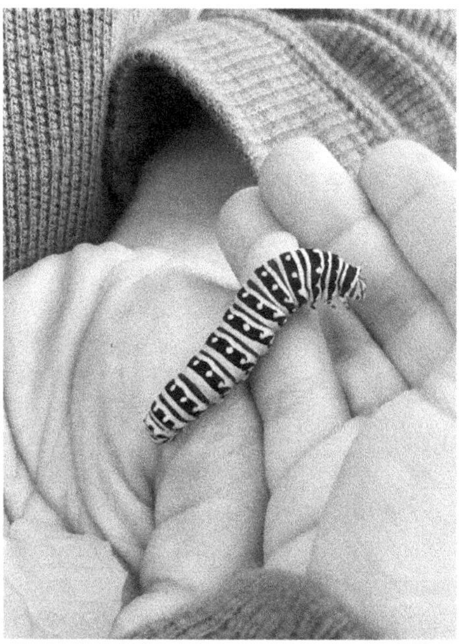

FIGURES 8.7 Butterfly larvae are often discovered by young children.

Source: Photograph by the author.

cucumbers, beans, or grapes that variously covered the wire structure. The perimeter of the South Dutch Garden was planted with raspberries, an espaliered pear tree, Western sand cherry, chokeberry, bayberry, herbs, native flowers, and other useful plants. A toddler meander was installed to break up a long rectangular edge where water runoff was a problem. A meander is a basic permaculture pattern of S-shaped curves that inherently slows down flows. The simple path converted a sloping, linear bed into an exploratory area for youngsters, away from the hot sidewalk. Children snaked naturally around the mounded beds while being able to step right next to the plants and touch them. Fragrant herbs, native flowers, vegetables, and berries were easy to see, as well as the insects that frequent them (See Figure 8.7).

Bio-intensive, no-till cultivation was a method we applied in the beds of the farm garden. Every niche of bed space was filled with annual and perennial plants. Hazelnut, bayberry, winterberry, and elderberry formed a protective hedge round the perimeter. Volunteers, frequent visitors, and DPR employees appreciated picking fresh produce and herbs there. We assisted in harvesting the plentiful raspberries, ground cherries, tomatoes, beans, greens, and other vegetables so the plants were not damaged. Harvesters were encouraged to check with us before picking, so that a substantial display always remained for educational purposes. Planting diverse crops in each bed furthered discussions, demonstrated intensive cultivation, decreased pest issues, protected the soil, reduced weed pressure, and increased interest. Most people respected the harvest guidelines even though the garden was never locked. The beds' shapes (heart, gourd, paisley, keyhole, and spiral) allowed groups to gather easily around them for conversation. Set roughly on contour, the beds were designed to slow the flow of rainwater from two directions (Figure 8.5). The recessed pathways held runoff, allowing it to infiltrate into the beds during storm events. This prevented erosion and mitigated water flows into the storm drains leading to the Gowanus Canal. The twisting, sunken pathways also slowed the movement of children flowing through the space.

Designing for Diverse Use

Disparate activities coexist frequently in well-defined areas of the landscape. The gardens are used for various purposes from film shoots to family meals. Many parks were closed during the one of the worst outbreaks of COVID-19 from April through June 2020, and their reopening provided much needed relief. At times, the elusive hope of finding tranquility in the city can be realized. Yoga, meditation, and tai chi are often practiced in secluded parts of the gardens. The shelter of plants provides a protective ease, within the community rather than separate from it, and a quiet space away from home. Mental health stresses became more prominent during the pandemic, and gardens were much sought after as therapeutic environments. A publication by the US Forest Service, *Restorative Commons: Creating Health and Well-Being Through Urban Landscapes*, documents precisely how greenspaces can provide physical and psychological benefits for urban residents (Campbell and Wiesen 2009).

FIGURE 8.8 Natural materials are used by children for imaginative play.

Source: Photographs by the author.

The theme of traditional use offers many plants, native and introduced, that thrive in difficult conditions. Seasonal interest, ease of care, abundant harvests, and multiple functions were characteristics we valued. Plants tolerant of salt, flooding, drought, and neglect, as well as alkaline soils, heat extremes, wind tunnels, pollutants, and compacted soil met our horticultural challenges. Washington Park often hosts 600 to 1,000 visitors a day, and many were not aware that stepping in beds could stress the plants. The landscapes near the skate park, ball courts, and playing field (one of 12 for a population of 3 million) were designed to withstand wayward balls and foot traffic. Children interacted with the abundant natural materials: stones, wood chips, leaves, flowers, berries, and sticks were used in their imaginative play (Figure 8.8). The bird bath became an unplanned outlet for expression when visitors floated flowers, berries, and leaves on the water. Classes on traditional herb use, crafting, forage, and dye gave purpose to the plantings. Adults used botanical material for contact printing, pressing, and studies. Flowers were given away or sold and were abundant in the landscape for all to enjoy. Sensory activities engage us with the natural world, sustaining our creativity, mental health, and sense of well-being.

The appearance of birds, bees, and butterflies followed years of planting appropriate shrubs and flowers to meet their breeding and feeding needs. Maintaining a butterfly population requires both nectar and larval support plants that are often specific to a species. At least ten kinds of butterflies, some reproducing on the site, were identified by the seventh year, whereas none were present in 2004. Toddlers are often the first to see the brightly striped larvae that find food and shelter in the gardens. Insect populations swelled to include rooftop honeybees, native bees, beneficial wasps, and pollinating flies. Praying mantises, grasshoppers, lady bugs, and damselflies also became common. Resident birds and migrating songbirds can now find food, shelter, and water in the gardens, and the number of species has increased dramatically. Though we started with pigeons, sparrows, and starlings, birdwatchers might now see visiting thrushes, warblers, vireos, kinglets, and finches, as well as red-tailed hawks, woodpeckers, robins, mockingbirds, and cardinals.

Gratitude to Supporters

The OSH Board and numerous agencies are the benefactors of our community commons. I am grateful to everyone who helped establish the gardens through their words, actions, and presence, especially to the New York City Department of Parks and Recreation for allowing me to demonstrate permaculture practice in such an important public space. Community service and religious organizations, therapeutic programs, and schools brought much needed help to the gardens. Neighbors and students who became familiar with the site mowed, weeded, and clipped on their own. The gardens became expansive and diverse because people dependably dedicated their time, skills, and energy.

Conclusion

Social and ecological actions are needed in response to the current pandemic, mass extinction, and climate upheavals. Ecological resilience can be improved by engaging citizens at every skill level to steward public space in partnership with community agencies and local government. Programs offering simple solutions, diverse yields, and skills training attract financial and physical support while providing significant ecosystem benefits. Installing park, school, and library landscapes to mitigate water flows during storm events and to buffer heat extremes also increases educational and material resources for neighborhoods. Led by community organizations, the stewardship of public landscapes can multiply, expand, and improve green infrastructure effectively. In alignment with permaculture principles, citizen engagement is necessary to revitalize urban ecosystems, increase their resilience, and support the well-being of urban residents.

Reference

Campbell, Lindsay, and Anne Wiesen. 2009. *Restorative Commons: Creating Health and Well-being through Urban Landscapes*. General Technical Report NRS-P-39, USDA Forest Service, Newtown Square, PA.

PART II
The Right to the City

Part II takes up Henri Lefebvre's idea of the "right to the city" (Lefebvre 1996) as a critique of the commodification and gentrification of urban spaces and a call to reclaim cities for those who built and inhabit them. Starting in the 1990s, the slogan "the right to the city" has facilitated alliances among human rights groups, environmentalists, NGOs, movements of the urban poor, local governments, and international bodies at United Nations conferences. The agenda of these grassroots movements is informed by the principle that the right to the social, economic, and cultural goods of the city depends on "greater democratic influence in shaping the provision, quality, and governance of those goods" (Cohen 2018).

The first cluster of chapters offers a variety of perspectives on the right to the city in relation to the ethos of sustainable cities, the role of liberatory education in achieving sustainability, and an ethic of care to guide urban dwellers, researchers, educators, and activists.

In "The Right *to* the City and the Rights *of* the City," Margaret Cuonzo examines the ethos of the sustainable city as a question of rights and what Rachel Carson famously referred to as "the obligation to endure" (Carson 1962/2002).

Deborah Mutnick presents a theory of urban ecosystems in "Making Our Own History: Urban Sustainability in a World in Crisis" as threatened by interrelated crises of climate change, species extinction, and racialized class inequality. Based on ecosocialist theory, she considers dystopian realities (Robinson 2017) in light of still available paths to achieving an equitable, sustainable city.

Scott Carlin's "Reconceptualizing the City in the Age of the Anthropocene" calls for a "radical reconceptualization of the city, its logic of agglomeration, and its role in global capital accumulation" based on an ethic of care.

The second cluster of chapters turns to case studies and histories focused on New York City's outer boroughs, a suburban city in Minnesota, and Rio de Janeiro, Brazil. Common themes range from the disproportionate impact of environmental

DOI: 10.4324/9781003217442-10

pollution on minority populations to urban plans that "green" some spaces while exposing others to pollution and climate change.

Melissa Checker examines troubling urban policies in "Industrial Gentrification and the Geography of Sacrifice and Gain." She depicts the impact of economic development and rezoning on low-income communities of color in Staten Island, in particular.

Focusing on a suburban Midwestern city, Nicole K. Ciulla reports in "Futurity and Equity in Local Climate Action" on the limitations of linking climate change policy to future generations. Based on her study, she finds that, in the public rhetoric on the future in local debates, an otherwise ambitious climate plan limits action as well as whom such action should serve.

We go from suburban Minnesota to Brooklyn with Jerome Krase's "Social Justice and Brooklyn Development: 500 Years of Struggle," an overview of the historical development of the city and its "periodically troubled history of growth as a consequence of conflicts and competition over Brooklyn spaces."

Closing out Part II, in "Participatory Action Research in Rio de Janeiro's Sustainable *Favela* Network," Leonel Lima Ponce assesses the impact of technical consultancy and academic research on local sustainability efforts by residents to create ecosystem services, build economic capacity, and enhance social resilience.

References

Carson, Rachel. (1962) 2002. *Silent Spring*. Boston, MA: Houghton Mifflin.

Cohen, Daniel Aldana. 2018. "Climate Justice and the Right to the City." In *Penn Current Research on Sustainable Urban Development*. https://penniur.upenn.edu/uploads/media/Cohen.pdf.

Lefebvre, Henri. 1996. *Writing on Cities*. Translated and Edited by Eleonore Kofman and Elizabeth Lebas. Hoboken: Blackwell Publishers.

Robinson, Kim Stanley. 2017. *New York 2140*. New York: Orbit.

9

THE RIGHT *TO* THE CITY AND THE RIGHTS *OF* THE CITY

Margaret Cuonzo

Introduction

Not everyone has equal or fair access to what the city has to offer. Factors such as race, class, gender, sexual orientation, and ethnicity influence the degree of access that different groups of people have to the good things that cities provide. This inequality of access has led to an important discussion of how human beings have a *right to the city*. David Harvey, in *The Right to the City*, has argued, "The freedom to make and remake our cities and ourselves is . . . one of the most precious yet most neglected of our human rights" (Harvey 2008, 23). In the present discussion, I'll compare the notion of the right *to* the city with that of the rights *of* the city. I'll argue for a particular conception of a city, one that treats the city as an ecosystem. As such, it includes both living (biotic) things and nonliving (abiotic) things as components. The city, I'll argue, can be thought of as having rights, in addition to being thought of as a thing to which others have a right. The idea of fair access to what the city provides can be fleshed out more fruitfully by conceiving of the city in this way rather than by thinking of the city as a "resource" to be used. Though the idea that cities have rights may seem paradoxical, the arguments I'll present will entail the extension of natural rights to cities.

This chapter will present what I take to be a key problem, the problem of access to the full array of urban life. It will look at the motivations for finding a solution to this problem in the Right to the City movement and turn the discussion more toward the nature of cities and, ultimately, what obligations we have toward them. The central argument I'll make begins with the premise that the city is an ecosystem, coupled with a second premise, namely, that ecosystems have rights. I will ultimately conclude that cities have rights. By conceiving of a city this way, we now have a way to think about cities that provides a strategy for solving the problem of access, namely, that accessibility to all aspects of urban life, such as public space and

DOI: 10.4324/9781003217442-11

healthy food, is a right *of* the city itself. In the same way that poisoning a stream within a forest harms the forest and violates its right to freedom from harm, the lack of access violates a city's inherent right to freedom from harm. Framing the discussion of access this way makes access a concern not only to the groups without the access and their allies but to all who are concerned about the well-being of the city in which they live.

A Key Problem of Cities: Access

One of the most fundamental problems of cities is that access to good things, such as education, healthcare, healthy food, clean air and water, safe and restful parks, cultural activities, transportation, employment opportunities, financial services, and legal representation,[1] is not equal. Wealthy and white city dwellers have greater access to these things than do the working class, poor, and people of color. The problem leads, I think, to other problems. Lack of access to education and employment opportunities leads to crime, lack of access to healthy food and clean air or water leads to poor health, and so on.

These facts regarding access have led activists and theorists alike to the Right to the City Movement. Two early, key thinkers behind this movement, Henri Lefebvre and David Harvey, see the problem of access as the result of cities being constructed without regard for the working class. Harvey, for example, notes that, "from their inception, cities have arisen through geographical and social concentration of a surplus product" (24) to the detriment of the working class, and he sees Georges-Eugene Haussmann's work in Second Empire Paris in the mid-19th century as a case in point. To solve the problem of surplus capital and unemployment, Haussmann planned and built the grand boulevards of Paris, including the eponymous Boulevard Haussmann, and displaced many working-class inhabitants. Haussmann, according to Harvey, "helped resolve the capital-surplus disposal problem by setting up a proto-Keynesian system of debt-financed infrastructural urban improvements" (Harvey 2008, 26).[2] Through urban development, the excess wealth is parked in physical locations like large building developments. In addition, future wealth can be used to pay for the projects through financing the cost of the projects. Unemployment, which brings the possibility of social unrest, is lessened through hiring workers to build these large projects, and spaces designed for the wealthy are the result. This calls to mind and explains frequent questions I find myself posing when confronted by a new residential development in my native Brooklyn. "But who is going to live there? How are the developers going to find so many people willing to pay that much money?" and "How will this impact the people who are already there?" These developments serve more than the purpose of generating wealth; they are also vehicles for storing it. Even if they don't make money, they serve the purpose of parking wealth. As for the question about impact, time and again the answer is that working-class people will suffer rising costs, displacement, and loss of community.

Lefebvre (1968) and Harvey (2008)

Turning now to the early Right to the City theorists will illustrate some key lessons from the movement and also reveal some understandable shortcomings. To begin, Lefebvre's discussion of the right to the city begins with a discussion of some uniquely human needs. For Lefebvre, the city is a place in which basic human needs for "encounters and simultaneity" and exchanges not based solely in commerce are satisfied. He also speaks of the city as the place where the "*oeuvre*" (participation in the work of creating the city) occurs and where needs for "information, symbolism, imaginary, and play" (Lefebvre 1968, 147) are met. According to Lefebvre:

> Would not specific urban needs be those of qualified places, places of simultaneity and encounters, places where exchange would not go through exchange value, commerce and profit?
>
> *(Lefebvre 1968, 147–148)*

This quote brings to mind the unique way in which city inhabitants are living quite different, simultaneous lives and, of necessity, must encounter others and become influenced by them. How the needs for things like simultaneity and encounters that go beyond commerce are specifically urban remains unclear, although Lefebvre does provide some reason for thinking that cities are well poised to provide these things. Speaking of the right that corresponds to these "urban needs," he claims:

> The *right to the city* cannot be conceived of as a simple visiting right or as a return to traditional cities. It can only be formulated as a transformed and renewed *right to urban life*. It does not matter whether the urban fabric encloses the countryside and what survives of peasant life, as long as the "urban," place of encounter, priority of use value, inscription in space of a time promoted to the rank of supreme resource among all resources, finds its morphological base, and its practico-material realization. . . . . . Only the working class can become the agent, the social carrier or support of this realization. . . . Who can ignore that the Olympians of the new bourgeois aristocracy no longer inhabit? They go from grand hotel to grand hotel, or castle to castle, commanding a fleet or a country from a yacht. They are everywhere and nowhere. That is how they fascinate people immersed into everyday life. They transcend everyday life, possess nature and leave it up to the cops to contrive culture.
>
> *(Lefebvre 1968, 158–159)*

The Right to the City is the right to urban life. Only the working class, the inhabitants of everyday life, can bring about a city that meets the needs for urban life because, as the ones who built and maintain the city, only the working class can know the city from the level of lived experience. The "new bourgeois aristocracy" does not "inhabit" the city in the same way and is thus blind to the texture of city life. In contrast to the Right to the City, Lefebvre criticizes a "right to nature" as a pseudo-right arising out of a faulty conception of nature as something commercial and to be sold to those seeking "leisure."

Four decades after Lefebvre, David Harvey also sees access as only a small part of the problem that a right to the city addresses. For Harvey:

> The question of what kind of city we want cannot be divorced from that of what kind of social ties, relationship to nature, lifestyles, technologies and aesthetic values we desire. The right to the city is far more than the individual liberty to access urban resources: it is a right to change ourselves by changing the city. It is, moreover, a common rather than an individual right since this transformation inevitably depends upon the exercise of a collective power to reshape the processes of urbanization. The freedom to make and remake our cities and ourselves is, I want to argue, one of the most precious yet most neglected of our human rights.
>
> *(Harvey 2008, 23)*

Not only do all people deserve access to the city, but they also need the power to transform the city as they see fit. In doing so, human beings will transform themselves. By changing the city, its citizens change the nature of the social ties that are formed, the relationship to nature that is available to its inhabitants, the kinds of technologies that city dwellers use, the aesthetic experiences and values that are developed. For Harvey, the right to the city is a collective right. That is, it is not the right of individuals to everything in a city but the right of a group of people, the city's inhabitants, to exercise its will in transforming the city. Just as a people collectively, not each individual member, has the right to choose those who govern the people, the collective group, in this case the group of inhabitants of a city, has a right to it, to make the city, to alter it, and to remake it along different lines. This notion of a collective right is important for the present discussion, as I'll argue that, since cities are ecosystems and since ecosystems have natural rights, cities have natural rights. The rights of the city for which I'll argue are collective in nature and go beyond any individual right or collection of rights.

Lessons from the Early Right to the City Movement . . . Along with Shortcomings

There are lessons to be learned from both Lefebvre and Harvey. These include the dangers and history of using cities as repositories for excess wealth and as engines for surplus profits, Lefebvre's critique of a misrepresentation of the "natural," and the need for the participation of the array of denizens of the city in its planning and actualization. Harvey puts the first lesson quite well, writing that "from their inception, cities have arisen through geographical and social concentration of a surplus product" (Harvey 2008, 24) and, again, writing that "urbanization has played a particularly active role . . . in absorbing the surplus product that capitalists perpetually produce" (Harvey 2008, 25) through "debt-financed infrastructural urban improvements" (Harvey 2008, 26), which have "dispossessed the masses of any right to the city whatsoever" (Harvey 2008, 37). If correct, and if the upcoming arguments about the rights of cities are successful, then it makes sense to claim that the cities have rights to self-determination. Large-scale projects often violate this right. The use of eminent domain to dispossess the inhabitants of their homes for construction of projects like sports stadiums is an example.[3]

A second important lesson to be had from the early Right to the City thinkers is that a commercialized/industrialized version of "nature" is not something to which we have an authentic right. The "right to nature" is a poor substitute for livable cities. Lefebvre writes, "This 'naturality' which is counterfeited and traded in, is destroyed by commercialized, industrialized and institutionally organized leisure pursuits" (Lefebvre 1968, 158). This quote from Lefebvre brings to mind highly manicured and chemically treated golf courses and lawns of suburbia in the United States, as well as the so-called naturalness of industrialized agriculture and its attendant chemically treated monocrop fields of corn, soybeans, and wheat. "Nature" in these contexts is anything but natural.

Another important lesson is that cities need the full participation of all of their constituent members. The "working class" inhabitants of cities must participate in the planning. According to Harvey, "The democratization of that right [to the city], and the construction of a broad social movement to enforce its will is imperative if the dispossessed are to take back the control which they have for so long been denied" (Harvey 2008, 40). This third lesson and the first are connected. There is room, in some sense, for wealth to be stored in cities in new urban developments, *if* the inhabitants of the city choose how this might happen and participate in the planning, construction, and evaluation of the success of these developments. The inhabitants, though, must have the freedom to refuse such developments as well.

Both Lefebvre's and Harvey's works contain common, understandable shortcomings. Both treat cities as products of our activities that stand outside of us, however much they influence how we live and who we become. I'd like to question this assumption. I will argue that cities are best thought of as ecosystems that we inhabit. In separating the people from the city, however connected they may be, these theorists conceive the relationship between humans and cities incorrectly, treating the city as a product or a resource to be used.

Also, quite understandably, the two pieces lack a deep environmental perspective that the times now require. Both bring up the issue of "nature" but drop it. Lefebvre dismisses a commercialized "right to nature" as a pseudo-right, and he does go into the nature of rural society but doesn't present an urgency or deep analysis of the environmental crises we now face. Harvey, too, speaks of our desired relationship with nature being part of how we must shape cities. However, he does not elaborate.

The City Is an Ecosystem

I believe that a city is more than the buildings and other physical infrastructure, more than the people and other living things, more than the laws, more than the major institutions, more than a history, a climate, or a culture. It is more than a port or marketplace. A *city* is the systemic interaction of all of the above (and possibly other things), with an underlying structure similar to that of other ecosystems, like forests, prairies, and so on. My reasons for believing this are twofold.

First, when we examine what an ecosystem is, we find that, surprisingly, the city conforms to the ecosystem category. On a rough, but useful definition, an *ecosystem* is:

> A community of living things (biotic component: humans, non-human animals, plants, fungi, and so on) interacting with nonliving things (abiotic component: energy, water, geology, and so on).

Cities have both biotic and abiotic components. A city's human and nonhuman inhabitants—its trees and plants, its fungal networks, and so on—form the biotic component of the urban ecosystem, while the physical structures—the air, the water, geology, and so on—form the abiotic component. In addition, both the biotic and abiotic components interact with each other in ways that are parallel to other ecosystems. Animals, part of the biotic component of an ecosystem, influence the abiotic part of this ecosystem. They tunnel through the soil and build dams and nests. They leave for periods of time, migrating to find food or to mate, and return. They take abiotic resources out of the system, such as water and air, to survive and produce waste. Also, their activities are not random but occur during certain times and for specific purposes. This can be said of the animals within the urban ecosystem, as well, including the humans. Their activities form a pattern. They use the surrounding abiotic elements, the air and the water, to survive. They enter and leave the ecosystem at times and according to patterns. That cities contain large human-constructed buildings and other structures is not an essential difference from the structures constructed by nonhuman animals in other ecosystems. Of course, human-constructed buildings will be more complex than structures created by nonhuman animals, but that is not a difference that needs a new category. It is a difference of level of complexity, not of kind. Thus all the features of an ecosystem are realized in a city: its biotic component, its abiotic component, and the interactions between them.

My second reason for believing that a city is an ecosystem can be illustrated by a thought experiment. If we engage in a little thought experiment, we can see that there is no "essence" to a city that can serve as a key component without other components.

Try to keep one of the following components of a city and subtract the rest:

living inhabitants
physical objects
cultural artifacts
laws
and so on

The end result is not a city; at best, it is only a former city. There is no one essential element to the city but only a fluid network of components, and this, again, is more readily thought of as being an ecosystem. Living things inhabit the city and interact with nonliving things, including the buildings and the laws. A bird and its nest and call are parts of an ecosystem. Humans and their cultural artifacts and the things that result from their activities are part of the urban ecosystem as well.

Here an analogy between a forest ecosystem and the urban ecosystem may be useful, as there are key similarities between ecosystems. For example, harm to one part of an ecosystem harms the system in general. As famously illustrated by writers such as Rachel Carson (2000), spraying a toxic chemical to destroy one species within an ecosystem can produce a "silent spring" in which a whole ecosystem is negatively affected. The same holds for the city as an ecosystem. "Solutions" to one problem within a city often give rise to others. There is a delicate balance between all the workings of ecosystems, and the city ecosystem is no exception. For example, adding jobs in a given neighborhood might lead to excess traffic, rent increases, and so on. Also, an ecosystem such as a forest is more than a collection of the living things in it. As the previous thought experiment showed, the same is true of cities.

Here, though, a potential breakdown in the analogy arises: forests have circular metabolisms, i.e., waste is recycled within the system, while cities, for now, have much more linear metabolisms, with food and other products entering and waste leaving. While this does introduce a difference in the organization of ecosystems, cities can progress to more circular systems, and I think it's fair to say that this would be a benefit to the city itself. The linearity of the metabolism of the city is not an essential feature, and indeed a more circular process in recycling waste and creating inputs is already in the works in many cities through composting systems and other activities that reduce the items that need to move out of cities. However, many ecosystems are in a disturbed or broken state, working through stages of regeneration, aspiring toward a sustainable balance between parts.[4]

Some Background on Expanding the Notion of Natural Rights to Ecosystems

To claim that a city has certain natural rights probably sounds quite strange, especially if we conceive of a city as composed solely of physical objects like buildings, streets, land, infrastructure, and other "inanimate" objects. If we think of a city as being composed of a set of laws that govern it, it may be equally strange to think that the laws themselves deserve or have a right to be obeyed.[5] If we think of a city as a collection of the individuals that reside near one another, it may be relatively easy to see that there might be a case for collective rights, as well as collective agency. Still, though, I believe that the rights of the city go beyond the collective rights of its citizens. If, as previously argued, a city is an ecosystem, which includes the living and nonliving things, then it becomes easier to see how a city is due certain consideration by those who live in and impact it and that to violate certain principles regarding this ecosystem is to violate its natural rights.

Conceptions of *natural* rights, rights which are accorded by nature or, as some philosophers[6] believed, from God, as opposed to *legal* rights, have expanded much since the time of the Enlightenment and Locke's (1980) *Second Treatise of Government*. Wollstonecraft (1782), who wrote *A Vindication of the Rights of Woman*, argued for the extension of natural rights to women. Bentham's famous defense of animal rights in 1789 made the case for the extension of natural rights to nonhuman animals. More recently, Stone (1972) argued for the legal rights of natural objects, and Holt (1974) argued for the natural rights of children. Also, in 1973, Baier (2010) argued for the rights of past and future persons. Raz (1986) provided an influential argument for collective rights, as did Cooper (1968) earlier, and Carlson and Kittay (2009) argued for the rights of the severely cognitively disabled.

Here I submit that the natural rights of an ecosystem are not very dissimilar to the collective rights that have already been proposed and accepted. If collective harm can be done to a people, then collective harm can be done to an ecosystem, given that it is already acknowledged that many of the living things within an ecosystem have natural rights, for example the animals that inhabit the ecosystem. Ecosystems therefore already include, as components, elements already considered to have natural rights. Would it be possible to deny rights to the entire ecosystem on the grounds that, because this whole contains nonliving elements, the whole must lack natural rights? If that were the case, then we would have to deny rights to any being or collective that contains nonliving components such as artificial organs, even water and oxygen. This seems implausible, as these are contained in humans and nonhuman animals.

In arguing simply for the *natural* rights of ecosystems, I have, until now, not focused on any *legal* or *state* rights they may be afforded. However, it should be noted that rivers, mountains, forests, and whole ecosystems have been afforded legal protections, and their allies have sued to guarantee that these protections are maintained (Stone 1972). For example, the Whanganui River in New Zealand has been granted legal rights (Dwyer 2017, 1), and the Ecuadorian constitution has a chapter on the "Rights of Nature," with its Article 71 stating, "Nature or Pachamama, where life is reproduced and occurs, has the right to integral respect for its existence and for the maintenance and regeneration of its life cycles, structure, functions and evolutionary processes" (Ecuador 2008). In addition, Ukraine and Vietnam have laws against ecocide, the destruction of an ecosystem (Boyd 2017, 214). As I write this, lithium-rich Chile, too, is considering extending constitutional rights to nature and possibly ecosystems like salt flats and lagoons that may be harmed by mining the lithium that is needed by newer, "greener" technology (Sengupta 2021).

If, as I have argued, cities are ecosystems, and ecosystems have natural rights, then we can only conclude that cities have natural rights, as well, ones that are distinct to them. What, then, might those rights be? Natural rights cannot be uniform for all things. Children, for example, may have the right to freedom but not to the extent of adults, owing to the nature of a child's emotional and cognitive development. Cities, then, will have rights based on their own natures. Of the various possible natural rights possessed by cities, the right to exist comes first to mind. Of the rights of collectives and individuals, the natural right to continued existence is primary. For cities, this means that the city itself has a right to continued existence, and this is not merely the agglomeration of the rights of its inhabitants. If a city becomes uninhabitable, as has happened and continues to happen, and the people need to flee, then a city can die the way a forest can die.

Another right, the right to be spared from harm, seems relevant as well. There are peculiar harms that can be done to cities. As mentioned earlier, the problem of access is harmful to the city as it underlies some of the other common problems that cities face, such as crime. A last one is the right to freedom and self-determination. A unique feature of the city ecosystem is that many of its inhabitants are humans, who can decide on the aesthetics, social structures, culture, and other aspects of the urban ecosystem. To deny the inhabitants of the city this freedom, as might happen in the case of a hostile takeover by another community, would be to violate the right of the city to self-determination. This speaks to Harvey's claim that "the right to the city is far more than the individual liberty to access urban resources: it is a right to change ourselves by changing the city" (Harvey 2008, 23).

Conclusion: Reconceiving the Problem of Unequal Access Within the City

The problem of unequal participation in urban life is an instance of the violation of the right of the city to remain free from harm. Cities are harmed the way forests are harmed by the restricted exchange between different elements within the ecosystem. When the natural flow of a river is stopped by a dam or when a highway is put where a particular species would normally travel to mating grounds, the flow of elements within the system is cut off, and this has a negative impact on the system as a whole. The same can be said of the city. When access to the good things that cities can offer is cut off to some groups, the city is harmed.

A natural corollary to a right is responsibility. If I have successfully shown that cities have rights, then doesn't it follow that cities would have responsibilities as well? While beyond the scope and space limits of this piece, I would agree that this is the case. One such responsibility would be to avoid harming other cities and ecosystems, which would also have rights, if my argument is successful. Cities that pollute surrounding environments, for example, violate the rights of others and have a responsibility not to do so.

Protecting the rights *of* cities is a worthy political enterprise and can begin with the construction of a "Bill of Rights for a City." Rectifying the violation of the rights of a city would entail studying the city as an ecosystem, noting its health, and curing the problems—beginning with access. In doing this, the city is not treated as standing outside of us and as a resource to be used and possibly exploited. Instead, it is an ecosystem, and we are a component of it. Our cities and our own destinies are more clearly connected in this way of viewing things, and this would hopefully remind us all that our fates are interdependent. If one part of the city ecosystem is unfairly prevented from access to another, then the city as a whole suffers.

Notes

1 See Bassetti (2018), Chawla (2021), Dutko and Farrigan (2012), and Giacoletti, Heimer, and Yu (2021) for further discussion.
2 This may sound counterintuitive, since surplus capital needing to be parked somewhere would not need to be financed, the whole point being the use of the surplus. However, the surplus belonging to one entity could finance another. Also, the scale of the projects provides an opportunity to park not only present capital but future capital as well.
3 The construction of the Barclays Center in Brooklyn (2010–2012) was an example of the use of eminent domain to force inhabitants of a neighborhood to move. Also, a project by the same developer displaced, for years, a well loved Brooklyn public library in the upscale neighborhood of Brooklyn Heights; as of this writing, the library has not yet returned to the newly developed tower, which was part of the agreement to allow construction.
4 Robin Wall Kimmerer's *Braiding Sweetgrass* (2013) provides a lovely illustration of this in the chapter "Old Growth Children" (269–283).
5 In Plato's *Crito*, Socrates imagines the laws of Athens speaking to him and saying that he has no right to destroy them by fleeing Athens to save himself from an unjust death sentence (Plato 2012).
6 Cicero, for example, in *De legibus*, holds that the combination of reason and speech separates human beings from other creatures and provides humans with inherent rights to life and liberty.

References

Baier, Annette. 2010. "The Rights of Past and Future Persons." In *Reflections on How We Live*. New York: Oxford. http://profs-polisci.mcgill.ca/muniz/intergen/Baier.pdf.

Bassetti, Michelle. 2018. "Academic Inequality in the Urban School Setting." *PATimes*. American Society for Public Administration. https://patimes.org/academic-inequality-urban-school-setting-funding-disparities-lead-educational-disadvantages/.

Bentham, Jeremy. 1789. *An Introduction to the Principles of Morals and Legislation*. www.econlib.org/library/Bentham/bnthPML.html.

Boyd, David. 2017. *The Rights of Nature*. Toronto: ECW Press.

Carlson, Licia, and Eva Kittay. 2009. "Introduction: Rethinking Philosophical Presumptions in Light of Cognitive Disability." *Metaphilosophy* 40 (3–4): 307–30.

Carson, Rachel. 2000. *Silent Spring*. New York Penguin Press.

Chawla, Ambika. 2021. "Green Space in Cities Can Bring Considerable Health Benefits for Communities, But Access is Unequal." *ENSIA*. University of Minnesota, March 19, 2021. https://ensia.com/articles/urban-nature-environmental-justice-maginalized-underrepresented-communities/.

Cooper, David. 1968. "Collective Responsibility." *Philosophy* 43 (165): 258–68.

Dutko, Ver Ploeg, and Tracey Farrigan. 2012. "Characteristics and Influential Factors of Food Deserts." *ERR-140 Economic Research Service/USDA*, August. www.ers.usda.gov/webdocs/publications/45014/30940_err140.pdf.

Dwyer, Colin. 2017. "A New Zealand River Now Has the Rights of a Human." *National Public Radio*, March 16, 2017. www.npr.org/sections/thetwo-way/2017/03/16/520414763/a-new-zealand-river-now-has-the-legal-rights-of-a-human.

Ecuador. 2008. "Rights of Nature." *Constitution*. Chapter 7. https://pdba.georgetown.edu/Constitutions/Ecuador/english08.html.

Giacoletti, Marco, Rawley Heimer, and Edison G. Yu. 2021. "Using High-Frequency Evaluations to Estimate Discrimination: Evidence from Mortgage Loan Officers." https://ssrn.com/abstract=3795547 or http://dx.doi.org/10.2139/ssrn.3795547.

Harvey, David. 2008. "The Right to the City." *New Left Review* 53 (September–October): 23–40.

Holt, John. 1974. *Escape from Childhood: The Needs and Rights of Children*. New York: Penguin.

Kimmerer, Robin Wall. 2013. *Braiding Sweetgrass*. Minneaopolis: Milkweed Editions.

Lefebvre, Henri. 1968. "The Right to the City." In *Writings on Cities*, translated by Eleonore Kofman and Elizabeth Lebas. Oxford: Blackwell.

Locke, John. 1980. *Second Treatise of Government*. Edited by C. B. McPherson. Indianapolis: Hackett.

Plato. 2012. *Crito*. Translated by Cathal Woods and Ryan Pack. San Francisco: Creative Commons. www.pitt.edu/~mthompso/readings/crito.pdf.

Raz, Joseph. 1986. *The Morality of Freedom*. Oxford: Clarendon.

Sengupta, Somini. 2021. "Chile Writes Its Constitution, Confronting Climate Change Head On." *New York Times*, December 28, 2021. Accessed January 2, 2022. www.nytimes.com/2021/12/28/climate/chile-constitution-climate-change.html.

Stone, Christopher D. 1972. "Should Trees Have Standing? Toward Legal Rights for Natural Objects" *Southern California Law Review* 45: 450–501.

Wollstonecraft, Mary. 1782. "Vindication of the Rights of Woman." www.gutenberg.org/ebooks/3420.

10

MAKING OUR OWN HISTORY

Urban Sustainability in a World in Crisis

Deborah Mutnick

In Kim Stanley Robinson's (2017) apocalyptic but plucky *New York 2140*, the city is underwater, squatters live in impoverished, half-submerged downtown neighborhoods, and the rich live uptown ensconced in skyscrapers. Everyone gets around in boats. The city survives the ecocatastrophe of the First Pulse and the Second Pulse caused by melting ice in Antarctica and Greenland and two surges from sea level rise that flood lower Manhattan. Inequality has intensified in a world in which the top 1% owns 80% of global wealth. Throughout the novel, one of many narrators known only as "the citizen" offers a running critique of 22nd-century capitalism, replete with bank bailouts between 5 and 15 trillion dollars after the Second Pulse and financialization and privatization of an antiterrorist surveillance state in which the rich hire private security armies to police hundreds of millions of rebellious refugees and squatters. It was all, explains the anonymous citizen, "just a result of the logical progression of the ordinary workings of capitalism, following its overarching operating principle of capital accumulation at the highest rate of return" (Robinson 2017, 205; see also Heise 2021).

Robinson's climate-conscious depiction of an urban future under water is bleak and menacing yet also a Beckettian expression of resilience in the face of catastrophe. Even as he describes how the forces and relations of capitalist production create intertwining environmental and socioeconomic crises and how spectacularly our century failed to stave off disaster, he also shows that material conditions arising from yet another crisis of capitalism give rise to new forms of resistance and hope. Teaching literature like this tragicomic epic of city life in the 2100s presents the educational challenge of discovering the essential relations of complex, discrete, yet interrelated scientific, political economic, and historical realities. I started to integrate environmental themes into my teaching in response to the climate crisis. Given that existential threat, my primary goal as a teacher of writing and literature is to foster multidisciplinary, critical reading, writing, and analytical skills needed

DOI: 10.4324/9781003217442-12

by nonspecialists and specialists alike to evaluate and act collectively to change the world before it is too late.[1] This chapter argues that education is necessary though insufficient in itself in struggles for environmental sustainability and an egalitarian world order that "reconnects our ecologies with our economies" (Wallace et al. 2020, 11).

Defining the Pedagogical Problem: "An Existential Threat to Humanity"

Devastating storms, sea level rise, drought, soaring temperatures, and species extinction, together with dire international scientific reports of catastrophic consequences if carbon emissions are not cut by at least 55% by 2030 and obscene levels of global inequality that put poor people, especially in the Global South, at higher risk of economic and environmental calamity—these conditions can seem overwhelming and disparate, threatening and abstract. Their totality can be hard to grasp; they may seem impossibly remote, theoretical, diffuse, unfathomable. Moreover, they can elude solutions in the widening gap between the rhetoric and the reality of efforts to address them, resulting in failures of action, especially threatening to those most vulnerable to the vagaries of capital. It is also challenging to parse the historical processes that produce these seemingly disparate problems, determine their underlying causes, and remedy them. The alternative is to feel terrified (or inured), swept up in the tides of history—and let the rich and powerful continue to determine the collective fate of humanity.

Let me begin with some questions. How can we go beyond symbolic support for cutting greenhouse emissions to achieve the reductions needed to avert the worst consequences of global warming? To what extent do mainstream sustainability programs perpetuate the very systemic forces that derail them? What are those forces? Once the structural barriers to social and environmental justice become visible, what can be done to overcome them? Crucially, who is the "we" with the capacity and will to bring about the needed change? Whose interests are at stake? Whose rights? What role do urban spaces play in sustaining ecosystems? As urban development accelerates worldwide, what does the slogan "the right to the city" mean? Who controls the city's resources? To what end?

This cascade of questions poses several pedagogical challenges. First, the critical evaluation of texts and contexts depends on basic comprehension of relevant scientific, economic, historical, and sociological facts, concepts, and processes (see Carlin, Chapter 11; Horning, Chapter 21). Second, our interpretation of reality is conditioned by the mode of production, a specific political economic system characterized throughout human history by class division and concealing, in our era, capitalist relations of production that exploit workers and destroy the environment. Third, such objective realities are mystified by class interests delimited by the ideology of neoliberal capitalism.

To invite critical reflection and action entails striking a balance between rigorous, historical materialist analysis and robust dialogue that engages multiple publics.

It means stepping outside the dominant discourse in order to critique it. In what follows, I unpack some contradictions and contingencies of mainstream, sustainable development policy. Rather than focus on specific curricula or methods, I propose a pedagogy of exigency—to occur in diverse social spaces—that can raise consciousness and build a mass movement for egalitarian, environmentally conscious societies needed to avert disaster and build a new world. I then delineate three key ecological and sociological concepts fundamental to a deeper understanding of urban ecosystems:

1. *Great Acceleration* refers to a rapid increase since 1945 in average global temperatures and other Earth trends that had remained relatively stable for half a millennium and that correlate with upward socioeconomic trends, including global inequality (Steffen et al. 2015).
2. *Metabolic rift* explains the rupture in the metabolism or interdependent processes of society and nature caused by the degradation of soil, carbon emissions, and other forms of human impact on the environment (Foster 2013; Foster and Clark 2018).
3. The *right to the city*, proposed by French sociologist Henri Lefebvre (1996) in response to urban inequalities produced by capitalism, has since become a worldwide slogan in urban movements for housing, education, and jobs, as well as climate action and environmental justice (see also, Harvey 2008).

Finally, I argue that the twin struggles for climate justice and social equality undergirding the right to the city (and the world) necessitate rebuilding a global left capable of challenging the capitalist state.

Contingencies of Sustainable Development and the Capitalist Mode of Production

The United Nation's goals for sustainable development aim to improve the world's economic and environmental prospects. However, as the intergovernmental body itself acknowledges, even prior to the COVID-19 pandemic, it was at risk of breaking its "promise to current and future generations" set forth in the 2030 Agenda (*Sustainable* 2020; United Nations 2021). The United Nations Climate Change Conference of the Parties known as COP26 in Glasgow, Scotland, in 2021, did little to allay the concerns of climate scientists. As one climate researcher put it, "COP26 has closed the gap but it hasn't solved the problem" (Masood and Tollefson 2021). And all but the most diehard climate change deniers now understand that time is running out.

As a brief history of sustainable development efforts reveals, this impasse is not the first time the program has stalled. In 1967, the World Commission on Environment and Development, led by former Norwegian Prime Minister Gro Harlem Brundtland, famously defined sustainable development as "development that meets the needs of the present without compromising the ability of future generations to meet their own needs." Following the UN Millennium Summit in 2000, eight Millennium Development Goals (MDGs) were formulated with a target date of 2015. But as early as 2004, at a joint meeting of the International Monetary Fund and the World Bank those goals were considered elusive by global leaders, who projected that Sub-Saharan Africa would not attain the MDGs until sometime in the 2200s.[2] Predictably, by 2015, the target date, progress on the goals was negligible—mocked as "Minimalist Development Goals" by some African aid workers (qtd. in Davis 2007, 200).

Jennifer L. Tucker and Manisha Anantharaman (2020) note that, despite elite policy makers' emphasis on eradicating poverty and formalizing the world's growing informal labor sector to foster economic growth and ensure more humane living conditions, the reality is moving in the opposite direction, accelerated by COVID-19. In *Planet of Slums*, Mike Davis (2007) traces the radical transformation in the 1970s of already existing informal workers into a permanent underclass. Instead of creating a pathway out of poverty for the global masses, neoliberal, structural adjustment programs (SAPs) imposed by the International Monetary Fund and the World Bank—both UN affiliates—have decimated public sector employment and the formal labor sector, resulting in "myriad invisible networks of exploitation" that are especially oppressive for women and children (181). According to Tucker and Anatharaman, these conditions point to a fundamental contradiction in the UN's goal of promoting both economic development and environmental sustainability:

> [T]he SDGs remain captive to "fairytales of eternal economic growth," a mindset decried by youth climate activist Greta Thunberg at the 2019 UN Climate Summit. The SDGs privilege technocratic planning and propose win-win scenarios, downplaying the trade-offs between economic growth, social development, and environmental protection. Furthermore, *the SDGs ignore compelling evidence that redistribution, not growth, is the key.*
>
> *(291, emphasis added)*

The impact of the UN programs on environmental matters has also been minimal, as can be seen most disturbingly in the slow pace of the international response to the accelerating rate and impact of climate change. Recalling Herman Daly's (1990) critique of the very notion of sustainable development as oxymoronic, Michael R. Redclift (2006) declared "the oxymoron had come of age" (65). As Daly predicted more than 30 years ago, "If the Brundtland Commission means multiplying the existing economy and its current matter-energy throughput by five to ten, then they are calling for an ecological holocaust" (401). While human life will always require agricultural and industrial production and rational infrastructural development, the profit model based on gross domestic product (GDP) embedded in the SDGs contradicts the attainment of sustainable social and Earth systems by continuing to plunder both labor and nature.

In a discussion of green strategy approaches to the climate crisis, Robert Pollin (2018) notes, "It is obvious that growth per se, as an economic category, makes no reference to the distribution of the costs and benefits of an expanding economy" (7). Nor does the GDP adequately measure environmental degradation, unpaid labor, mostly done by women, or income and wealth distribution. Even so, Pollin argues that the climate crisis can be solved in 40 years by eliminating fossil fuel pollution, which constitutes 70% of all greenhouse gas emissions, while investing heavily in clean, renewable energy that meets global needs and supports economic development worldwide through job creation and programs for displaced energy workers. Driven by the imperative of acting now to stop global warming, his plan assumes that capitalist states can realize these objectives—an appealing assumption given the planet's rapidly worsening conditions, the weakness of the global left, and the absence of any semblance of revolutionary conditions.

Yet the stubborn fact remains that the capitalist mode of production has a "historically specific, systemic and class character, which must be taken into account seriously if we are to effectively face the problem" (Liodakis 2010, 2608). Since the 1970s energy crisis, U.S. capitalism has suffered from repeated economic shocks, including the long-lasting effects of the 2008 financial crash. Its neoliberal policies have enriched the 1% and increased global inequality and environmental destruction, only to be redoubled in 2020 by the COVID-19 pandemic, which left tens of millions of Americans unemployed as the stock market soared. The return to Keynesian policies to achieve the environmental reforms promised by the Biden Administration is thus tenuous given the worsening crisis of capitalism as it exhausts both its markets and the surplus it can extract from workers and nature. Moreover, as the legislative impasse in 2021 over Biden's already much diminished Build Back Better plan persists, hope ebbs for the administration's commitment to a 50% reduction from 2005 levels of greenhouse gas pollution by 2030. As Pollin acknowledges, fossil fuel companies like Exxon-Mobil and Chevron have "massive interests at stake" and will have to be defeated. Along with coauthor Noam Chomsky (Chomsky and Pollin 2020), he believes the most effective lever for climate action is mobilization of masses of people.

Urban spaces play a key role in such struggles insofar as they contain more than half the world's current population with 70% projected by 2050 and 90% by 2100 (UN GA 2015; Sanderson, Walston, and Robinson 2018), thrusting millions of people together in conditions conducive to movement building. Additionally, according to Eric Sanderson, Walston, and Robinson (2018), cities have the potential for a renaissance of nature a century from now—a breakthrough from the current bottleneck of pollution and carbon emissions—produced by "long-term sociological change" driven by lower fertility rates, less poverty, higher levels of education, and more job opportunities, particularly for women. These social processes tend to reduce population and social inequalities, increase per capita income, and foster greater ecological consciousness. At the same time, the requirements for more compact greenspaces and public transportation inherent in city life contribute to lowering the per capita carbon footprint through denser housing, mass transit, accessible education, and falling birth rates. All of these factors benefit working- and middle-class urban dwellers as well as plant and animal life in and beyond city borders by reducing planetary global warming.

While Sanderson et al. acknowledge the global conservation movement's "brave and worthy but often futile struggle" so far to protect the environment, they nonetheless predict that long-term, positive socioeconomic trends will mitigate climate change and create conditions for a recovery of diverse plant and animal life. Thus, even as cities bear traumatic histories of dispossession, migration, and brutal living conditions, they hold the key to our collective survival insofar as the interdependence of society and nature is never fixed in time or space but always in the process of becoming, a dialectic that could lead to catastrophic conditions like those we currently face but also to a radical transformation of society and nature. However, like the gap between the rhetoric and the reality of sustainability development goals, Sanderson proposes a set of future scenarios mired in what-if contingencies—such as the elimination of extreme poverty and investments in health, education, and affordable housing. While declines in global population will happen faster than previously predicted, sparking new concerns about worker shortages and proliferating retirees, global poverty is far from ending and climate change mitigation is nowhere near sufficient to stop, much less reverse rising greenhouse gas emissions (Cave, Bubola, and Sang-Hun 2021; Plumer 2021).

Sanderson's predicted renaissance of nature is contingent not only on population stabilization but also on substantial reductions in poverty and a "shared belief" that it is best "to care for—rather than destroy—the natural bases of life on Earth" (413). This imagined breakthrough to social equality and natural recovery assumes that the same global market forces that have resulted in half the globe living on less than $5 a day and "unprecedented destruction of the environment" will invest in "human welfare" and rational environmental policies. In one sense, the argument that the destruction of nature is creating the "necessary conditions" for transformative change suggests a dialectic that echoes the logic discovered by Marx and Engels in the rise of capitalism and modern urbanized life, a historical process that drove peasants off the land into teeming, polluted cities and factories. Being thrust together in crowded, unsanitary, disease-ridden quarters inflicted misery on masses of people but also created the context for collective struggle for better working and living conditions.

However, as Marx and Engels (1893) declare at the outset of *The Manifesto of the Communist Party*, class division defines human history, assuming a specific character in different historical epochs—a reality that is both obvious and obscured by the dominant ideology of capitalism and absent from Sanders et al.'s breakthrough theory. Since 1973, the contradictions of capitalism have intensified as it struggles to recover from deepening financial crises of overaccumulation and a falling rate of profit through global restructuring and neoliberal programs of austerity, privatization, deregulation, and financialization that have, in turn, resulted in deprivation for the many and obscene concentrations of wealth for a very few. It is precisely the analysis of class and late capitalism that is missing from neoliberal projects like the 2030 Agenda and Sanders' "breakthrough" forecast for urban ecological recovery.

Even Pollin's left-of-center proposal for a global green new deal as a necessarily immediate response to climate change underestimates the intransigence of capitalism. What, then, is to be done? As Chomsky put it in a 2020 virtual forum: "Make more trouble. Come with the pitchforks. That will make a difference. They [banks and corporations] have to listen."[3]

Phenomenal Forms of Essential Relations: Key Concepts for Urban Sustainability

The role of educators in this pivotal moment of history is not only to produce the next generation of discipline-specific researchers, scholars, and practitioners but also to educate nonspecialists about climate science and promote critical analysis for specialists and laypeople alike of 21st-century political economic and environmental realities. A useful pedagogical tool in this regard lies in Marx's (1867) theory of the phenomenal forms, related to ideology and to commodity fetishism but pertaining more broadly to the tendency for reality to appear in distorted or inverted form. As Marx (1845) wrote in *The German Ideology*, "If in all ideology men and their circumstances appear upside-down as in a camera obscura, this phenomenon arises just as much from their historical life-process as the inversion of objects on the retina does from their physical life-process." Similarly, the phenomenal forms or surface manifestations of reality conceal its "hidden substratum" or essential relations. But unlike Marx's emphasis on how ruling class ideology masks or distorts reality, the phenomenal forms can be applied more broadly to any human activity, including scientific discovery, in which there exists "a gap between how processes really develop and how they end up being perceived by human observers" (Villacañas de Castro 2016, 6).

For urban sustainability educators, the theory can be applied to phenomenal forms of climate change such as intensifying wildfires. While wildfires are indisputable facts, the underlying physicochemical processes of greenhouse emissions that produce them have been distorted by disinformation campaigns as well as by attribution to human impact in general rather than the specific impact of capitalism. Moreover, students may struggle with complex scientific concepts and schemas that are difficult to grasp. The educator's role is thus to present the essential relations of scientific and sociological processes that may be obscured by learners' resistance to new, unsettling, or challenging knowledge or by their susceptibility to dominant ideologies that resist critique and reflect their social conditioning.

Three key concepts for the critical study of urban ecosystems are the *Great Acceleration* of climate change and other destructive environmental and socioeconomic trends, often seen as broadly anthropogenic rather than largely produced by capitalism; the *metabolic rift* in the metabolism or interdependence between nature and humanity; and the *right to the city* vis-à-vis forces of disinvestment, dispossession, accumulation of surplus capital, and development that have for centuries rebuilt urban spaces to accommodate the needs of capital. Given space constraints, I will briefly define each concept in relation to its phenomenal form as a prelude to some final thoughts about making history.

The Great Acceleration

Though it has yet to become official, the Anthropocene is widely invoked as a new geological epoch marking significant human impact on Earth's geological and ecological systems. Researchers have disagreed on the start date, with suggestions ranging from the advent of agricultural production 7,000 to 10,000 years ago to the rise of industrialism in the late 18th and 19th centuries to the massive increase in industrial and weapons production after World War II. Numerous studies, however, locate a sharp rise in greenhouse gas emissions starting in 1945 as shown by the "hockey stick graph" of multiple measures of CO_2 over the past two millennia (Mann, Bradley, and Hughes 1998; Steffen, Crutzen, and McNeill 2007).[4] Hard put to dismiss the reality of increasingly violent weather events, climate change deniers, including companies like Exxon, have long disputed the evidence of human impact on ecosystems, claiming that the myriad phenomenal forms of a hotter planet result from normal variations in climate rather than chemical interactions of CO_2 and infrared radiation that heat up the Earth's surface.

While massive disinformation campaigns by Exxon and other fossil fuel companies sought to convince Americans that global warming was a hoax, many climate change believers attribute it to inexorable historical development rather than the rapid capitalist expansion that triggered the Great Acceleration in the postwar era. For example, *The New York Times Magazine* devoted an entire issue to Nathaniel Rich's (2018) dramatic history of climate science findings from 1979 to 1989 and the political failure to take action as economic concerns driven by oil lobbyists trumped scientific evidence. Yet even as he exposes the demonic role of the fossil fuel industry and the complicity of the state, Rich ascribes the lost momentum of the 1980s to human nature rather than the "neoliberal crusade" that coincided with climate scientist James Hansen's congressional testimony in 1988 that the planet was on the verge of heating up to a point of no return. Pedagogically, this distinction between human nature and unfettered global capitalist accumulation of wealth through the extraction of natural and human resources is fundamental to identifying the primary source of the destruction of life on the planet to which we are bearing witness.

The Metabolic Rift

Underlying the discovery of the Great Acceleration of environmental depredation is the concept of a metabolic rift of interactions between humanity and nature. Just as the denial of climate science and the attribution of the climate crisis to "human nature" conceal the essential relations and geological-ecological consequences of the global spread and intensification of capitalism since 1945, the idea that nature is a force to be mastered or that capitalism is infinitely adaptable and can surmount the ecological crises of the 21st century conceals the rift between humanity and nature that objectively endangers our existence. John Bellamy Foster's (2013) recovery of Marx and Engels' extensive explorations of the dialectic between nature and society forms the basis of his elaboration of a historical materialist analysis that the deepening alienation of humanity from both the labor process and from nature in the Anthropocene is in direct proportion to capitalism's extraction of wealth from workers and the Earth.

Through this materialist lens, Marx identified an "irreparable rift in the interdependent process of social metabolism, a metabolism prescribed by the natural laws of life itself" (qtd. in Foster). In this process, labor mediates the metabolic interaction between humanity and nature according to the objective conditions of existence. This view of human and natural history rests on two related points: first, human beings are part of nature yet distinct from it in their capacity to "*produce* their means of subsistence" (Marx and Engels 1845); second, capitalism extracts surplus value from labor *and* nature in processes that produce calamitous but also, at least potentially, revolutionary effects. As a key term, "metabolic rift" can thus help learners conceptualize the link between biophysical and socioeconomic processes, the interaction of nature and human beings, and the extraction of surplus capital from nature and from workers as a rupture of the social metabolism.

The Right to the City

An emancipatory slogan coined by French sociologist Henri Lefebvre in 1968, "the right to the city," has become a critical concept and slogan for leftwing urban social movements as the process of urbanization continues to expand and accelerate (for other perspectives on "the right to the city," see Cuonzo, Chapter 9; Carlin, Chapter 11). Unlike established rights such as freedom of speech or freedom of assembly, the right to the city is a contested, amorphous, yet powerful call to action in response to social inequality, racism, poverty, gentrification, dispossession, segregation, environmental injustice, and, more recently, catastrophic climate events. Bound up in 18th-century Enlightenment thought—the Age of Reason that embraced the scientific method and forged new principles of liberty, fraternity, and equality—today's established freedoms comprise both bourgeois rights like free speech and capitalist rights to property and expropriated surplus. As such, the right to the city is a concept whose meaning is up for grabs depending on the uses to which the slogan is put.

Two urban planners notorious for their massive redesign of cities—George-Eugène Haussmann who rebuilt Paris during the Second French Empire and Robert Moses who transformed the New York Metropolitan Area in the Depression and postwar periods—obliterated whole working-class neighborhoods to make cities more amenable to the bourgeoisie and absorb surplus labor and capital. Harvey (2008) explains how the dialectic of overaccumulation and expansion resolves periodic crises of capital by spurring development and displacing working-class residents from their neighborhoods. Urban restructuring, he observes, is always a "class phenomenon since surpluses are extracted from somebody, while the control over their disbursement typically lies in a few hands" (24). The right to the city means democratic control over the surplus from urban processes to be used for the collective good, now including whatever steps can be taken to mitigate and adapt to climate change. For urban dwellers, this concept has immediate applications and relevance yet also poses formidable challenges. Daniel Aldana Cohen (2018) argues that "climate justice is the right to the city" in its focus on housing, transit, and land use, with intelligent densification at the center of urban climate politics and the struggle for affordable housing and climate justice (12).

In his discussion of a Marxist aesthetics of cognitive mapping, Fredric Jameson (1988) brilliantly explains the imperialist basis of monopoly capitalism, which expands its geographical reach to the far corners of the world, locating the "truth" of individual experience in London or Rio de Janeiro or elsewhere and giving rise to modernist approaches to literature that seek to close the cognitive gap created by a local reality that is overdetermined by elusive, typically invisible global forces. He writes: "Without a conception of the social totality (and the possibility of transforming a whole social system), no properly socialist politics is possible" (347). Thus, to assert their right to the city, urban dwellers must not only confront the hopelessness, atomization, and alienation that the unemployed, the homeless, the aged, the young, the marginalized, and others may experience but also the limits of electoral politics and liberal solutions to the pressing local problems inscribed by global forces that increasingly demarcate city streets.

★ ★ ★

At the end of Robinson's novel *New York 2140*, in the summer of 2142, Hurricane Fyodor strikes New York City, wiping out all the trees in Central Park and leaving "Harlem a bedraggled bay" (489).[5] Refugees camp in the park in the wake of the devastating storm, and Amelia Black, a pilot who transports endangered animals to safety and broadcasts "Assisted Migration" episodes from her plane, calls for a householders' strike, a jubilee, "Because at this point it's democracy versus capitalism. We the people have to band together and take over. We can only do that by mass action. . . It's a case of all for one and one for all" (528). General strikes and rioting spread throughout New York and the world, causing a housing bubble in the finance market that pops. "The house of cards had fallen again, and the whole world was left standing in the rubble of the crashed economy" (533). When the Rad Dems get elected to Congress in 2143, they nationalize the banks, pass a "Picketty tax," and propose universal healthcare, free education through college, a living wage, and full employment to be paid for by a "healthy budget surplus" (602). The citizen concludes, "History happened. It does not stop happening. Seemingly frozen moments are transient, they break up like the spring ice, and then change occurs" (603).

"Men make their own history, but they do not make it as they please," Marx (1852) wrote in *The Eighteenth Brumaire of Louis Bonaparte*. "[T]hey do not make it under self-selected circumstances, but under circumstances existing already, given and transmitted from the past." As terrifying as they are, the cataclysmic effects of global warming are ripening the potential for radical change. The ever-deepening crisis of capitalism has rendered a stark divide between the tiny fraction of the world's billionaires and its vast poor, low-wage, and precariously middle-class majority, intensified by COVID-19 and an environmental collapse already underway. These conditions have created what Arundhati Roy (2020) calls, writing about the impact of the pandemic, a "portal, a gateway between one world and the next." Like Beckett (1953/2010), who famously declares, "I can't go on. I'll go on," Roy insists: "We can choose to walk through it, dragging the carcasses of our prejudice and hatred, our avarice, our data banks and dead ideas, our dead rivers and smoky skies behind us. Or we can walk through lightly, with little luggage, ready to imagine another world. And ready to fight for it." In defining the "right to the city," Harvey remarks on sociologist Robert Parks's observation that " 'in making the city man [sic] has remade himself.' The right to the city is therefore "the right to change ourselves by changing the city" (23). Like the dispossessed in Robinson's novel, urban (Earth) dwellers can exercise that right by rising up to demand a more egalitarian, regenerative social and natural world.

Notes

1 Here, I am conjuring the words of Karl Marx—"The philosophers have only interpreted the world in various ways. The point, however, is to change it"—inscribed on his tombstone.
2 UK Chancellor of the Exchequer Gordon Brown stated, "Sub-Saharan Africa will not achieve universal primary education until 2130, a 50 percent reduction in poverty in 2150 and the elimination of avoidable infant deaths until 2165" (qtd. in Davis 2007, 18–19).
3 See "Noam Chomsky and Robert Pollin on the Climate Crisis, Capitalism, Trump, and the Green New Deal." www.youtube.com/watch?v=a5rflVwfEP4.
4 Hockey stick graphs reflect similar transformations since the mid-20th century in measurements of all socioeconomic and Earth systems trends (Steffen et al. 2015).
5 This future is rapidly unfolding in the present in increasingly deadly weather events. In September 2021, remnants of Hurricane Ida struck the New York Metropolitan Area, dropping 3.15 inches of rain in Central Park in one hour, smashing a record set the previous week of 1.94 inches and killing at least 48 people, many of whom died in flooded basement apartments in New York City (Levenson and Barnard 2021).

References

Beckett, Samuel. (1953) 2010. *The Unnamable*. London: Faber & Faber.

Cave, Damien, Emma Bubola, and Choe Sang-Hun. 2011. "Long Slide Looms for World Population, With Sweeping Ramifications." *New York Times*, May 22, 2011. www.nytimes.com/2021/05/22/world/global-population-shrinking.html.

Chomsky, Noam, and Robert Pollin. 2020. *Climate Crisis and the Global Green New Deal: The Political Economy of Saving the Planet*. New York: Penguin.

Cohen, Daniel Aldana. 2018. *Climate Justice and the Right to the City*. Penn Current Research on Sustainable Urban Development. https://penniur.upenn.edu/uploads/media/Cohen.pdf.

Daly, Herman E. 1990. "Sustainable Growth: A Bad Oxymoron." *Journal of Environmental and Science Health* C8 (2): 401. https://doi.org/10.1080/10590509009373395.

Davis, Mike. 2007. *Planet of Slums*. London: Verso.

Engels, Friedrich. 1893. *The Dialectics of Nature*. Marx Engels Archive. www.marxists.org/archive/marx/works/1883/don/index.htm.

Foster, John Bellamy. 2013. "Marx and the Rift in the Universal Metabolism." *Monthly Review* 11 (7). https://monthlyreview.org/2013/12/01/marx-rift-universal-metabolism-nature/.

Foster, John Bellamy, and Brett Clark. 2018. "The Robbery of Nature: Capitalism and the Metabolic Rift." *Monthly Review* 70 (3). https://monthlyreview.org/2018/07/01/the-robbery-of-nature/.

Harvey, David. 2008. "The Right to the City." *New Left Review* 53 (September–October): 23–40.

Heise, Ursula. 2021. "Urban Narrative and Climate Change." In *Transcultural Ecocriticism: Global, Romantic, and Decolonial Perspectives*, edited by Stuart Cooke and Peter Denney. London: Bloomsbury Academic.

Jameson, Fredric. 1988. "Cognitive Mapping." In *Marxism and the Interpretation of Culture*, edited by Cary Nelson and Lawrence Grossberg, 347–60. Urbana and Chicago: University of Illinois Press.

Lefebvre, Henri. 1996. *Writings on Cities*. Translated by Elenore Kofman and Elizabeth Lebas. Oxford: Blackwell.

Levenson, Michael, and Anne Barnard. 2021. "Scenes from New York as Ida Paralyzes Region." *New York Times*, September 2, 2021. www.nytimes.com/2021/09/02/nyregion/flash-floods-new-york.html.

Liodakis, George. 2010. "Political Economy, Capitalism, and Sustainable Development." *Sustainability* 2: 2601–16. doi:10.3390/su2082601.

Mann, Michael E., Raymond S. Bradley, and Malcolm K. Hughes. 1998. "Global-Scale Temperature Patterns and Climate Forcing over the Past Six Centuries." *Nature* 392 (April): 779–87.

Marx, Karl. 1852. *The Eighteenth Brumaire of Louis Bonaparte*. Marx Engels Archive. www.marxists.org/archive/marx/works/1852/18th-brumaire/ch01.htm.

———. 1867. *Capital: A Critique of Political Economy*, Vol. 1. Marx Engels Archive. www.marxists.org/archive/marx/works/1867-c1/index.htm.

Marx, Karl, and Friedrich Engels. 1845. *The German Ideology*. Marx Engels Archive. www.marxists.org/archive/marx/works/1845/german-ideology/index.htm.

———. 1893. *Manifesto of the Communist Party*. Marx-Engels Archive. www.marxists.org/archive/marx/works/1848/communist-manifesto/.

Masood, Ehsan, and Jeff Tollefson. 2021. "COP26 Hasn't Solved the Problem: Climate Scientists React to UN Climate Deal." *Nature* (November 15, 2021). www.nature.com/articles/d41586-021-03431-4.

Plumer, Brad. 2021. "Carbon Dioxide in Atmosphere Hits Record High Despite Pandemic Dip." *New York Times*, June 7, 2021. www.nytimes.com/2021/06/07/climate/climate-change-emissions.html.

Pollin, Robert. 2018. "De-Growth Vs. a Green New Deal." *New Left Review* 112 (July–August): 5–25.

Redclift, Michael R. 2006. "Sustainable Development (1987–2005): An Oxymoron Comes of Age." *Horizontes Antropológicos* 12 (25): 65–84.

Rich, Nathaniel. 2018. "Losing Earth: The Decade We Almost Stopped Climate Change." *New York Times Magazine*, August 1, 2018. www.nytimes.com/interactive/2018/08/01/magazine/climate-change-losing-earth.html.

Robinson, Kim Stanley. 2017. *New York 2140*. New York: Orbit.

Roy, Arundhati. 2020. "The Pandemic Is a Portal." *Financial Times*, April 3, 2020.

Sanderson, Eric W., Joseph Walston, and John G. Robinson, 2018. "From Bottleneck to Breakthrough: Urbanization and the Future of Biodiversity Conservation." *Bioscience* 68 (6): 412–26.

Steffen, Will, Wendy Broadgate, Lisa Deutsch, Owen Gaffney, and Cornelia Ludwig. 2015. "The Trajectory of the Anthropocene: The Great Acceleration." *The Anthropocene Review* (March): 1–18. doi:10.1177/2053019614564785.

Steffen, Will, Paul J. Crutzen, and John R. McNeill. 2007. "The Anthropocene: Are Humans Now Overwhelming the Great Forces of Nature?" *Ambio* 36 (8): 614–21.

Tucker, Jennifer L., and Manisha Anantharaman. 2020. "Informal Work and Sustainable Cities: From Formalization to Reparation." *One Earth* 3 (3): 290–99. https://doi.org/10.1016/j.oneear.2020.08.012.

United Nations. 2021. *The Sustainable Development Goals Report 2021*. New York: United Nations Publications. https://unstats.un.org/sdgs/report/2021/.

United Nations, General Assembly. 2015. "Transforming our World: The 2030 Agenda for Sustainable Development." October 21, 2015. A/ RES/70/1. Sustainable Development Knowledge Platform. https://sustainabledevelopment.un.org/post2015/transformingourworld/publication.

Villacañas de Castro, Luis S. 2016. *Critical Pedagogy and Marx, Vygotsky, and Freire: Phenomenal Forms and Educational Research*. London: Palgrave Macmillan.

Wallace, Rob, Alex Liebman, Luis Fernandez Chaves, and Rodrick Wallace. 2020. "Covid-19 and the Circuits of Capital." *Monthly Review* 72 (1). https://monthlyreview.org/2020/05/01/covid-19-and-circuits-of-capital/.

11

RECONCEPTUALIZING THE CITY IN THE AGE OF THE ANTHROPOCENE

Scott Carlin

The world faces a cascade of rolling crises in coming decades as global corporations orchestrate the movement of capital around the world and acquire more wealth and power through their investments and their control of governments. Cities are at the center of this storm. As capitalism transforms and degrades the world's lands, oceans, world biomes, and climate, scientists are forced to wrestle with this as a geologically significant process. This new era, the Anthropocene, is also a call for planetary management, but for whom and by whom? What is urgently needed is not just a radical reshaping of energy systems away from fossil fuels but a deeper transformation of economies and aspirations toward a culture of peace developed through the right to the city.

Cities remain the most likely geographic spaces for innovating practices of social and ecological justice. Today's metropolitan areas are home to just over half of the world's population. This chapter outlines four themes important for reconceptualizing the city in the Anthropocene. First is the right to the city, which expands and reconceptualizes urban democratic practices. Second is building a culture of peace through an ethic of care. Third is a call for metabolic praxis: using the right to the city and a culture of peace to reimagine and reconstitute the metabolic functions of the city—its material flows and their transformations. Urban metabolism is a socioecological process that creates and distributes the vital nutrients that support human health and productive ecologies. Lastly, I confront the grim realities of existing urban growth dynamics and their destructive logic.

Capitalism has revolutionized our world, but it reduces human and ecological relationships to economic calculations. Its central engine is profits; whenever profits falter, the system slips into crisis. Cycles of both boom and bust are inherently violent. The concentration of power into the hands of the few is inherently corrupting. The global scale of poverty and environmental degradation now requires us to struggle for a postcapitalist future. What conditions and economic models

DOI: 10.4324/9781003217442-13

would permit urban movements to build healthy, socially just, and peaceful societies? Three different ways to conceptualize this future are restructuring business enterprises, municipalism, and ecological democracy.

Participatory economics offers one example for radically restructuring business enterprises. Michael Albert (2021) advocates for a world with "No Bosses," where the responsibilities and benefits of work are shared fairly. In her survey of worker-owned cooperatives, Virginie Pérotin (2020) concludes that they "are more productive than conventional businesses" (3). Whereas Albert envisions a nonhierarchical economy, Richard Wolff (2017, 2012) advocates for workers' self-directed enterprises, where workers fully and democratically control their organization's core economic decisions. These decisions include what is produced, how it is produced, and how profits are shared, but organizations could still retain hierarchical structures, such as managerial jobs and complex divisions of labor.

At the urban scale, municipalism offers many examples of creative engagement. Matthew Thompson (2021) surveys municipalism's visions and initiatives and calls to radically rescale political economies through projects such as the Fearless Cities (2021) network, "an informal global movement . . . to radicalize democracy, feminize politics and drive the transition to an economy that cares for people and our environment."

A third example is Randolph T. Hester's (2006) short summation of ecological democracy:

> Ecological democracy, then, is government by the people emphasizing direct, hands-on involvement. Actions are guided by understanding natural processes and social relationships within our locality and the larger environmental context. This causes us to creatively reassess individual needs, happiness, and long-term community goods in the places we inhabit. Ecological democracy can change the form that our cities take creating a new urban ecology.
>
> *(4)*

The rise of authoritarian political movements is one of many practical difficulties with building economies that support cultures of peace. As capitalism propels us deeper into a planetary climatic and ecological crisis, state-sanctioned violence will likely increase. The challenge is to expand networks of locally based international networks of egalitarian practices.

The Right to the City

Cities are hierarchically organized into global, national, and regional centers. A defining aspect of economic globalization is its organization through these urban networks (Taylor and Derudder 2016). Profits fuel cycles of interurban reinvestment and growth. The financial system is critical to capitalism's ability to expand credit and rates of growth. In the right to the city, David Harvey (2008) provides a compelling narrative for a geographic reinterpretation of human rights: "Since the urban process is a major channel of surplus use, establishing democratic management over its urban deployment constitutes the right to the city" (37). He further notes that this process of establishment "has to be a global struggle, predominantly with finance capital" (39).

Harvey recounts how the process has previously worked under capitalism. One early example is Georges-Eugène Haussmann who "transformed the scale at which the urban process was imagined" in mid-1800s Paris, in effect, "setting up a proto-Keynesian system of debt-financed infrastructural urban improvements" (26). After World War II, Robert Moses transformed the New York metropolitan region, using similar methods. This became a template for cities across the country and "entailed a radical transformation in lifestyles" for American consumers and the creation of "new financial institutions and tax arrangements that liberated the credit to debt-finance urban expansion" (27). Now there is an urgent need to design climate-friendly eco-cities. Finance capital is already aggressively shifting resources to fund these projects. In 2021, the Glasgow Financial Alliance for Net Zero (2021) pledged to develop and implement voluntarily defined net zero carbon emission goals to decarbonize the global economy. As of November 2021, Alliance members managed $130 trillion in assets among its "banks, insurers, pension funds, asset managers, stock exchanges, credit rating agencies, and audit firms" (Alderman and Nelson 2021). In other words, finance capital is building financial alliances that can be directed toward international urban restructuring. How can democratic coalitions demand more public accountability and control over this funding? Architects, designers, planners, scientists, and activists have already created scores of models for progressive, new urban designs. Those who control the money, however, will also control the designs that are built and replicated. National green new deal campaigns have the potential to legally require new funding streams for urban climate justice projects. But this will not fundamentally change the dynamics of capitalism and wealth inequality. An urgent opportunity exists for urban coalitions to assert democratic controls over these investments that will define the future of cities.

Corporations have created innovative new frameworks to address new ecological demands. Circular economics, for example, is a framework to engineer eco-industrial parks to achieve zero waste. In Kalundborg, Denmark, industrial companies pipe waste from one facility into neighboring plants that use those materials as raw materials. This cooperative model is known as "Kalundborg Symbiosis" (Faria, Caldeira-Pires, and Barreto 2021). China plans to develop over 200 eco-industrial parks. Kalundborg is innovative, but its worldview remains driven by commodities, not ecology.

The greening of some corporations, however, is not a viable model of reform. The expansion of unbridled corporate power has dramatically degraded democratic politics. The power and wealth corporations wield is fueling authoritarian movements. Sheldon Wolin (2008) suggests the specter of a new postdemocratic form of power: inverted totalitarianism. It is the "*political* coming of age of corporate power and the *political* demobilization of the citizenry" (x). Joel Bakan (2004) asked psychologist Dr. Robert Hare to apply a diagnostic checklist of psychopathic traits to the modern corporation, and the modern corporation failed that test. Corporations are irresponsible, grandiose, manipulative, lack remorse and empathy, and often fail to take responsibility for the problems they generate. In striving to achieve their goals, "everybody else is put at risk" (57).

Marek Nowak (2020) finds that " 'urban social movements' and 'right to the city' initiatives are becoming an increasingly important subject for researchers representing various disciplines" (8). Researchers find that these movements often focus on divergent goals. Anna Domaradzka (2018, 615) categorizes mobilization efforts into economic and social, housing, civic and political, environmental, cultural, and ownership rights to the city. Paul Raskin (2016) scripted a variety of future world scenarios to assess what life will be like if we choose a more democratic or a more authoritarian future. The democratic, equitable path best meets human needs. Today, it is not only possible, it is absolutely necessary that we challenge and restructure what Harvey (2008) calls the "hegemonic liberal and neoliberal market logics" (23).

An Ethic of Care

It is the exploitive city, with its spatial density of capital, workers, and creative talent, that nurtures oppositional social movements that strive to universalize human rights and protect ecological systems. If the right to the city is a global urban struggle to reorganize finance capital to support economic democracy, it also requires a new geographic politics of care that expands our collective capacities to develop our potential to live meaningful lives. Care is essential to the emancipatory project of cultivating human freedom.

Carol Gilligan (2011) defines care as "a relational ethic, grounded in a premise of interdependence. . . . Within a *patriarchal* framework, care is a feminine ethic. Within a *democratic* framework, care is a human ethic." Riane Eisler (2017) promotes caring economics built upon democratic, postcapitalist, partnership-based institutions as a framework that can replace contemporary urban systems of dominance. And Vandana Shiva (2015) urges us to join "in an amazing uprising of love and care where we act as one heart, as one mind and one consciousness to say no to this ecocide and genocide." Shiva's own work focuses upon rejecting the corporate privatization of public goods such as water and genetically engineered farm seeds. Rejecting the commodification of life, Shiva promotes transformative social change based on Earth Democracy, a democracy that supports the interconnectedness of all life and "globalizes peace, care and compassion" (Shiva 2005, 11).

Care stands in stark contrast to the instrumental rationality of capitalism, which sees workers and nature as factors of production in a profit-driven system. From this perspective, the employment wage contract not only deprives workers of their just pay and alienates them from their labor product, it undermines their right to achieve self-actualization. Abraham Maslow's (1943, 1971) hierarchy of human needs articulates a universal human desire to achieve our full potential, our self-actualization and self-transcendence. The exploitive and hierarchical nature of social relationships thwarts our collective capacity to achieve these higher aims of life.

Self-transcendence is fundamentally an awakening to interdependence as an ontological and scientific reality. Revolutionary consciousness requires the personal and social cultivation of self-transcendence. Dr. Martin Luther King Jr. (1963) eloquently expressed his own sense of self-transcendence in explaining why he was sitting in the Birmingham jail. "Injustice anywhere is a threat to justice everywhere" (1).

The emerging field of planetary health and the Green New Deal offer two contexts for exploring the necessity for global governance, new modes of urbanization, and the relevance of an ethic of care. In his review of Earth's planetary health, Samuel Myers (2017) concludes:

> [A]chieving planetary health will require a renaissance in how we define our place in the world. A new narrative will reject the one streaming into our homes—that happiness comes from relentlessly acquiring more things—and embrace what we know: that what truly makes us happy is time spent with those we love, connection to place and community, feeling connected to something greater than ourselves, taking care of each other.
>
> *(2867)*

Myers confines his renaissance to the realm of reproduction. Myers acknowledges that the economy has mortgaged our future but fails to address that this is a necessary outcome of capitalist social relations. Pierre-Marie David, Nicolas Le Dévédec, and Anouck Alary (2021) critique the planetary health scientific project as an effort to strengthen the technofinancial resilience of capitalism through the development of biotechnology and the erosion of local democracy. Their own model of planetary health is "one that puts the establishment of democratic and ecological post-capitalist health above all else" (1151).

A potential example of this is the Green New Deal. The Paris Agreement is built around nationally determined voluntary contributions to reduce carbon emissions. The Green New Deal is therefore a national economic policy framework to spur investments in renewable energy production, among other goals. A key issue is urban transportation. Will the automobile continue to dominate urban landscapes or will cities reclaim roadways for innovative new uses? It is likely that climate change will force cities to undertake more radical transformations in urban infrastructure and social relations than currently envisioned.

Green New Deal initiatives have cultivated important new urban coalitions focused on jobs and infrastructure, but they have not examined the problematic nature of urban property relationships. In land relationships, Native American tribes see the natural world as an integral part of their family relations; they reject an instrumental understanding of nature. They did not understand the European construct of land as property. Their cultural ontology was interdependence (Cajeta

2000). In developing his land ethic, Aldo Leopold (1949) urged us to give up the role of conqueror. The conqueror knows "what makes the community clock tick, and just what and who is valuable, and what and who is worthless" (204). This certainty in war, ecology, and commerce always ends in self-defeat. In contrast, a land ethic can only flourish in a context of love and respect (223).

The Green New Deal can also learn from the Poor People's Campaign, which holds a mirror up to American society and asks how much cruelty, violence, poverty, and injustice should a moral society accept? As a leader of the revived Poor People's Campaign, Rev. William Barber III (2020) believes that:

> [t]he rejected must lead, caring for the poor and living wages and health care and environmental protection. It's the rejected who are going to call us to study war no more. It's the rejected who will demand equal protection under the law. It is the rejected who say that the government must be a government of conscientious aspiration and not authoritarian destruction.
>
> *(74)*

Barber is calling each of us to embrace self-transcendence; the rejected will lead because they have already embraced a worldview of care and service. Care is essential for human well-being, for creating an equitable urban economy, and for democratizing self-realization. Joanna Macy (1991) poetically highlights a self-transcendent stance by declaring that our world "can appear to us now both as self and as lover" (14).

Lastly, education is critical to developing an ontology of interdependence and an ethic of care. For example, Durryle Brooks (2017) calls for (re)conceptualizing love. Critically reflecting on a social justice professor's unprofessional aim to "*colonize us*" (102), Brooks concludes:

> I wonder if she had spent time deeply exploring love and its role in social justice education if she would have stumbled across bell hooks's work on love and read that even if the reasons for committing such acts of violence were to teach survival within systems of domination that it is still is [sic] violence and creates *soul wounds*. Maybe she would have stumbled across Erich Fromm's work that suggested that love is indeed an art, which would require her to develop new ways of being that do not rely on control and domination.
>
> *(110)*

I read this as an invitation for teachers and students to collectively transform educational experiences from colonizing and soul wounding to emancipatory. This work cannot be the exclusive province of elite institutions of learning. Noliwe Rooks (2017) challenges readers to acknowledge and transform policies across America's cities that have systematically produced "caste-creating apartheid schools," what she calls segrenomics (207).

The sustainability of cities—and, by extension, the planet—depends in part on an ethic of care for humanity and for nature that has been delineated by philosophers, psychologists, ecologists, and social justice activists. This relational ethic is rooted in the rejection of dominating or colonizing others and in egalitarian, democratic principles of care and love for one another and for the land. Achieving a sustainable planet will first require a reallocation of trillions of dollars of military expenditures to create a global, clean energy infrastructure that cuts carbon emissions by half by 2030. Urban labor and social movements must rise up to make these demands. They must also push for a more radical restructuring of capitalism's global ecology of material flows with its catastrophic reliance on social and economic exploitation and wholesale destruction of the earth's living systems.

Metabolic Praxis

Much like planetary health, scientists and engineers have developed complex models of urban metabolic activity that includes flows of energy, metals, water, food, forest products, chemicals, construction materials, and pollutants (Baccini and Brunner 2012). Quantitative models of urban metabolism include flow analysis, energy assessment, carbon and ecological footprints, input/output, life cycle assessment, network analysis, and integrated studies (Beloin-Saint-Pierre et al. 2017). Similar to the planetary health concept, these quantitative models treat class relationships as exogenous; it is assumed that material flows can be optimized without attention to exploitive social relationships.

In contrast, Marxism sees class relationships as central to capitalist production and profit making. Karl Marx explained that the central profit dynamic of capitalism can be summarized with a simple shorthand formula: M-C-M. Capital (M) is converted into a commodity through the labor process (C); a commercial sale converts that commodity back into capital (M), with the goal of earning and growing net profits. As previously discussed, corporations are amoral because they have no ethical precept beyond completing M-C-M transactions. Furthermore, this formula is also an implicit expression of global metabolic transformations; as natural resources are converted into commodities, ecological systems are degraded. Cities function as complex hierarchical global economic networks to maximize profits. Urban and economic activity produces significant local and global impacts on human health, natural resource systems, air and water quality, and the climate. Cities in the Global North have been importing raw materials from around the world for hundreds of years beginning with European mercantilism. These circuits of capital have also created enormous global inequalities.

These economic dynamics have placed us all into a new era of social crisis and mass extinction. Metabolic praxis is the transformative work of rebalancing the world's complex socioecological metabolic processes through new and revolutionary forms of consciousness, political economy, and urbanization. This transformative work is necessarily experimental; there is no working template for how to do this successfully.

Can cities rise to this challenge? What resources would labor and social movements draw upon? What can eco-industrial parks look like if they champion caring, postcapitalist economics? Should their factories produce for the world or local markets? What new systems of finance will support healthy metabolic processes? How do we ensure that eco-industrial production systems meet social needs and regenerate biologically diverse natural landscapes? John Bellamy Foster, Brett Clark, and Richard York (2010) offer a socialist conception of ecologically rational metabolic relations delineated by:

> (1) social use, not ownership of nature, (2) rational regulation by the associated producers of the metabolism between human beings and nature, and (3) the satisfaction of communal needs—not only of present but also future generations.
>
> *(417)*

Cities should have better designed, publicly financed, and cooperatively run eco-industrial parks because these will offer cities far greater value than those designed by corporations.

Regenerative designs offer new opportunities to create ecologically diverse urban landscapes as part of the practice of cultivating successful economic enterprises. Regenerative design offers new opportunities for creating nonexploitative models of coevolving mutualism among human and biotic communities that allow people and nature to flourish. Jaime Lerner (2014), former mayor of Curitiba, Brazil, calls his regenerative approach to urban revitalization urban acupuncture. For example, Lerner famously provided the city's *favelas* with fresh food in exchange for their trash. Regenerative designs include urban food production and permaculture, urban wetlands and water bodies, green infrastructure, and urban forests. The goal is to build places that are more symbiotic, diverse, complex, and creative (Mang, Haggard, and Regenesis 2016, xiv). Building designs and landscapes based upon mutualism and symbiosis can open opportunities for more peaceful and caring cultures. What is urgently needed is conversations and project developments that bring together urban designers with labor leaders to explore opportunities for new urban partnerships and coalitions.

Metabolic processes at the urban scale impact the global and vice versa. The 1970 Limits to Growth study simulated future human–environmental system dynamics at the global scale and concluded that industrial societies would collapse mid-21st century under several model scenarios. Branderhorst (2020) recently confirmed those same findings. In other words, the era of growth-driven development is ending either through our conscious choices or systemic collapse. Crutzen and Stoermer (2000) state:

> Mankind will remain a major geological force for many millennia, maybe millions of years, to come. To develop a world-wide accepted strategy leading to sustainability of ecosystems against human induced stresses will be one of the great future tasks of mankind.

In their opinion, we are no longer living in the Holocene. Geologists continue to debate this issue, but they may conclude that the Earth system has entered a new geologic epoch, the Anthropocene.

Climate change is one of the crucial global issues facing cities. If cities are to successfully eliminate greenhouse gas emissions, they need national governments to set socially and ecologically rational energy prices and policies. The European Commission (2020) has set a draft goal of 100 cities reaching carbon neutrality by 2030. The International Carbon Neutral Cities Alliance (2021) has 21 city members. Only a handful of countries have pledged to meet this goal by 2030. The European Union and China's carbon neutrality goals are respectively 2050 and 2060. In addition to meeting the challenge of climate mitigation—eliminating its causes—cities must also adapt to changing weather patterns and extreme events, including forest fires, flooding, drought, and extreme heat.

Urban Agglomeration

The core reason for why cities have been successful for hundreds of years is that they have been and remain extraordinary incubators of creativity and innovation, drawing upon each city's unique history, natural resources, entrepreneurial talent, and labor skills. Alan Scott et al. (1994) summarized these processes in high-tech centers as:

> Networks of crisscrossing relationships that run through the multifaceted and localized complexes of universities, research laboratories, private industrial firms, venture capitalists, service providers, and so forth, and out of which there (sometimes) emerges a virtuous circle of industrial innovation and self-transformation.

The pursuit of higher profits drives urban competition and innovation. As industries mature and innovation slows, rates of profit decline. Tensions over wages also play a critical role in shaping urban dynamics. Cities are essentially labor markets, each with its own unique skill set. Learning and innovation in one business or sector spills over to other local economic activities, accelerating urban growth. In Silicon Valley "job-hopping between companies creates the likelihood of knowledge spillovers because human capital acquired in one firm is employed in another" (Fallick, Fleischmann, and Rebitzer 2005).

Shared infrastructure is essential to sustaining propulsive growth, and in larger cities these systems are massive. New York City pipes in water from 125 miles north of the city. Under the city are nearly 7,000 miles of water mains (Barron 2020). The water distribution system for Los Angeles includes the 242-mile Colorado River Aqueduct (CRA). Economists refer to the economic benefits of shared infrastructure as "agglomeration economies"; they are the economic value that accrues "when firms and people locate near one another. . . . These benefits all ultimately come from transport costs savings" (Glaeser 2010, 1).

Historically, corporations were anchored to particular cities for skilled labor or natural resource requirements. Globalization strengthened global supply chains, transportation, and communications systems and gave corporations more latitude to rethink their locational choices. Cities in China and other Asian countries became magnets for corporate investment. Leslie Chang (2008) vividly describes Dongguan, China's growth: "Everywhere is construction and motion, jackhammers and motorcycles, drills and dust; at street level the noise is deafening. . . . The city is built for machines not people" (19).

Agglomeration also confers disadvantages. The high concentration of workers, materials, and machinery increases traffic, pollution, taxes, and high rents. Facing rising costs, marginal economic activities must move or cease operations. Historically, cities have attracted new immigrants and people of color; these working-class populations are subject to lower-quality housing conditions, higher rates of wage exploitation, and criminalization by police.

All industries go through cycles of growth, maturation, and decline. Under capitalism, the process of decline can be highly destructive to local workers and communities. The once indomitable Northeast and Midwest manufacturing belt of the United States turned into a rustbelt as industries closed or moved to new locations, such as Sunbelt cities. Detroit long enjoyed high rates of profit and employment as the global center of the automobile industry. Rising labor costs and oil prices, mismanagement and labor conflicts, and foreign innovation squeezed profits. Companies shifted production lines toward trucks and sport utility vehicles and moved production south and overseas (Galster 2012).

Conclusion

Detroit offers a cautionary tale. For all of the damage created by profit and growth-driven economics, the solution cannot be the chaotic and destructive abandonment of existing production and distribution systems. That would drive cities and the world into a depressionary spiral of unemployment and social distress. The economic systems in place today are complex and scalar; they involve an array of legal and institutional networks that operate from the local to the global scale. The right to the city offers us an agenda for not only envisioning but realizing radical changes at a time of global crisis using the tools of social protest, legal reform, urban design, and economic restructuring. For example, cities can incentivize cooperative practices among an array of actors—labor unions, workers, business cooperatives, social movement organizations, health and environmental professionals, teachers, researchers, and entrepreneurs. Their vision can inspire progress toward a postcapitalist, egalitarian economy and culture. The terrain of engagement should be the physical city itself with an emphasis on contesting investment decisions that focus exclusively on profits rather than human needs and ecological wealth. Rather than assuming that an ethic of care must be divorced from economics, it is time to assume the opposite.

References

Albert, Michael. 2021. *No Bosses: A New Economy for a Better World*. Hampshire, UK: Zero Books.

Alderman, Liz, and Eshe Nelson. 2021. "Global Finance Industry Says it has $130 trillion to Invest in Efforts to Tackle Climate Change." *New York Times*, November 3, 2021. Accessed November 6, 2021. www.nytimes.com/2021/11/03/world/europe/cop26-climate-change-finance-industry.html.

Baccini, Peter, and Paul H. Brunner. 2012. *Metabolism of the Anthroposphere: Analysis, Evaluation, Design*. Cambridge: MIT Press.

Bakan, Joel. 2004. *The Corporation: The Pathological Pursuit of Profit and Power*. New York: Free Press. www.google.com/books/edition/The_Corporation/IniiDwAAQBAJ?hl=en&gbpv=1&printsec=frontcover.

Barber II, William J. 2020. *We Are Called to Be a Movement*. New York: Workman Publishing.

Barron, James. 2020. "Water Mains Are Bursting All Over New York. Can They Be Fixed?" *The New York Times*, February 12, 2020. www.nytimes.com/2020/02/12/nyregion/nyc-water-mains.html.

Beloin-Saint-Pierre, Didier, Benedetto Rugani, Sébastien Lasvaux, Adélaïde Mailhac, Emil Popovici, Galdric Sibiude, Enrico Benetto, and Nicoleta Schiopu. 2017. "A Review of Urban Metabolism Studies to Identify Key Methodological Choices for Future Harmonization and Implementation." *Journal of Cleaner Production* 163: S223–40. https://doi.org/10.1016/j.jclepro.2016.09.014.

Branderhorst, Gaya. 2020. "Update to Limits to Growth: Comparing the World3 Model with Empirical Data." Master's Thesis. Harvard Extension School, Cambridge, MA. https://dash.harvard.edu/handle/1/37364868.

Brooks, Durryle N. 2017. "(Re)Conceptualizing Love: Moving Towards a Critical Theory of Love in Education for Social Justice." *Journal of Critical Thought and Praxis* 6 (3): 102–14.

Cajeta, Gregory. 2000. *Native Science: Natural Laws of Interdependence*. Santa Fe: Clear Light Publishers.

Chang, Leslie T. 2008. *Factory Girls: From Village to City in a Changing China*. New York: Random House Publishing Group.

Crutzen, Paul J., and Eugene F. Stoermer. 2000. "The 'Anthropocene.'" *IGBP Global Change Newsletter* (41): 17–18.

David, Pierre-Marie, Nicolas Le Dévédec, and Anouck Alary. 2021. "Pandemics in the Age of the Anthropocene: Is 'Planetary Health' the Answer?" *Global Public Health* 16 (8–9): 1141–54. https://doi.org/10.1080/17441692.2021.1893372.

Domaradzka, Anna. 2018. "Urban Social Movements and the Right to the City: An Introduction to the Special Issue on Urban Mobilization." *VOLUNTAS: International Journal of Voluntary and Nonprofit Organizations* 29 (4): 607–20. https://doi.org/10.1007/s11266-018-0030-y.

Eisler, Riane. 2017. "Roadmap to a Caring Economics: Beyond Capitalism and Socialism." *Interdisciplinary Journal of Partnership Studies* 4 (1): Article 3.

European Commission. 2020. "100 Climate-Neutral Cities by 2030—by and for the Citizens." Report of the Mission Board for Climate-Neutral and Smart Cities. https://doi.org/10.2777/46063.

Fallick, Bruce, Charles A. Fleischmann, and James B. Rebitzer. 2005. "Job Hopping in Silicon Valley: Some Evidence Concerning the Micro-Foundations of a High Technology Cluster." Working Paper 11710. Working Paper Series. National Bureau of Economic Research. https://doi.org/10.3386/w11710.

Faria, Emilia, Armando Caldeira-Pires, and Cristiane Barreto. 2021. "Social, Economic, and Institutional Configurations of the Industrial Symbiosis Process: A Comparative Analysis of the Literature and a Proposed Theoretical and Analytical Framework." *Sustainability* 13 (13): 7123. https://doi.org/10.3390/su13137123.

Fearless Cities. 2021. "Fearless Cities: The Global Municipalist Movement." https://fearless-cities.com/en/about.

Foster, John Bellamy, Brett Clark, and Richard York. 2010. *The Ecological Rift: Capitalism's War on the Earth*. New York: Monthly Review Press.

Galster, George. 2012. *Driving Detroit: The Quest for Respect in the Motor City*. Philadelphia: University of Pennsylvania Press.

Gilligan, Carol. 2011. "Looking Back to Look Forward: Revisiting in a Different Voice." In *Classics@9*. Washington, DC: Center for Hellenic Studies. http://nrs.harvard.edu/urn-3:hlnc.essay:Gilligan.Looking_Back_to_Look_Forward.2011. https://archive.chs.harvard.edu/CHS/article/display/4025.

Glaeser, Edward L. 2010. "Introduction to 'Agglomeration Economics.'" In *Agglomeration Economics*, 1–14. Chicago: University of Chicago Press. www.nber.org/books-and-chapters/agglomeration-economics/introduction-agglomeration-economics.

Glasgow Financial Alliance for Net Zero. 2021. "GFANZ Progress Report." www.gfanzero.com/progress-report/.

Harvey, David. 2008. "The Right to the City." *New Left Review* 53 (September–October). https://newleftreview.org/issues/ii53/articles/david-harvey-the-right-to-the-city.

Hester, Randolph T. 2006. *Design for Ecological Democracy*. Cambridge: MIT Press.

International Carbon Neutral Cities Alliance. 2021. Accessed December 1, 2021. https://carbonneutralcities.org/cities/.

King Jr., Dr. Martin Luther. 1963. "Letter from Birmingham Jail." www.csuchico.edu/iege/_assets/documents/susi-letter-from-birmingham-jail.pdf.

Leopold, Aldo. (1949) 1989. *A Sand County Almanac and Sketches Here and There*. Special Commemorative ed. New York: Oxford University Press.

Lerner, Jaime. 2014. *Urban Acupuncture: Celebrating Pinpricks of Change that Enrich City Life*. Washington: Island Press.

Macy, Joanna. 1991. *World as Lover, World as Self*. Berkeley: Parallax Press.

Mang, Pamela, Ben Haggard, and Regenesis Group. 2016. *Regenerative Development and Design: A Framework for Evolving Sustainability*. Hoboken: John Wiley & Sons.

Maslow. Abraham H. 1943. "A Theory of Human Motivation." *Psychological Review* 50 (4): 370–96.

———. 1971. *The Farther Reaches of Human Nature*. New York: Viking Press.

Myers, Samuel S. 2017. "Planetary Health: Protecting Human Health on a Rapidly Changing Planet." *The Lancet* 390 (10114): 2860–68. https://doi.org/10.1016/S0140-6736(17)32846-5.

Nowak, Marek. 2020. "The 'Right to the City' on Various Scales." *Society Register* 4 (4): 7–48. https://doi.org/10.14746/sr.2020.4.4.01.

Pérotin, Virginie. 2020. "What Do We Really Know about Worker Co-operatives? | Co-Operatives UK." www.uk.coop/resources/what-do-we-really-know-about-worker-co-operatives.

Raskin, Paul. 2016. *Journey to Earthland: The Great Transition to Planetary Civilization*. Boston: Tellus Institute.

Rooks, Noliwe Rooks. 2017. *Cutting School: The Segrenomics of American Education*. New York: The New Press.

Scott, A. J., A. Thornley, U. Collier, A. C. Pratt, A. R. Townsend, J. Simmie, and R. B. Potter. 1994. "Review: Technopoles of the World: The Making of Twenty-First Century

Industrial Complexes, the British Planning System: An Introduction, This Common Inheritance: The Second Year Report 1992. Britain's Environmental Strategy, Restructuring Rural Areas, Volume 1. Constructing the Countryside, Enterprise and Human Resource Development: Local Capacity Building, Urban Sociology, Capitalism and Modernity, Retail Change: Contemporary Issues." *Environment and Planning B: Planning and Design* 21 (6): 769–78. https://doi.org/10.1068/b210769.

Shiva, Vandana. 2005. *Earth Democracy: Justice, Sustainability, and Peace*. Cambridge: South End Press.

———. 2015. "We Care for Life—The Power in Caring." *Regeneration International*, October 13, 2015. https://regenerationinternational.org/blog-five/we-care-for-life.

Taylor, Peter J., and Ben Derudder. 2016. *World City Network: A Global Urban Analysis*, 2nd ed. London: Routledge.

Thompson, Matthew. 2021. "What's so New About New Municipalism?" *Progress in Human Geography* 45 (2): 317–42.

Wolff, Richard D. 2012. *Democracy at Work: A Cure for Capitalism*. Chicago: Haymarket Books.

———. 2017. "Start with Worker Self-Directed Enterprises." *The Next System Project*. https://thenextsystem.org/sites/default/files/2017-08/RickWolff.pdf.

Wolin, Sheldon S. 2008. *Democracy Incorporated: Managed Democracy and the Specter of Inverted Totalitarianism*. Princeton: Princeton University Press.

12

INDUSTRIAL GENTRIFICATION AND THE GEOGRAPHY OF SACRIFICE AND GAIN[1]

Melissa Checker

Introduction

In September 2021, Beryl Thurman, president of the North Shore Waterfront Conservancy of Staten Island (NSWC), e-mailed several high-level officials at the US Army Corps of Engineers, New York District (ACOE). Thurman lives on the North Shore of Staten Island, and for nearly two decades she has led NSWC in an ongoing struggle for environmental justice in that area. For her constituents, this meant access to more greenspaces and the mitigation of local pollution, contamination, and flooding. That day, Thurman and the ACOE officials were discussing a new blasting and dredging project to take place in the Kill Van Kull, a narrow tributary that runs between the North Shore of Staten Island and Bayonne, New Jersey.

Roughly every ten years, the Corps dredged the Kull to make room for ever larger container ships serving the Port of New York, on the North Shore's western edge. Previous blasting and dredging projects had created intolerable noise and air pollution as well as vibrations that cracked windows and the foundations of people's homes. Unsurprisingly, Thurman's exchange with the ACOE officials grew tense. In one message, she wrote,

> The waterfront Environmental Justice communities are already underserved and considered undesirable because of contaminated sites and industrial sites and their truck traffic at the waterfront that runs through them. They are considered undesirable due to flooding issues from the rivers and poor storm water management due to neglect. . . . None of these things listed above have any benefit to the Environmental Justice communities, none of it, they are all burdens. Environmental burdens, health burdens, financial burdens, burdens.[2]

DOI: 10.4324/9781003217442-14

For local residents, the noise and air pollution from container ships and the stream of diesel trucks traveling to and from the port were bad enough. But they also contended with approximately 18 contaminated sites clustered into a 5.2-mile stretch of waterfront including two private waste transfer stations, a Department of Sanitation garage, a sewer treatment plant, an industrial salt company, and several bus depots. Only a two-lane road separated these sites from residential neighborhoods. Making matters worse, the area flooded frequently. Residents worried that the effects of climate change would intensify these floods, possibly dislodging waterfront contaminants and distributing them across the neighborhood. Their concerns, however, often fell on deaf ears—a situation they attributed to the following demographics.

Across the North Shore's approximately 15 neighborhoods, racial composition varied. In high-density residential zones, approximately 47% of residents were Hispanic and as few as 16% were white, non-Hispanic. But in low-density, residential zones, up to 72% of households were white, non-Hispanic (see Mehrotra et al. 2018). Approximately 13% of households on the North Shore lived below the poverty line (about the same percentage as in New York City). Although per capita incomes averaged 20% below the city average, household incomes averaged slightly higher. This suggests that individual households on the North Shore contained higher than average numbers of earners (US Census Bureau 2018). Regardless of racial background or income levels, North Shore neighborhoods shared the sense that they had little to no political clout and that they lived in the most forgotten part of "the forgotten borough."

Drawing on the case of the North Shore, this chapter examines the arrangement of industrial space in the industrial and postindustrial city and its impact on the environmental health and well-being of low-income communities of color. Specifically, I describe how zoning codes charted a course for economic development that protected affluent residents and relegated out-of-the-way, minority communities to live side by side with polluting facilities. Zoning decisions thus mapped out a path for the city's shifting economic agendas and structured the kinds of opportunities available to different groups of people (see Angotti and Morse 2016). They dictated whether neighborhoods would be adjacent to a waste transfer station or factory, to a park, or to a commercial area. Zoning decisions set property values, transportation options, the quality of schools, and health outcomes. In short, they demarked which parts of the city would be protected from environmental harms and which would be sacrificed to them.

City Meets Industry

When Henry Hudson sailed into the New York Bay in the early 1600s, he spotted the northern edge of an island he later named *Staaten Eylandt*. Like most of the region, the land was inhabited by Lenape Indians. After three attempts, European settlers drove the Lenape from their lands and planted their own farms, which would supply food to the rest of the city for several centuries (Papas 2007). As the rest of New York City grew over the next couple of centuries, the Island remained accessible only by boat. This relative remoteness made it both a popular vacation destination for the city's hoi polloi and a common location for institutions that served the city's poor, infirm, mentally ill, and developmentally disabled citizens (Lam 2013). By the end of the 1800s, Staten Island's northern shore came to play a crucial—if largely unseen—supporting role in the city's economic system. Its port became one of the East Coast's largest, and its waterfront a busy hub for the industries that would propel New York into the 20th century.

The opening of the Panama Canal in 1914 provided passage for ships traveling from the Pacific Ocean to the East Coast, almost tripling import/export traffic to the Staten Island port (La Rocco 2004). Some of the city's largest manufacturers took their places along the North Shore waterfront, including the Jewett White Lead Company, US Gypsum, Proctor & Gamble, and Archer Daniels Midland (ADM) (Bayles 2010; NSWC 2007). Those companies provided factory work for a steady stream of European migrants, who settled in the neighborhoods closest to the waterfront. By the early 20th century, the North Shore comprised only one-third of Staten Island's land mass but was twice as populated as the rest (Kramer and Flanagan 2012).

Across the New York Harbor, Manhattan was bursting at the seams. Thanks to new feats of engineering, dozens of skyscrapers pierced the sky. A growing number of factories sprawled throughout the lower part of the island, and an increasing immigrant population arrived to work in them. In addition to expanding upward, advances in transportation allowed the city to expand outward. Wealthy families fled the downtown crowds for more pristine environs on the city's outskirts. But, they worried that both industries and workers would infiltrate their neighborhoods. Eventually, local elites pressured municipal leaders to take action to create a more organized and use-specific plan for the city.

In 1916, the city of New York established the nation's first set of zoning codes. Tall buildings now had to begin narrowing at a certain height and continue to narrow as they climbed. This "wedding cake" design (made famous by the Chrysler and Empire State Buildings) mitigated skyscrapers' blocking of light. Zoning regulations also divided the city into three types of use districts: residential, commercial, and unrestricted Going forward, residential districts would be restricted to apartments, single- and two-family homes, hotels, clubs, schools, churches, and other cultural and institutional uses. Although not officially affiliated with income, residential districts mostly encircled already wealthy areas like Fifth and Park Avenues on the east side of Manhattan, Murray Hill, and parts of the Upper West Side. Commercial districts included a range of commercial functions but excluded any industrial use involving noxious fumes. Most of these were located in downtown and midtown Manhattan (Willis 1993).

Unrestricted districts, however, could include any kind of use from industrial to commercial businesses to residential. In fact, city leaders deliberately placed housing for factory workers in unrestricted districts.[3] At the time, factory workers did not complain since many of them considered such proximity to be critical. For instance, garment workers made up nearly half of Manhattan's entire workforce, and more than 70% of them were women (Freeman 2001). On top of working 12 to 14 hours per day, seven days a week, female factory workers had childcare and domestic duties, which made long-distance commutes impossible (Orleck 1995, 23–30). Unrestricted districts became vibrant ethnic enclaves that eventually accommodated multiple income levels. Lack of resources or restrictive covenants excluded upwardly mobile, ethnic workers from living in affluent neighborhoods. But they were able to move into new, middle-income housing developments in unrestricted districts. By the 1940s, over half of the city's inhabitants lived in one of these zones (Sze 2006, 45).

The new zoning codes reinforced a stratified city. Affluent residents lived in quiet, low-density neighborhoods surrounded by commercial and higher-density residential zones. These acted as buffers, further insulating residential neighborhoods from the noise, pollution, and overcrowding of industrial districts. At the same time, industrial zones themselves were protected from being overtaken by commercial and residential development. This allowed property values in manufacturing zones to remain relatively low.

The new zoning codes thus provided more than a spatial blueprint for the city. They also preconfigured property values and established which New Yorkers had access to which resources, including greenspace and clean, healthy air.

City "Loses" Industry

Manufacturing in New York City peaked at the end of World War II. But after 1947, multiple factors led to its sharp decline (Freeman 2001). For instance, the city's antiquated and cramped factory buildings could not accommodate advances in machinery, especially in the city's two largest manufacturing sectors—printing and garment. In addition, the invention of refrigerated trucks allowed food manufacturing businesses to move away from the city to cheaper environs, just as containerization made overseas manufacturing more viable. Around the late 1950s, city leaders recognized these trends and reversed course when it came to the retention of manufacturing businesses. The city pulled back its tax incentives and loans to manufacturers, and redirected its resources to courting corporate growth of the nonmanufacturing variety (Levinson 2006).

The vision was to create a city that relied almost exclusively on finance, insurance, and real estate (aka "FIRE") sectors, along with "producer services" (law, accounting, advertising). As historian Robert Fitch argues, "New York rid itself of everything that blocked its potential to become the biggest and best FIRE and producer services city in the world" (Fitch 1993, 12–13). However, the old zoning codes presented a major obstacle to achieving this new agenda. Unrestricted zones were now taking up much needed space that could be used to grow FIRE industries.

In 1961, the City Council passed a new set of zoning resolutions, dividing New York City into residential, commercial, and manufacturing areas. Unlike unrestricted districts, the new manufacturing, or "M-zones," were subdivided into "light," "heavy," and "mixed-use" designations. Heavy M-zones were moved to the outer boroughs, while most of Manhattan was rezoned for commercial and residential uses, with light manufacturing businesses along its edges. For residential and commercial districts, the codes offered more flexibility in building heights by allowing developers to add extra floor space in exchange for incorporating public plazas into their projects (NYC Department of City Planning 2019).

The reforms of 1961 charted a course forward for the borough's economic future. Its rearrangements of space further diminished New York City's manufacturing sector and concentrated it in low-income neighborhoods and the outer boroughs. Between 1967 and 1976, the city lost a fourth of its factories and one-third of its manufacturing jobs, while FIRE industries continued to grow (Levinson 2006, 99).

As blue-collar jobs left the city, so did many of its middle-class residents. Post-World War II–era federal programs like the GI Bill and the Federal Homeowners Association were making homeownership possible for working- and middle-class white families. Federal subsidies also enabled the expansion of the auto industry, which in turn allowed these families to move to any number of new suburban developments (Jackson 1985; Massey and Denton 1993; Sugrue 2014). On Staten Island, which was still relatively undeveloped (except for the North Shore), city leaders saw an opportunity. They could mitigate "white flight" by creating a suburb within the city limits. Already, Robert Moses's most ambitious project, the Verrazano Bridge, was well underway. By connecting southern Brooklyn to Staten Island, the bridge would allow upwardly mobile, white, ethnic families easy access to their former Brooklyn communities while also offering them the single-family homes, driveways, shopping malls, and homogeneity that were luring them to the suburbs. New zoning codes thus divided two-thirds of Staten Island into a bedroom community.

But the remaining third—the North Shore—retained most of its mixed-use, high-density zoning. While white families filled the Island's residential zones, black and brown families were most often denied mortgages and housing insurance. As a result, many remained stuck in unrestricted neighborhoods. Throughout the fiscal crisis of the 1970s, city leaders cut services to these neighborhoods, both creating and intensifying crime, arson, and dilapidated housing. Discrimination in employment, education, healthcare, and criminal justice multiplied the effects of residential segregation, curtailing opportunities for black and brown families even further (Maantay 2002; see also Maantay and Moroko 2018).

Meanwhile, municipal facilities replaced a number of the area's abandoned factories. Environmental reforms of the early 1970s facilitated this process. For instance, the Clean Air Act included opportunities for local communities to oppose the siting of new facilities in or near their neighborhoods. While these opportunities were open to anyone, they took time, money, and political clout (Angotti 2008). This meant that wealthy areas like Manhattan's Upper West Side blocked hazardous facilities (such as waste treatment plants and transfer stations, incinerators, and waste storage facilities), which were then clustered in poor, minority neighborhoods. In addition, the 1961 zoning regulations included an "as of right" clause that allowed owners of properties in existing M-zones to expand their facilities without City Planning approvals. An owner could also sell his or her property to another industrial facility without seeking zoning approvals.

As a result, along the North Shore waterfront, one factory replaced another without concern for the contaminants left behind. For instance, the Jewett White Lead Paint works became Sedutto's Ice Cream factory; an oil tank storage facility became the Port Richmond Sewer Treatment Plant; and soap production facilities at Port Ivory were acquired by the Port Authority. As layers of contaminants accumulated, so did the environmental, economic, and social risks that threatened local residents (NSWC 2007).

Economist Yale Rabin refers to zoning codes that allowed and even encouraged the clustering of industrial facilities in neighborhoods of color as "expulsive zoning." For Rabin, this term refers to zoning decisions that "permit—even promote—the intrusion into black neighborhoods of disruptive, incompatible uses that have diminished the quality and undermined the stability of those neighborhoods" (Rabin 1989, 101). Rabin's formula emphasizes the degree to which industrial facilities can be a cause, not just a symptom of neighborhood degradation. In the case of New York City, expulsive zoning played an active role in determining the fate of the city's industrialized areas and the people who lived in them.

City Wins Industry Back

In his 1978 inaugural speech, Mayor Ed Koch announced his intention to promote gentrification and rescue neighborhoods previously targeted by austerity measures. To that end, Koch offered private developers tax abatements in certain formerly industrial areas, where land values had bottomed out (Soffer 2012). Redevelopment quickly became the cornerstone of the FIRE economy. Gleaming new office towers helped lure global corporations, luxury developments housed their employees, and commercial districts supplied residents and visitors with places to eat, drink, and be merry. Over the next several decades, mixed-use manufacturing districts in neighborhoods close to downtown Manhattan were rezoned for commercial and residential uses, transforming them into affluent enclaves for white-collar professionals.

This process got a significant boost with the election of Mayor Michael Bloomberg in 2002. On a mission to bring New York City out of the post-9/11 economic slump, Bloomberg vowed to recreate New York as a "luxury city." He ramped up the rezoning process and stepped up efforts to push the city's remaining industrial uses to the margins of the outer boroughs or out of the city altogether. A year into his first term, Bloomberg famously told a reporter from the *Financial Times*, "If you are a pharmaceutical company or a steel company, you do not need to be here. . . . New York City should not waste its time with manufacturing" (qtd. in David 2012, 138). Between 2002 and 2014, over one-third of the city was rezoned (Mistry and Byron 2011, 20).

Then, during his second run for office, Bloomberg reversed his rhetorical course. Facing criticism from land use and progressive advocates, he announced that the city would welcome and make room for industrial businesses. In a press release describing the new industrial incentives, Bloomberg said, "Over the past half century, the City's industrial base has declined, along with many other American cities, but industry remains a powerful engine of our economy and its 500,000 jobs represent about 15 percent of our workforce" (Skyler and Falk 2005). Within four years, Bloomberg's narrative about industry had gone from seeing it as "a waste of time" to celebrating it as "a powerful engine of [the] economy" (Mandel 2013, 2).[4]

During his campaign, Bloomberg launched the Made in NYC program, which provided 5% tax credits on production costs for qualifying film and television productions filmed in New York City (NYC.gov 2006). Using similar tax credits, as well as small business loans and subsidies for office space and broadband, Bloomberg also made a significant effort to court the technology sector from start-ups to tech giants like Google, Yahoo, and LinkedIn. As part of this effort, he established the Industrial Business Zone (IBZ) program. The IBZ set aside 16 safe havens that protected industrial areas from being rezoned for residential uses, and it created a tax credit for the one-time relocation of businesses—many coming from rezoned areas. Additionally, the city would dedicate several of its publicly owned properties to manufacturing (NYCEDC 2013a).

To kick off the IBZ program, the city pledged $200 million to expand and update the Brooklyn Navy Yard. To oversee this development, the NYCEDC installed as CEO Andrew Kimball, a rising, young urban planner who had just spent three years trying to bring the Olympic Games to New York City. From the get-go, Kimball made sustainability integral to the Navy Yard's new brand. Its website devoted an entire page to listing the green manufacturers who leased space at the site. Another highlighted all of the environmental initiatives undertaken at the Yard, including LEED-certified buildings, hybrid vehicles, the largest rooftop farm in New York City, solar-powered trash compactors, and solar- and wind-powered streetlamps (Brooklyn Navy Yard Development Corporation 2019).

Industrial Gentrification and the Green Growth Machine

The Navy Yard's emphasis on green manufacturing and energy signals another of Bloomberg's new visions for the city. In 2007, he launched PlaNYC 2030, a sweeping plan for achieving a robust urban economy while also cleaning up environmental pollution and reducing greenhouse gas emissions. Bloomberg's zeal for such "win-win" scenarios attracted worldwide attention and accolades, and his name became almost synonymous with urban sustainability. In this way, Bloomberg rebranded himself and the city at the same time. In her study of how PlaNYC 2030 rebranded New York, sociologist Miriam Greenberg writes, "Greening, with its connotations of quality of life, transparency, innovation, and resilience, was the ideal antidote to image crises" (Greenberg 2015, 122). Old-school New

York liberals feared that "the luxury city" was a regressive attempt to expunge the city's low-income populations. "The sustainable city," however, was aimed directly at liberals' concerns for the environment and climate change, and it resonated with their proenvironmental identities.

A revamped industrial sector fit neatly into the mayor's sustainability—and redevelopment—agendas. The Navy Yard development coincided with the renovation of Brooklyn Bridge Park into a state-of-the-art park featuring repurposed and recycled materials and native plants and habitats. Both helped catalyze a host of upscale housing complexes, storefronts, bars, and restaurants.

Building on the success of the Navy Yard and other industrial incentives, in 2011, Bloomberg launched Manufacturing 2.0. This initiative funded the public purchase and renovation of vacant or underused industrial properties. Small manufacturers could then rent space in these buildings at subsidized rates (Business Programs n.d.). Manufacturing 2.0 targeted local crafters, woodworkers, artisans, and food and beverage manufacturers, as well as technology start-ups, film studios, visual artists, and fashion designers (NYCEDC 2013a). Importantly, the program's promotional materials featured market studies showing that upwardly mobile, professional New Yorkers favored locally sourced, specialty products and high-end, handcrafted items. In short, both the producers and consumers of these new manufacturing businesses synced up with Bloomberg's vision for a sustainable/luxury city.

Indeed, the location of Manufacturing 2.0 properties was highly correlated with gentrification. All but one of the new spaces created by the program were located in neighborhoods undergoing high-end redevelopment. This proximity was no accident. Small-scale, green manufacturing businesses had become an amenity that facilitated neighborhood gentrification. At the same time, they marked the gentrification of industry itself.

The Other Makers

Politicians and developers promoted green manufacturing as a way to provide thousands of well paying, secure jobs to unskilled, noncollege-educated workers. But studies show that jobs in the new manufacturing sector were dramatically different from unionized jobs in the old manufacturing sector. Between 2001 and 2012, the total number of manufacturing jobs in New York City decreased by nearly 50%. Yet, after 2006, the average number of manufacturing businesses stabilized. During the same period, industrial output declined at a far slower rate than industrial jobs (NYCEDC 2013b). These numbers indicate that manufacturing businesses were hiring fewer workers and relying heavily on technology.

Manufacturing 2.0's high-tech and small-scale businesses follow this trend. In 2014, for instance, the 330 companies at the Navy Yard averaged 19 people per business. The Brooklyn Army Terminal in Sunset Park averaged 34 workers per company. A report by the City of New York's Independent Budget Office found that industrial subsectors such as biotech, computer, and electronics manufacturing, broadcasting, and motion picture and sound hired relatively small numbers of workers, especially without college degrees, relative to traditional industrial firms. Moreover, the report found that "many of the industrial subsectors that did employ a large number of workers without college degrees, like food manufacturing or warehousing, pay wages that are comparable to the average wages for workers in nonindustrial [service] sectors" (New York City Independent Budget Office 2014, 3). These statistics demonstrate that in practice, small-scale manufacturers provided few opportunities for low-skilled workers.

Traditional manufacturing jobs have also undergone irrevocable change. Automation and technology dramatically have altered the scale and type of jobs available to industrial workers, demanding highly specialized and skilled expertise. Only some of those jobs offer the livable wages and benefits attributed to them. Anti-union legislation, globalization, and outsourcing have all diminished the power of unions, which lowered both wages and benefits. In 2020, only 8.6% of US manufacturing workers belonged to labor unions (Bureau of Labor Statistics 2020, Table 3). Over the years, industrial manufacturers have increasingly turned to informalized, or "day" labor, often filling low-tech jobs with undocumented workers (Constable 2015).

The recent explosion of online retail platforms (especially since the start of the COVID-19 pandemic) has given rise to a new category of manufacturing job—"last mile" laborers. "Last mile" describes the last leg of the supply chain—where trucks drop off goods for sorting and loading onto other trucks or cars for distribution to customers. The need for last mile workers during the pandemic called attention to the high-pressure and often dangerous labor conditions they endure. In fact, New York's largest Amazon warehouse, located on Staten Island's western shore, made news early in 2020 when its workers staged a walkout due to unsafe conditions. In December 2021, a coalition of those workers filed an official vote request with the National Labor Relations Board (De Vynck 2021). These efforts spread to Amazon warehouse workers in other parts of the country, who have also held walkouts and issued demands for higher wages and safer working conditions (Siberling 2021).

Despite their questionable labor practices, last mile facilities are booming in New York City. In addition to Staten Island's west shore, between 2019 and 2021, more than six fulfillment centers appeared in low-income sections of Brooklyn's Sunset Park and Red Hook neighborhoods. Currently, these facilities are classed as manufacturing uses and allowed to locate in M-zones without needing to undergo environmental review. The truck traffic generated by these last mile facilities compounds local pollution problems and exacerbates the health risks faced by the low-income communities of color living nearby (see Kuntzman and Cuba 2021).

Dynamics of Sacrifice and Gain

Like any city, New York needed space within municipal boundaries for certain kinds of industrial facilities, such as waste transfer stations, sewage treatment plants, truck and bus depots, and power stations. In recent years, it also needed an increasing amount of space for last mile warehouses and fulfillment centers. But as more and more neighborhoods were rezoned for residential, commercial, and green industrial uses, there were fewer and fewer places to locate such facilities. As a result, out-of-the-way neighborhoods that were not gentrifying became prime targets for housing these facilities.

NSWC members constantly found themselves challenging new industrial projects. During my research period (roughly 2009–2018), these projects included two new cement manufacturing plants, a natural gas pipeline, a container port expansion, a waste-to-energy plant, and a facility to treat dredged soils. All the while, NSWC continued to fight against existing contamination and pollution.

For instance, in the case of the dredged soils facility, city planners had proposed locating a similar site in two different Brooklyn neighborhoods, both of which were redeveloping and upscaling at a rapid pace. Officials proposed the facility first in one neighborhood, where residents raised significant opposition, and then in a second, with the same result. Eventually, they announced that they were abandoning plans to locate the facility anywhere in Brooklyn. Just a few months later, residents of the North Shore received notice that a local waste transfer station had applied for permits to host a dredged soils facility that was almost identical to the one Brooklynites had rejected.[5]

The gentrification and greening of certain neighborhoods led to the displacement of heavy industry from those areas. As a result, such industries were placed in nongentrifying neighborhoods, even if they were already saturated with polluting sites. This dynamic was not lost on North Shore activists. As one NSWC member wrote in an e-mail, "Creating all the wonderful new playgrounds and high-value waterfront residential areas pushes HEAVY industrial use, into parts of Queens and Staten Island."[6] This statement epitomizes the dynamic relationship between gentrification and environmental degradation. In other words, the greening of some parts of the city appeared to accelerate the browning of other parts.

Conclusion

In 1916, the institution of zoning codes drew protective boundaries around wealthy areas, insulating them from industries and immigrant workers. Simultaneously, the elite players protected industrial zones from the commercial and residential real estate market. In 1961, zoning reforms rearranged urban space to recreate New York City as a worldwide center for the new finance and real estate–based global economy. In both cases, low-income communities and communities of color were relegated to living near and among industrial facilities, serving as human buffers for the waste such facilities generated.

Rather than correcting the systemic inequities of industrial and real estate capitalism, the 21st-century reboot of the industrial economy remained embedded in those very same inequities. High-tech and small-scale, trendy manufacturing businesses raised the cultural capital of a neighborhood and emitted significantly fewer pollutants. But this reinvention (and celebration) of industrial businesses also concealed growing inequities in the manufacturing sector. It also shifted the burden for all of the city's heavy manufacturing onto the shoulders of neighborhoods that were not slated for revitalization and that were already saturated with polluting facilities. Industrial gentrification thus became another mechanism for reinforcing the same old dynamics of sacrifice and gain that benefited certain neighborhoods at the expense of others.

Notes

1 Much of the information contained in this chapter can also be found in Checker 2020, especially Chapter 3, and Checker 2021.
2 This quote is drawn from an e-mail received by the author on September 9, 2021. It is currently on file with the author.
3 NYC Planning Commission, Division of Master Plan 1940, 5, as quoted in Maantay 2002, 70.
4 This report was commissioned in advance of the Second Annual Bloomberg Technology Summit, held in New York City in September 2013.
5 Although the permits for this facility were approved, it was never built.
6 This quote is drawn from an e-mail received by the author in August 2012. It is currently on file with the author.

References

Angotti, Tom. 2008. *New York for Sale: Community Planning Confronts Global Real Estate*. Cambridge: MIT Press.

Angotti, Tom, and Sylvia Morse, eds. 2016. *Zoned Out! Race, Displacement and City Planning in New York City*. New York: Terreform.

Bayles, Richard Mather. 2010. *History of Richmond County (Staten Island), New York from Its Discovery to the Present Time*. Charleston: Nabu Press.

Brooklyn Navy Yard Development Corporation. 2019. *A History of the Yard*. Brooklyn Navy Yard. https://brooklynnavyyard.org.

Bureau of Labor Statistics. 2020. *Occupational Employment Statistics*. United States Department of Labor. www.bls.gov.

Business Programs. n. d. "NYCEDC." Accessed March 22, 2020. https://edc.nyc/business-programs.

Checker, Melissa. 2020. *The Sustainability Myth: Environmental Gentrification and the Politics of Justice*. New York: NYU Press.

———. 2021. "Environmental Justice and Gentrification in New York City." *Environment: Science and Policy for Sustainable Development* 63 (2): 16–27.

Constable, Pamela. 2015. "Majority of Undocumented Immigrants Work in Low-Skill Jobs, Report Finds." *Washington Post*, March 26, 2015. www.washingtonpost.com/local/majority-of-undocumented-immigrants-work-in-low-skill-jobs-report-finds/2015/03/26/dada9f2a-d3bc-11e4-a62f-ee745911a4ff_story.html.

David, Greg. 2012. *Modern New York: The Life and Economics of a City*. New York: St. Martin's Press.

De Vynck, Gerrit. 2021. "Amazon Warehouse Workers in New York Are Taking Another Stab at Forming a Union." *Washington Post*, December 22, 2021. www.washingtonpost.com/technology/2021/12/22/amazon-warehouse-union/.

Fitch, Robert. 1993. *The Assassination of New York*. New York: Verso.
Freeman, Joshua B. 2001. *Working-Class New York: Life and Labor since World War II*. New York: New Press.
Greenberg, Miriam. 2015. "'The Sustainability Edge': The Post-Crisis Promise of Eco-City Branding." In *Sustainability in the Global City: Myth and Practice*, edited by Cindy Isenhour, Gary McDonough, and Melissa Checker, 105–30. New York: Cambridge University Press.
Jackson, Kenneth T. 1985. *Crabgrass Frontier: The Suburbanization of the United States*. New York: Oxford University Press.
Kramer, Daniel, and Richard Flanagan. 2012. *Staten Island: Conservative Bastion in a Liberal City*. Lanham: University Press of America.
Kuntzman, Gersh, and Julianne Cuba. 2021. "Advocates: Change Zoning to Regulate 'Last-Mile' Trucking Facilities Flocking to Low-Income Communities." *Streetsblog*, December 15, 2021. https://nyc.streetsblog.org/2021/12/15/advocates-change-zoning-to-regulate-last-mile-trucking-facilities-flocking-to-low-income-communities/.
La Rocco, Barbara. 2004. *Going Coastal New York City*. Brooklyn: Going Coastal, Inc.
Lam, Joyce. 2013. "The Staten Island Farm Colony's Disturbing Secrets." *Untapped Cities*, October 15, 2013. https://untappedcities.com/?s=The+Staten+Island+Farm+Colony's.
Levinson, Marc. 2006. *The Box: How the Shipping Container Made the World Smaller and the World Economy Bigger*. Princeton: Princeton University Press.
Maantay, Juliana A. 2002. "Zoning Law, Health, and Environmental Justice: What's the Connection?" *Journal of Law, Medicine & Ethics* 30 (4): 572–93.
Maantay, Juliana A., and Andrew R. Maroko. 2018. "Brownfields to Greenfields: Environmental Justice Versus Environmental Gentrification." *International Journal of Environmental Research and Public Health* 15 (10): 2233.
Mandel, Michael. 2013. "Building a Digital City: The Growth and Impact of New York City's Tech/Information Sector." Report prepared for the Bloomberg Technology Summit, New York City. September 30, 2013. https://southmountaineconomics.com.
Massey, Douglas S., and Nancy A. Denton. 1993. *American Apartheid: Segregation and the Making of the Underclass*. Cambridge: Harvard University Press.
Mehrotra, Apurva, Bijan Kimiagar, Marija Drobnjak, and Sophia Halkitis. 2018. *The North Shore of Staten Island: Community Driven Solutions to Improve Child and Family Well-Being*. Citizens' Committee for Children of New York. https://cccnewyork.org/data-publications/the-north-shore-of-staten-island-community-driven-solutions-to-improve-child-and-family-well-being/.
Mistry, Nisha, and Joan Byron. 2011. *The Federal Role in Supporting Urban Manufacturing*. What Works Collaborative. Accessed January 13, 2022. www.urban.org/sites/default/files/publication/26966/1001536-The-Federal-Role-in-Supporting-Urban-Manufacturing.PDF.
New York City Independent Budget Office. 2014. *Budget Options for New York City*. New York: New York City Independent Budget Office. https://ibo.nyc.ny.us/iboreports/options2014.pdf.
NSWC (North Shore Waterfront Conservancy of Staten Island). 2007. "Staten Island's Gold Coast 5.2 Miles from St. George to Arlington." On file with author.
NYCEDC (New York City Economic Development Corporation). 2013a. "NYCrafted: State of Local Manufacturing, October 2013: A NYCEDC Special Report." Report on File with the Author.
———. 2013b. "Industrial Business Zone Boundary Commission: Staff Recommendations." Report on File with the Author.

NYC Department of City Planning. 2019. *City Planning History*. NYC Planning. https://www1.nyc.gov/site/planning/about/city-planning-history.page.
NYC.gov. 2006. "Mayor Bloomberg Announces $30 Million Annual Increase in City Financing for Film Production Tax Credit Program." New York City, Office of the Mayor, May 10, 2006. On File with the Author.
Orleck, Annelise. 1995. *Common Sense and a Little Fire*. Chapel Hill: University of North Carolina Press.
Papas, Phillip. 2007. *That Ever Loyal Island: Staten Island and the American Revolution*. New York: NYU Press.
Rabin, Yale. 1989. "Expulsive Zoning: The Inequitable Legacy of Euclid." In *Zoning and the American Dream: Promises Still to Keep*, edited by C. Haar and J. Kayden. Chicago: Planners Press and American Planning Association.
Siberling, Amanda. 2021. "Amazon Workers at Two Chicago Warehouses Walk out to Demand Better Treatment." *TechCrunch*, December 22, 2021. https://techcrunch.com/2021/12/22/amazon-workers-at-two-chicago-warehouses-walk-out-to-demand-better-treatment/.
Skyler, Edward, and Jennifer Falk. 2005. "Mayor Michael R. Bloomberg Introduces New Initiatives to Support New York City's Industrial Sector." January 19, 2005. On File with the Author.
Soffer, Jonathan. 2012. *Ed Koch and the Rebuilding of New York City*. New York: Columbia University Press.
Sugrue, Thomas. 2014. *The Origins of the Urban Crisis: Race and Inequality in Postwar Detroit—Updated Edition*. Princeton: Princeton University Press.
Sze, Julie. 2006. *Noxious New York: The Racial Politics of Urban Health and Environmental Justice*. Cambridge: MIT Press.
US Census Bureau. 2018. "American Community Survey 1-Year Estimates." Census Reporter Profile page for NYC-Staten Island Community District 1-Port Richmond, Stapleton & Mariner's Harbor PUMA, NY. https://censusreporter.org/profiles/79500US3603903-nyc-staten-island-community-district-1-port-richmond-stapleton-mariners-harbor-puma-ny/.
Willis, Carol. 1993. "A 3D CBD: How the 1916 Zoming Law Shaped Manhattan's Central Business Districts." In *Planning and Zoning New York City: Yesterday, Today and Tomorrow*, edited by Todd Bressi. New Brunswick, NJ: Center for Urban Policy Research.

13

FUTURITY AND EQUITY IN LOCAL CLIMATE ACTION

Nicole K. Ciulla

Introduction

In the early 2000s, climate action efforts rose significantly at the local level in response to a lack of sustained federal interest. For three years, I was a participant and observer of local climate action work emanating from the city government of St. Louis Park, Minnesota (SLP), a midsize first-ring suburb of Minneapolis, Minnesota. During this time, the city enacted a climate action plan with some of the most ambitious emissions reduction targets in the state and in the country. My observations focused on how the city and its resident liaison arm, the Environment and Sustainability Commission, spoke and wrote about the motivations and goals for local climate action and how these conversations and writings shaped the actions of the city.

I was serving as a member of the city's Environment and Sustainability Commission from 2017 to 2020. During this time, I often found myself embroiled in discussions about how to communicate climate action: how to write about it, promote events, and discuss it with fellow residents. The stakeholders involved shared a concern: how we spoke about climate action would affect how it was taken up. It would also have a significant impact on what that action ended up looking like. While the climate action plan provided overarching goals and emissions reduction targets, the finer details of the city's climate action were hammered out on a daily basis through interactions between city staff, the commission, the city council, and other dedicated participants, such as local youth, who were involved.

A complex ecology of events and actions shaped climate action in SLP. While I began by analyzing the climate action plan document, officially enacted in 2018, I quickly realized I had to record the perspectives and practices of those involved in the climate action work. Over two years' time (2018–2020), I was acting as both a highly engaged *participant* and *researcher*, and, for my research, I was recording

DOI: 10.4324/9781003217442-15

meetings, writing memos and observation notes, conducting semistructured interviews, and studying texts (the climate action plan, primarily, but also news articles, press releases, photographs, and any documents publicly distributed about the climate action work in SLP). I collected data using methods common to participant-observation research in sociology, anthropology, and—more recently—rhetorical fieldwork (McKinnon et al. 2016; Rai and Gottschalk Druschke 2018). In all, I attended over 30 meetings, taking notes and recording over 20 of those meetings. I conducted eleven 30- to 60-minute, semistructured interviews with fellow commissioners and relevant city staff, along with close analysis of the climate action plan itself and corresponding documents. My approach to analyzing texts may be best described as moving between inductive and deductive analysis, along the lines of what Tavory and Timmermans (2014) have recently termed "Abductive Analysis."

In the following chapter, I trace temporal rhetorics, specifically rhetorics of futurity, as defining mechanisms of local climate action in SLP. The following account presents the partial results of textual analysis, meeting and event observations, and semistructured interviews, all collected over two years' time. The results selected to present in this chapter include those that most relate to the theme of temporal rhetorics. This theme emerged prominently from the data, and then the data analysis was further focused around it. An underlying assumption of the following chapter is that *how* we talk and write about climate change will necessarily affect the shape and outcomes of climate action taken, and vice versa. Perspective and practice are inextricably linked. As environmental communication scholar Danielle Endres (2016) says, rhetoric "chang[es] the ways we collectively organize, construct, make sense of, and engage with our socio-material worlds" (Endres 2016). In collecting the perspectives of climate action participants via discussion at meetings and interviews, I developed a sense of how participants discussed and communicated climate action at this site. By observing the unfolding of climate action, I observed connections between perspective and practice, what actions followed from preliminary conversations and what spoken ideas preceded policy. Here, I attempt to recreate some of the relationships observed between rhetoric and action at this site, not in order to come to clear conclusions about causality but rather to emphasize the importance of increased attention to how we discuss the temporality of climate change in local governance.

Temporal rhetorics and—more specifically—rhetorics of futurity were constantly present in these discussions and played a significant role in shaping climate action in SLP. It is common and arguably necessary for calls to climate action—whether local, national, or global—to be articulated through our imagined future and supported as much as possible through scientific research. Hence, not surprisingly, when it comes to public policy, instead of focusing on present populations, climate action in the United States is often construed to benefit future generations. In the present, calculating and explaining to what extent *this* fire or *this* coastline building collapse should be attributed to climate change can be difficult. That said, individual factors contributing to climate change, such as air pollution, have clear and immediate effects on the health and well-being of present populations. Additionally, there have been notable efforts to measure present global health outcomes. As early as 2005, the World Health Organization estimated that anthropogenic climate change causes 150,000 deaths annually due to related health conditions (Patz et al. 2005). To frame climate change as a *future* issue can limit our understanding of both the contributing factors to climate change and the present impacts of environmental injustice.

"For Future Generations": A Climate Action Commonplace

Commissioners, city staff, and other key participants inflected futurity at this site through appeals to the welfare of future generations. The vast majority of appeals presented in meetings, interviews, and key texts couched the benefits of climate action for "our children." We see this starting with the founding documentation of climate action in SLP, the city's Climate Inheritance Resolution, which reads:

> NOW THEREFORE BE IT RESOLVED that the city of St. Louis Park City Council commits to working constructively, using ingenuity, innovation, and courageous determination to create a St. Louis Park Climate Action Plan for consideration that significantly reduces St. Louis Park's greenhouse gas emissions to levels that would protect our community's children and grandchildren from the risk of climate destruction.

This section of the climate inheritance resolution was eventually included in the city's climate action plan (CAP). The passage frames the threats of climate change as future threats, limiting the concern to the children and grandchildren of the local community. The CAP resists mention of threats to other communities or nations across the globe.

Given that participants frequently referenced future generations and given the lack of explicit rationale surrounding its use in public deliberation and in the CAP document itself, I read "future generations" as a commonplace of climate action occurring at this site. While I have focused my inquiry on understanding its use in this particular context, anyone who studies or actively participates in climate action has likely heard and/or uttered the refrain "for future generations," and we should consider the implications of its use in wider contexts.

While an understanding of commonplaces, common topics, or *topoi* in rhetoric extends back to Aristotle, contemporary environmental rhetoricians, such as Derek Ross (2013) and Denise Tillery (2018) have come up with expanded definitions of the term "commonplace" that reflect present contexts. Ross (2013), in his study of environmental commonplaces, defines them as ideas that activate a shared understanding among audience members. He argues that commonplaces are necessary for establishing understanding among diverse audiences (expert and layperson) on complex scientific and environmental topics: "Topics shape an audience's perceptions and influence decision making because of the way they catalyze frames of understanding" (96). Similarly, Tillery (2018) describes how commonplaces serve to "collapse or abbreviate arguments that would otherwise need to be fully explained" (17). For example, every time a participant in climate action says "for future generations," this is shorthand for a longer argument that would require an explanation of how climate change will affect future generations if left unchecked *and* why present populations should be acting with the primary goal of protecting future generations' welfare (one side of a debate that is played out *at length* in economic discussions of climate change (see Rendall 2011; Weisbach and Sunstein 2009).

Audiences are left to accept or reject commonplaces depending on whether their values are in accordance with that framework (Tillery 2018, 17). Uttering the shorthand "for the children" allows stakeholders to frame their appeals in a way that persuades through a widely held value (the welfare of children), while simultaneously paving the way for certain policy choices to take precedence. This commonplace serves an important role in encouraging audiences to think about vulnerable future populations whose livelihoods are at stake in the movement to address climate change. Tillery (2018) sees "future generations" as part of the melodramatic scientist-as-hero framework, which allows "environmental activists to cast their heroes [scientists] as warriors fighting against oppressive polluters on behalf of victims who cannot speak for themselves"; these victims include animals, the Earth itself, and future generations. Drawing from Schwarze's study of environmental melodrama (2006), Tillery argues that melodrama can serve as a useful and productive frame for environmental issues by "drawing clear moral distinctions" in order to "clarify issues of power" (157). However, as Ross says, "Because environmental argument is strongly shaped by sociopolitical climate . . . motivating an audience to action by framing an argument based on common shared narratives has an ethical component" (122). Following from this concern of ethics, I considered how the appeal to climate action on the behalf of future generations, when used as the sole or primary motivator for climate action, may obfuscate specific types of mitigation policy and civic engagement, as opposed to what could be encouraged by alternative or diversified frameworks. With this concern in mind, I dug further into the factors that motivated climate action participants at this site and how those motivating factors are connected to the CAP itself and the actions it outlines.

Motivating Factors of Climate Action in SLP

Upon asking interview participants about their motivations for involvement in local climate action, seven of 11 participants framed their motivation in terms of the futures of their own children, other children they know, or "future generations" more broadly. Only two participants explicitly framed their motivations in terms of present conditions and impacts. Six of the seven participants who were motivated by the livelihoods of future generations talked about the futures of the children in their family. One drew on a saying heard elsewhere, "The Earth does not belong to us. . . . We're borrowing it from our children." One strongly expressed that "climate work in general is the future of my kids." This participant described changing careers in order to pursue work geared toward this aim because, before the switch, thinking about the future of their children had become "terrifying."

City staff chose to open the climate action plan with two letters, each from a group of key participants in climate action in SLP. The first letter, from city youth, asks readers to think of their future livelihoods. In this letter, the word "future" appears in nearly every paragraph at least once. Each appeal is framed through the idea of insurance; in other words, how the CAP is "demonstrating the city's commitment to protecting our future and the lives of generations to come." The letter mentions presently felt suffering once, in the following passage:

> The climate crisis will have a huge impact on our future. An inadequate response now will cause dangerous economic and environmental disruptions, many of which are already being felt around the globe. But we see this as an opportunity to rethink our current actions and imagine a better, more sustainable future.

Presently experienced, global suffering is acknowledged; however, the beginning and end of this paragraph once again emphasize taking action for insurance purposes, returning the lens to yet-to-be-seen-or-felt, local conditions. The youth letter ends on three quotes, all of which employ the language of futurism. One reads, "The time is now to come together as a united force, to ensure a livable future for us, for your children, for your grandchildren, and for the generations to come. The time is now to create a sustainable future."

Notably, future outlooks were not the only motivation present. Two participants explicitly framed their motivations in terms of present conditions and experiences. One was motivated primarily by the present state of air pollution and public health, along with their own personal health. This participant cares about the present quality of life and wants to live in a healthy environment. The current state of pollution is enough to warrant change: "You know, when I walk to work, when there's a big diesel truck, it feels unpleasant. I mean . . . It's very immediate. There's nothing abstract about it. It's not about, well maybe the sea levels will rise in 10 . . . No, it's right now, I'm light-headed, I don't feel well." The other participant took a global perspective on presently felt climate conditions. "People think of it as the future, but it's now. It's already happening in places," they stated, and they went on to talk about refugees as an example of how climate change is already upending people's lives. These two participants' viewpoints stood out among the majority perspective that local climate action in St. Louis Park is motivated by the welfare of local families' future generations.

In US politics, we often hear the appeal "for the children" used as an inarguable rationale for a large variety of issues from education reform to abortion rights, and strangely the appeal can be seen employed in defense of multiple, opposing sides of an argument. The reason: people generally care about the welfare of children, a vulnerable population without much voice in our society as they do not have the right to vote or make any number of their own choices. Children are left to inherit whatever world the preceding generations leave to them. For this reason, young people have been some of the most impactful climate activists led by icons such as Greta Thunberg and, in the United States, the Sunrise Movement. While calls to consider *future generations* do work toward the welfare of one important vulnerable population, we must consider how the use and dependence on these calls as a commonplace also shapes climate action in potentially undesirable ways. When enacting local climate action at this site, participants often had the welfare of future generations in mind, a specific group who stands to benefit. This focus operates at the expense of those currently suffering among the local population; it also avoids a global mindset that could beneficially shape local action.

Notably, when the perspective shifted to a global one, either in interviews, meetings, or in the youth letter, the repercussions of unmitigated climate change became much more *presently* imagined. For instance, the youth letter mentioned environmental disruptions, "many of which are already being felt around the globe," and one interview participant talked about their motivations in terms of current refugee crises. However, aside from vague "cobenefits" discussed in the CAP itself and from the one interview participant who discussed current air quality in the area, benefits of climate action at this site were largely discussed in terms of mitigating future damages and ensuring the welfare of generations to come. Employing this commonplace regularly steered action here in ways that led to ambitious emissions targets but that lacked attention to important cobenefits that would ensure salient, equitable action.

Our Economic Present and Future

Many believe that the ethical questions and dimensions of climate change have not been addressed adequately in policy debates or in the scientific and economic literature on the topic. Invoking the welfare of future generations as a motivation to enact climate action indirectly takes a stance on a long-standing and ongoing economic debate concerning whether to "discount" the future. In these debates, the futurity of climate change is put into stark focus, some arguing that we should not act drastically as this is an issue for future generations to deal with. Economists differ on how quickly climate mitigation policies, like a carbon tax, should be applied based on how they interpret present generations' ethical obligation to future generations. As Weisbach and Sunstein (2009) succinctly put it, "At the core of the climate change problem . . . is an ethical debate involving the allocation of resources across generations" (434). Economists who advocate for slower emissions reduction measures argue that, since the beneficiaries of emissions reduction measures will be future generations, not present ones, and since future generations will be richer than present ones, climate mitigation policy should not be sudden or drastic. They believe it is likely that future generations will have more wealth, and "it is necessary to ensure consistent comparisons of resources spent in different time periods" (435). Economists who argue for more immediate and significant emissions reduction measures believe we must use a principle of "intergenerational neutrality" and cast doubt on the idea that future generations will be wealthier, especially in light of climate change's impacts. Further, they suggest that discounting the future "ensures that future generations will receive less attention, and perhaps far less attention, than those now living" (Weisbach and Sunstein 2009, 435; Rendall 2011).

The debate over "discounting the future" clarifies some of the tensions with climate mitigation policy decisions. In these decision-making processes, stakeholders must make choices, whether explicitly or not, about the relationship between present and future populations and weigh the needs of these generations. I primarily use the concerns of this debate as a way to further explore how reliance on the commonplace of future generations may shape local climate action in particular ways. In its hyperfocus on establishing our economic responsibility to future generations, the economic debate can fail to address present suffering. Instead, it assumes a present population with sufficient means and a future population that is at risk. This is the same assumption made, I would argue, by climate action stakeholders who are solely concerned with the futures of people yet to be born. When it comes to considerations of public policy, acknowledging present suffering can provide us clarity in both (1) our need to take immediate actions and (2) specific types of action that work across the needs of generations, present and future.

Imagining the Peoples and Places of Climate Action

While temporal rhetorics abounded in SLP, spatial considerations were conspicuously absent from discussions, even though the two are inextricably linked. Ursula Heise (2008), in *Sense of Place and Sense of Planet*, contends that conceptualizing climate change challenges our imaginations. Similarly, the theorist Timothy Morton (2013) calls climate change a "hyperobject," an incomprehensible thing radically distributed across time and space, and Amitov Ghosh (2016) calls it the "unthinkable," citing our inability to conceptualize climate narratives beyond sci-fi. As a professor, Heise also approaches this conceptualization issue from the perspective of how literature can contribute to the challenge. She argues, "Like other processes of global systemic transformation, ecological or not, climate change poses a challenge for narrative and lyrical forms that have conventionally focused above all on individuals, families, or nations, since it requires the articulation of connections between events at vastly different scales" (205). It is challenging to represent climate change through narrative forms given its vast distribution across time and space, and this difficulty extends to how we communicate and legislate on the issue as well. We must *imagine* climate change (in one way or another, and there are multitudinous ways of doing so) in order to write laws and ordinances regarding it. Heise traces how our imaginations are challenged by the globally distributed nature of the environmental issues we face, and the tensions between local, national, and global identities that are catalyzed by issues such as climate change.

In this case, future generations discourse shows a privilege for local outcomes for specific populations, while the climate action at this site still draws on the global nature of the issue in order to forward a sense of urgency. In "Climate Crisis Made

Manifest: The Shift from a *Topos* of Time to a *Topos* of Place," Nielsen (2017) argues that when climate change is framed primarily through time-based topics such as future consequences and tipping points, which often are accompanied by apocalyptic narratives, the arguments for action tend to become abstract and, at the same time, "infused with a sense of inevitability" (91). This abstraction can be seen at this site through the lack of motivational clarity in the CAP itself and the tensions over *who* the climate action will benefit. Nielsen argues that shifting emphasis more toward *place* in framing climate change can offer a more concrete sense of climate change's realities and better motivate grounded action.

Along similar lines, Navroz Dubash (2020), a professor at the Centre for Policy Research, posits that highly contextualized climate policy (in place-based values and ideas) can be both more motivating and farther-reaching in its short-term and long-term outcomes. Dubash encourages "learning by doing" over a "target-driven approach" to climate action, saying "there are more and less productive ways for ambition (or urgency) to be interpreted" and, in particular, drawing "a distinction between urgency for deeper and broader action and urgency for enhanced quantitative emissions limitation and reduction targets" (2). Citing the importance of embedding climate priorities into other national political narratives, Dubash says the United States failed to integrate climate policy with other national priorities and, for this reason, could not sustain climate policy across administrations. Since climate change is a politically divisive topic in the United States and since the work has not been done to convincingly wed climate action to other developmental targets in the United States, the country has been set back in its climate goals *significantly* through four years by an administration (2016–2020) that refused to acknowledge the risks (or existence) of anthropogenic climate change. In other words, without recognition of present, tangible issues or short-term economic outcomes, there was no saliency for climate policy. Dubash focuses on national governments in his analysis and approach. However, on the level of local municipalities, recognizing cobenefits has also been found to be an important factor leading to climate change policy adoption (Salon, Murphy, and Sciara 2014, 73).

One of Dubash's key points is that when climate policy is reduced to the need for "more aggressive quantitative emission targets," the approach then signals strong *future intentions and outcomes* as opposed to focusing on what can be done in the present. Such reasoning can even lead, in some cases, to a *gamification* of climate policy that leads countries to inflate their costs or actions or to reformulate benchmarks or the like to suit the perception of their accomplishments. "Far more productive than this gaming," Dubash says, is to recognize as ambitious the actions "that seek to bring together climate and other objectives in the service of transforming key emitting sectors, even if quantitative emission targets cannot be reliably attached, up front, to this transformation" (5). Rhetorics of futurity in climate action rely on the communication of tipping points, future intentions, and future results, potentially at the expense of present conditions, short-term solutions and outcomes, and cobenefits across sectors.

Equity in Local Climate Action

The commonplace *for future generations*, in addition to obscuring the present suffering caused by climate change around the globe, also risks climate action that ignores the hardships of local communities. The SLP Climate Action Plan makes reference to addressing racial equity one time in the document:

> The city continues to advance race equity [sic] and inclusion through activities, outreach, programs, policies, projects, initiatives, contracting, budgeting, hiring, and all other areas of work and service delivery. The Climate Action Plan will impact everyone in the community, and it is important that racial equity considerations are made and used as part of achieving the targets and goals outlined in the document.
>
> *(9)*

This mention can at best be seen as a placeholder for future considerations of how climate action and racial equity intersect on the local level. In the years I was a participant at this site, expounding on how racial or class equity would motivate local climate action was not a significant part of public conversation and deliberation. St. Louis Park's CAP is not alone in this regard. Schrock, Bassett, and Green (2015) found that, while the majority of local climate and sustainability action plans (around 90%) refer to equity, only about one in three of the plans they studied gave equity prominence as a cobenefit in their plan and described specific action steps to achieve equity as a goal (286). The researchers also observed that a climate or sustainability action plan that had undergone revisions or an "update" significantly increased the likelihood that it meaningfully engaged equity in its goals (287). Climate action activists and climate policy makers across the country are having issues imagining how climate action intersects with equity locally, despite the long-lived presence of the environmental justice movement. Toward the end of my tenure on the Environment and Sustainability Commission in SLP, three years after the climate action plan had been enacted there, participants began more earnestly engaging in discussions to address equity in the city's climate action goals. A number of converging factors could have contributed to the increase in equity discussions: recent staffing and commissioner changes, the hire of a racial equity coordinator, and public outrage and mobilization in response to the murder of George Floyd in neighboring Minneapolis.

The future-looking focus of the CAP is strongly connected to its initial, exclusive focus on ambitious emissions reduction targets. The CAP document primarily focuses on how to keep more greenhouse gases out of the atmosphere to reduce the increase in global warming that will be especially detrimental to future generations. This focus led to a climate action plan that was highly ambitious (indeed, at the time, the most ambitious in the state) in terms of energy targets but somewhat narrow in its goals and objectives. In addition to fostering ambitious energy targets, a future-looking CAP can be very motivating for some. As shown by the interviews, many CAP participants were motivated by the thought of the welfare of their families and others' futures. Unlike some more suspect uses of "for the children" in politics, the push to envision the future is intrinsically tied to the imaginative work of understanding and responding to climate change. Considering the welfare of future residents of the Earth is also part of an intergenerational justice-based approach to climate change. In addition, without ambitious emissions reductions made by rich, postindustrial nations, those who will suffer the most and the most quickly are the poorest of the world who have done the least to contribute to global warming.

That being said, if taken as the sole or primary motivator of climate action, this focus on future generations can result in a type of climate action that ignores synergies that acknowledge and respond to present suffering. As Nielsen (2017) argues, it can create an image of climate change as a future and abstract issue, as opposed to one with concrete ramifications in the present. In addition to ignoring the needs of the present, futurism can also foster public apathy toward the issue, making it harder to enact ambitious climate policy. This may alienate some publics, including those who see the future as still in question or, more importantly, those who are presently suffering or who witness present conditions of oppression that continue to go unaddressed. A future-focused CAP framework can suggest that policy makers are more concerned about the offspring of the privileged than the present suffering of the marginalized.

A question that should always be considered in governance is who stands to benefit and, potentially, at whose expense? To turn, urgently, our efforts to the well-being of future generations suggests that the currently living are already well-off or even OK. Climate change has present consequences, and the pollution contributing to it has already been wreaking havoc on oppressed communities for many generations. Futurity in climate action leads to additional concerns, including the apathy generated through framing effects as future and uncertain. At the same time, we know that without significant emissions reduction, there will be increased suffering among the world's most vulnerable populations. Urgent and future-focused action must be coupled with a concern for all those suffering in the present *and* in the future, and this concern should be made manifest in both our rhetoric and our actions. Adopting clearer justice-based and place-based frameworks along with temporal drivers for climate action is key to working toward these goals. Seeing climate change as a future issue in the United States is a point of privilege. In the words of Ayana Elizabeth Johnson and Katharine Wilkinson, "Equity is not secondary to survival, as some suggest; it is survival" (2020, xix). We must temper our concerns about the future with a critical eye toward our present and the uneven distribution of pollution, wealth, and resources in order to act equitably on climate.

References

Dubash, Navroz. 2020. "Revisiting Climate Ambition: The Case for Prioritizing Current Action Over Future Intent." *WIREs Climate Change* 11: 1–7. https://doi.org/10.1002/wcc.622.

Endres, Danielle. 2016. "Rhetoric, Climate Change, and Social Justice: An Interview with Dr. Danielle Endres." Interview with Michael DuPont. *Journal of Critical Thought and Praxis* 5 (2). https://doi.org/https://doi.org/10.31274/jctp-180810-63.

Ghosh, Amitav. 2016. *The Great Derangement: Climate Change and the Unthinkable*. Chicago: University of Chicago Press.

Heise, Ursula K. 2008. *Sense of Place and Sense of Planet: The Environmental Imagination of the Global*. New York: Oxford University Press.

Johnson, Ayana Elizabeth, and Katharine K. Wilkinson. 2020. "Begin." In *All We Can Save: Truth, Courage, and Solutions for the Climate Crisis*, edited by Johnson and Wilkinson. New York: One World.

McKinnon, Sara L., Robert Asen, Karma R. Chávez, and Robert Glenn Howard, eds. 2016. *Text + Field: Innovations in Rhetorical Method*. University Park: Penn State University Press.

Morton, Timothy. 2013. *Hyperobjects: Philosophy and Ecology after the End of the World*. Minneapolis: University of Minnesota Press.

Nielsen, Esben B. 2017. "Climate Crisis Made Manifest: The Shift from a Topos of *Time* to a Topos of *Place*." In *Topic-Driven Environmental Rhetoric*, edited by Derek G. Ross, 87–105. Taylor & Francis Group.

Patz, Jonathan, Diarmid Campbell-Lendrum, Tracey Holloway, and Jonathan A. Foley. 2005. "Impact of Regional Climate Change on Human Health." *Nature* 17 (438): 310–17. doi:10.1038/nature04188.

Rai, Candice, and Caroline Gottschalk Druschke. 2018. *Field Rhetoric: Ethnography, Ecology, and Engagement in the Places of Persuasion*. Tuscaloosa: University of Alabama Press.

Rendall, Matthew. 2011. "Climate Change and the Threat of Disaster: The Moral Case for Taking Out Insurance at Our Grandchildren's Expense." *Political Studies* 59: 884–99. doi:10.1111/j.1467-9248.2010.00877.x.

Ross, Derek. 2013. "Common Topics and Commonplaces of Environmental Rhetoric." *Written Communication* 30 (1): 91–131. doi:10.1177/0741088312465376.

Salon, Deborah, Sinnott Murphy, and Gian-Claudia Sciara. 2014. "Local Climate Action: Motives, Enabling Factors, and Barriers." *Carbon Management* 5 (1): 67–79. doi:10.4155/cmt.13.81.

Schrock, Greg, Ellen Bassett, and Jamaal Green. 2015. "Pursuing Equity and Justice in a Changing Climate: Assessing Equity in Local Climate and Sustainability Plans in U.S. Cities." *Journal of Planning Education and Research* 35 (3): 282–95. doi:10.1177/0739456X15580022.

Schwarze, Steven. 2006. "Environmental Melodrama." *Quarterly Journal of Speech* 92 (3): 239–61. http://dx.doi.org/10.1080/00335630600938609.

Tavory, Iddo, and Stefan Timmermans. 2014. *Abductive Analysis: Theorizing Qualitative Research*. Chicago: University of Chicago Press.

Tillery, Denise. 2018. *Commonplaces of Scientific Evidence in Environmental Discourses*. New York: Routledge.

Weisbach, David, and Cass R. Sunstein. 2009. "Climate Change and Discounting the Future: A Guide for the Perplexed." *Yale Law & Policy Review* 27: 433–57.

World Health Organization. 2005. "Climate Change." *Health and Environmental Linkages Initiative*. www.who.int/heli/risks/climate/climatechange/en/.

14

SOCIAL JUSTICE AND BROOKLYN DEVELOPMENT

500 Years of Struggle

Jerome Krase

The Inequitable Distribution of Public Goods in New York City

As a consequence of conflicts and competitions over the centuries among and between groups who have come to Brooklyn, there has been an unequal distribution of both public and private goods. *Public goods* are commodities or services provided by government, offered for the benefit of all members of that society. *Private goods*, as we might expect, are the opposite of public ones, being inherently scarce, whose ownership is restricted to only certain individuals or groups (Fernando 2021). Examples of public goods include public safety, clean air and drinking water, and services such as education and health. Governments at all levels in the United States provide a wide variety of public goods. New York State and City, having progressive political traditions, have offered its residents more in the way of public goods than other municipal and state governments.

Distributive Social Justice

For a society to be called "just," it must guarantee its members equal access to the liberties, rights, and opportunities it offers. This requires acceptance of a social contract freely entered into by its members, to which John Rawls (1971) believed free rational people would ascribe. The principles of justice in the contract "specify the basic rights and duties to be assigned by the main political and social institutions, and they regulate the division of benefits arising from social cooperation and allot the burdens necessary to sustain it" (Rawls and Kelly 2003, 7). David Harvey offers three principles addressing the uneven distribution of resources and rights: need, contribution to the common good, and merit (1973, 100–101). Harvey's most important principle was "Need—Individuals have rights to equal levels of

DOI: 10.4324/9781003217442-16

benefit which means that there is an unequal allocation according to need" (100). Over the centuries, Black Brooklynites consistently have gained the least and lost the most despite having contributed much to Brooklyn's prosperity. Why, then have they not received their fair share?

It is worth noting that one of the United Nations'17 Sustainable Development Goals (SDG) is "Reduced Inequalities" and the guiding principle for all the Goals: "Leaving no one behind" (https://sdgs.un.org/2030agenda). The most relevant SDGs here is 11, ensuring access to adequate, safe and affordable housing and basic services and enhancing capacity for participatory, integrated and sustainable human settlement planning and management. Equally important is Goal 16's promotion of inclusive societies and providing access to justice for all. As will be made clear, for Black Brooklynites these goals remain "devoutly to be wished."

Black Brooklyn: 500 Years of Growth and Change

Brooklyn's first inhabitants were the Lenapes' Canarsie tribe, whose lands were expropriated during Dutch and then British colonization. A census taken in 1698 counted 2,017 people in Kings County (Brooklyn). About half of these early settlers were Dutch. Others came from Germany, England, France, and Scandinavia, and a large number were enslaved Africans. Since initial invasions of Europeans in the 16th and 17th centuries, a trickle and then a torrent of migrants and immigrants arrived from every corner of the nation and the globe in the 19th and 20th centuries. Of especial importance for Brooklyn were African Americans, most of whom first came to Dutch New Amsterdam and British New York as slaves. African slavery increased and became harsher under British rule as the institution flourished in Brooklyn's rich farmlands during the 18th century. After the American Revolution, New York State passed a Gradual Emancipation Act of 1799 that freed those born after July 4, 1799. In 1817, a new law freed slaves born before 1799 but not until July 4, 1827. Just before the Revolutionary War, slaves composed nearly one-third of the population of Kings County. Even before legal slavery ended in New York, a small number of free Black Americans had settled in Brooklyn, and over time, settlements such as Weeksville grew and even prospered within the limitations of racial discrimination. Schools, associations, and businesses were established in the confines of such small, racially segregated, communities, which also contributed greatly to antislavery movements and the Underground Railroad.

As New York City flourished, so did Brooklyn, as development spread from the waterfront toward the interior. At the turn of the 19th century, the first large-scale immigration began with Irish immigrants. When the Erie Canal, which led through the Hudson and Mohawk Rivers to the Great Lakes, was connected to Brooklyn in 1825, the borough's economy exploded. By 1860, the City of Brooklyn, with more than 200,000 residents, was America's third largest city. By 1880, Brooklyn was a major national manufacturing center in need of workers. Later in the decade came a huge wave of immigrants from Eastern and Southern Europe. In 1900, Brooklyn's population was 1,166,582, almost a third of which was foreign-born; however, the Black population of 18,367 remained in a small number of settlements in Central Brooklyn.

During the early 20th century, Brooklyn's population grew from 1,634,351 (1910) to 2,018,356 (1920). At the same time, the Black population grew from 22,708 to 31,912 as the urbanization of Brooklyn intensified and spread. By 1930, Brooklyn's population exploded by almost a million to 2,560,400, and, as part of the Great Migration of Black Southerners to Northern cities, the borough's Black population grew to 68,921. However, Black residential confinement to Central Brooklyn continued, giving rise to a majority Black population in downtown Brooklyn, Fort Greene, Clinton Hill, Prospect Heights, Bedford Stuyvesant, and Crown Heights (Connolly 1977). The growth of the borough continued more slowly into 1940 and 1950 with 2,698,285 and 2,738,175, respectively, while the Black population almost doubled again from 107,263 to 208,478. Unfortunately for Black Brooklynites, the borough's economic engines, which had fostered upward mobility for legions of im/migrants, stalled after World War II.

The period 1950 to 1970 was the most crucial for Black Brooklynites. All poor and working-class im/migrants suffered discrimination in housing, employment, and electoral politics, as detailed by Daniel Patrick Moynihan (1963). Unfortunately, his optimism that Black Americans would also equally achieve prosperity in only another decade or two, was misplaced. As others, such as Elijah Anderson, later argued, racism was a much more difficult bias than religion and ethnicity to overcome (Anderson 2000).

From 1950 to 1980, Brooklyn's total population decreased, primarily due to "white flight," spurred by government-sponsored Federal Housing Administration loans for housing outside the central city and the construction of new highways that helped white working- and middle-class residents flee deteriorating Brooklyn neighborhoods. Black and other people of color moved into these vacated spaces in crumbling northern and Central Brooklyn neighborhoods.

Themis Chronopoulos (2020) argues that racial discrimination had, until the 1980s, confined most African Americans to what he called "Black Brooklyn" in Central and Eastern Brooklyn, where Craig Steven Wilder (2000) showed how and why between 1930 and 1953, "a vast black ghetto stretched across Brooklyn and was becoming the largest concentration of its kind" (177). Also, according to Wilder, the federal government's New Deal's Home Owners' Loan Corporation contributed to this ghettoization by grading neighborhoods in color-coded maps. Race was a negative for coding and the lowest graded (red-colored) Brooklyn areas usually the most Black. On the 1938 map, almost all of Central and Eastern Brooklyn was thus "redlined" and generally excluded from mortgage markets, which continued until the 1980s.

In the 1960s, a bleak future for American cities was predicted because of the influx of nonwhites. This self-fulfilling prophecy of urban decay was due in part to an assumption that nonwhite "invaders" were of lower socioeconomic status than those they replaced. However, as Taeuber and Taeuber (1969) found, invading Black residents were of higher educational and occupational status than their white counterparts (163–164). Black Brooklynites also paid higher prices for homes and, in general, were "acceptable" to dominant whites in the area. Despite the turmoil of the 1970s and early 1980s, Black people continued to develop community institutions that were instrumental to Brooklyn's survival. Fraternal associations, political clubs, arts and cultural associations were critical for the making of modern Brooklyn (Manoni 1973). During this period, Black Brooklyn also produced leaders in many fields such as Shirley Chisholm, Spike Lee, Kimberly Denise Jones (Lil' Kim), and Louis Gossett Jr.

Declining Brooklyn

Crime and arson rates soared in minority neighborhoods in the 1960s and 1970s. These problems, universally blamed on the victims, also had other culprits. John Barracato who served as the New York City Fire Department's Deputy Chief Fire Marshal from 1960 to 1976 wrote: "landlords who saw a bleak future for their investments in Brownsville were hiring torches to burn their buildings for insurance" (Barracato and Michelmore 1976, 18). Similarly, with respect to Black Brooklyn crime, the *Knapp Commission Report* (1973) revealed widespread New York City Police Department corruption in minority communities. There, police officers were directly and indirectly involved in narcotics trafficking, illegal gambling, and prostitution.

This period was also punctuated by urban riots in many major cities (National Advisory Commission on Civil Disorders 1968). In the summer of 1977, a power failure led to widespread rioting, looting, and arson in predominantly Black Brooklyn neighborhoods. Several blocks of the main Broadway shopping district in Bushwick were torched, and a third of the remaining stores closed immediately. More than 40% of Bushwick's commercial and retail operations went out of business within a year (Malanga 2008). The city's nadir coincided with the mayoralty of Abraham Beame and the New York City fiscal crisis that forced a virtual bankruptcy. The future looked so bleak that rather than cut citywide services, Housing and Development administrator Roger Starr suggested that a "planned shrinkage" policy be implemented with nonwhite neighborhoods in the Bronx and Brooklyn to be cut off from city services (Fried 1976).

As early as 1940, more than 90,000 out of almost 110,000 Black Brooklynites lived in a small redlined portion of Black Brooklyn. This concentration became more pronounced as large numbers of Black Southerners and Afro-Caribbeans moved to Black Brooklyn, replacing whites. White residents also engaged in various practices of "neighborhood defense" against Blacks and other people of color (Chronopoulos 2020; DeSena 2006; Krase 1982). Real estate agents and landlords employed a range of more and less obvious tactics to deny apartment rentals or housing sales to nonwhites. According to Chronopoulos, neighborhood defense "resulted in the hoarding of benefits and resources by white populations through the denunciation of Black advancement and the embrace of political entrepreneurs from the right" (560). Joel Schwartz added that the building of public housing projects contributed to the racial segregation of neighborhoods in that subsidized housing projects accepted only white or Black tenants, depending on where they were built (Schwartz 1993). Almost totally Black, Brownsville became one of many minority neighborhoods in New York to be neglected and suffer from the maldistribution of municipal services (Chronopoulos 2020, 560).

There were also some unanticipated consequences of liberal policies in the 1960s, a decade framed by civil rights movements, during which "restrictive" covenants and "neighborhood preservation" groups were stigmatized by liberal-minded publics and political authorities. The effect was the elimination of both real and imagined racial discrimination. Subsequently, many city neighborhoods, previously defended by social class hurdles, became open to Black residents, and the white middle class flight from the city accelerated. According to Thabit, East New York turned from white to Black in only six years because about 200 real estate firms resorted to blockbusting, " 'Ripe' blocks were flooded with scare literature; brokers and speculators paraded black families up and down the streets to frighten whites into selling" (2003, 1).

During the two-term mayoralty of John Lindsay (1966–1973), the NYC Planning Commission implemented several different experiments in decentralized government. These efforts stimulated the proliferation of local community and neighborhood associations in Brooklyn. In the 1970s, government agencies focused on problems in minority and changing neighborhoods as evidenced in Central Brooklyn Model Cities, the Office of Neighborhood Government in Crown Heights, and the Ocean Hill–Brownsville Community School District. In the 1970s and 1980s, local groups in Central Brooklyn aggressively competed for the limited number of various federal "poverty area" designations in order to be eligible for increased funding opportunities and programs.

In 1968, the Federal Civil rights Act was passed. Title VIII of the Act was the first comprehensive federal open housing law. It banned discrimination on the basis of race, color, religion, or national origin in the sale, lease, and financing of housing and in the furnishing of real estate brokerage services. It also resulted in numerous public and voluntary organizations seeking to enforce its mandate. Fair housing was a major concern of the Neighborhood Stabilization Program of New York City's Commission on Human Rights. The goal of their housing unit was to provide minorities with housing options in areas from which they were excluded as well as to prevent brokers from steering whites away from integrated neighborhoods. Organizations and activists acted as fair housing "checkers" or "testers."

In the 1970s, a broad spectrum of civic, business, and political forces tried to reverse the decline. For example, local activists joined in the fight against related real estate practices such as block-busting and racial steering, which are real estate practices that encourage white owners to sell their homes by exploiting fears of racial change. As a result, Black Brooklyn neighborhood associations became party to a 1977 "antisolicitation" order by then New York State Secretary of State Mario M. Cuomo.

New York City's Second Fiscal Crisis

Commenting on a Center for an Urban Future study, *New York in the World*, Patrick McGeehan (2011) wrote that New York has benefited and suffered more from globalization "than any other state in the nation. . . . Within New York State, manufacturing has disappeared fastest in New York City, leaving few options for the rising tide of immigrants needing to support their families." The median income for individuals in Brooklyn, Queens, and the Bronx, as a percentage of the nation's median income, fell significantly from 1970 to 2008 (von Nostitz and Bowles 2011, 5).

The crisis also created opportunities. Sassen (2009) argues that the securitizing of mortgages created a new channel for extracting household income, bundling it up with other types of debt and selling it off to financial investors. These "subprime" mortgages were often marketed to low-income households without disclosing the risks and possible changes in interest rates. The data shows that African Americans and low-income neighborhoods had a disproportionately higher incidence of subprime mortgages from 2000 to 2007 leading up to the 2008 financial crash. A further breakdown by neighborhoods (Community Districts) showed that the ten worst hit were those where between 34 and 47% of mortgages were subprime (Bloch and Roberts 2010). The foreclosure data for Bedford Stuyvesant and Bushwick in Brooklyn was startling.

Political Disempowerment

Black Brooklynites' political potential suffered greatly during the 20th century from discrimination. As a result, the Federal Voting Rights Act, associated with racial discrimination in the Deep South, applied to Brooklyn's Congressional Districts. According to Doug Muzzio (2021), the Act finally fulfilled "the promise of the Fifteenth Amendment, enabling blacks to exercise the franchise unhindered by exclusionary devices such as literacy tests, white primaries, poll taxes and grandfather clauses." As a result of a suit brought under the Act in 1967, Black voters who were spread among several, mostly white congressional districts were concentrated in one that enabled Afro-Caribbean Shirley Chisholm to win in 1968 (Muzzio 2021).

Immobility: Race, Poverty, and Stop-and-Frisk

According to Sheller and Urry (2006), "[S]patialities of social life presuppose, and frequently involve conflict over, both the actual and the imagined movement of people from place to place, event to event" (6). Ethnic population maps of Brooklyn since the 1920s show the remarkable immobility of Black enclaves. Black New Yorkers' limited residential mobility was due to discrimination as well as poverty (Krase and Shortell 2016). In 2008, the Community Service Society and United Way (2008) found that poverty and low incomes, unemployment, subsidized housing, emergency rent assistance, housing court actions, emergency feeding sites, disconnected youth, and low educational attainment were all highly concentrated in Upper Manhattan, the South Bronx, and Central Brooklyn. According to Tesfahuney (1998) "Differential mobility empowerments reflect structures and hierarchies of power and position by race, gender, age and class, ranging from the local to the global" (501). Therefore, it is not surprising to discover that race and poverty-induced immobility produce other patterns such as those for crime and arrests. If authorities wish to make arrests, the geography of race and class is like shooting fish in a barrel. Gelman, Fagan, and Kiss (2007) looked at claims of racial bias against the New York City Police Department's Stop-and-Frisk Policy. Blacks and Hispanics represented 51% and 33% of the stops but were only 26% and 24% of the New York City population.

The Gentrification of Black Brooklyn

Today's displacement of Black people due to gentrification has a parallel in the 1960s and 1970s. Frances Fox Piven and Richard A. Cloward (1971) noted then that federal urban renewal ("Negro removal") programs were undertaken in most big cities to deal with the Black invasion through slum clearance (241–242). According to Chronopoulos (2020), gentrification by Black renters and home buyers took place with the growth of the middle class in the 1970s. Due to limited choices, middle-class Black people moved into low-income northwest Black Brooklyn and eventually southward and eastward in the borough. The final decade of the 20th century witnessed a revival in Brooklyn's fortunes, and neighborhoods in Black Brooklyn, like Bedford Stuyvesant, Bushwick, Fort Greene, and Clinton Hill, improved as the outflow of capital in the city reversed. North Brooklyn gentrification was also pioneered by a new generation of artists, fleeing from the high rents in Manhattan, northern Williamsburg, and Greenpoint (Krase and DeSena 2016).

A seemingly benign form of gentrification in the 1970s was the so-called Brownstone Revival Movement that was part of the national Back to City movement. These movements aimed to recreate and preserve traditional, urban, middle-class residential values in the face of urban decline. Unfortunately, this activism changed to the celebration of what Neil Smith (1998) would call urban neighborhood "revanchism," which "blends revenge with reaction" (1). In northwest Black Brooklyn, whites became the numerical majority after 2013 (Chronopoulos 2020, 562). A 2019 study by Kristen Lewis and Sarah Burd-Sharps (2019) found, "The five neighborhoods with the largest increase in White residents accompanied by a decrease in residents of another racial or ethnic group were all found in Brooklyn—Bedford-Stuyvesant, Williamsburg, Clinton Hill, Park Slope and Gowanus, and Crown Heights." Consequently, between 1990 and 2020, as the total population of Brooklyn increased by 347,788, the Black population decreased by 38,622.

Racial Segregation and Environmental Hazards

According to Peggy Shepard, housing segregation has created "sacrifice zones," or "throwaway communities" in which environmental quality is low (2016). Blacks,

Latinos, Asians, and Native Americans and, secondarily, low-income communities are the most afflicted. Shepard noted that deficiencies in low-income buildings invite allergens that exacerbate asthma. Indoor chemical hazards and the extreme cold, heat, and moisture conditions lead to mold and compromised immune systems. While 15% of all New York City households report three or more such hazards, the rate is almost doubled for Black and Latino neighborhoods. Citywide, one in six children under age 13 is diagnosed with asthma. The rates are twice as high in the South Bronx North/Central Brooklyn and East/Central Harlem. These predominately Black and Latino neighborhoods also have the highest poverty, morbidity, and mortality rates. Finally, Shepard reported that the Coalition to End Lead Poisoning found that the city has failed to enforce the 2004 Local Lead Law requiring landlords to abate apartments and to notify tenants with children under the age of 6. Over 90% of all new cases of childhood lead poisoning involve children of color.

In New York City, residents of Harlem, the South Bronx, Greenpoint/Williamsburg, and communities of color in Brooklyn and Staten Island reside in these sacrifice zones. The causes of environmental injustice and health disparities are biased land use and zoning policies, as well as discriminatory enforcement of protective environmental laws. Morello-Frosch and Bill M. Jesdale (2006) examined links between residential segregation, ambient air toxin exposures and cancer risks. They found cancer risks were highest in tracts located in highly segregated areas, also associated with increased risks of adult and infant mortality. The research also showed that individual health in these neighborhoods was negatively impacted by lack of access to quality foods, such as fresh fruits and vegetables, affordable supermarkets, and crucial health services. The areas also have a dearth of public parks and open space, which are essential for healthy living.

As to the COVID-19 pandemic in New York City, Pereira, et al. reported on access to COVID testing and treatment (2021; also, Bergin and Cruz 2020). Overall, they found Black and Latino residents have received half their fair share expected based on demographic data. The data reveal that the COVID-19 vaccinations reflect the biased realities of New York City, with Black and Latino residents suffering the highest death rates, which they say are attributable to inequitable access to healthcare and housing.

Environmental racism has caused Black Brooklynites to suffer disproportionately from the flooding effects of Hurricanes Sandy in 2012 and Ida in 2021. This is especially true in regard to the New York City Housing Authority (NYCHA) low-income housing developments in which the vast majority of tenants are Black. As noted by Devin Gannon (2019), seven years after Hurricane Sandy hit, a majority of New York City's storm-damaged public housing developments have not been repaired. Of the 35 NYCHA complexes wrecked in 2012, totaling roughly 200 buildings, upgrades have been completed at only two. Brooklyn's Red Hook Houses, in a sea-level neighborhood, was deluged. Senior citizens and the disabled were trapped in cold and dark upper floors for three weeks, as elevators were inoperable. While work was scheduled to begin in April 2016, it did not begin until

August 2017, and officials estimated completion of the work in 2022. After Sandy deluged NYCHA developments, it scrambled to restart wrecked basement boilers and electrical rooms. In 2019, Brooklyn's Red Hook Houses were still using a temporary boiler put in place after the storm.

Summary

In this chapter I have employed a distributive social justice perspective developed by John Rawls (1971) and David Harvey (1973) to describe and discuss how the long history of conflicts and competitions over Brooklyn spaces and resources have negatively impacted the lives of Black Brooklynites. I have tried to frame this uneven distribution of urban resources and rights within the principles of the most relevant United Nations Sustainable Development Goals. Principal among these have been ensuring access to adequate, safe, and affordable housing and basic services, enhancing capacity for participation, promotion of inclusive societies, and access to justice for all. As noted early in this chapter, for a society to be called "just," it must guarantee its members equal access to the liberties, rights, and opportunities it offers. From my brief analysis presented here, I believe the answer to the question of whether Black Brooklynites have received "equal access" during their half millennium of involuntary and voluntary residence in and their many contributions to the borough is clearly no.

References

Anderson, Elijah. 2000. "Beyond the Melting Pot Reconsidered." *The International Migration Review* 34 (1): 262–70.

Barracato, John, and Peter Michelmore. 1976. *Arson!* New York: W.W. Norton and Company.

Bergin, Brigid, and David Cruz. 2020. "As Data Shows Hispanics and Blacks Hit Hardest by Coronavirus, Elected Officials Ask Why City Didn't Plan Better." *Gothamist*. https://gothamist.com/news/hispanics-blacks-covid-19-nyc-planning-coronavirus.

Bloch, Matthew, and Janet Roberts. 2010. "Mapping Foreclosures in the New York Region." *New York Times*, May 30, 2010. www.nytimes.com/interactive/2009/05/15/nyregion/0515-foreclose.html.

Chronopoulos, Themis. 2020. "What's Happened to the People? Gentrification and Racial Segregation in Brooklyn." *Journal of African American Studies* 24: 549–72.

Community Service Society and United Way. 2008, January. *Mapping Poverty in New York City: Pinpointing the Impact of Poverty, Community by Community*. New York: Community Service Society.

Connolly, Harold X. 1977. *A Ghetto Grows in Brooklyn*. New York: New York University Press.

DeSena, Judith N. 2006. *Protecting One's Turf: Social Strategies for Maintaining Urban Neighborhoods*. Revised ed. Latham, MD: University Press of America.

Fernando, Jason. 2021. "Public Good." www.investopedia.com/terms/p/public-good.asp.

Fried, Joseph P. 1976. "City's Housing Administrator Proposes 'Planned Shrinkage' of Some Slums." *New York Times*, February 3, 1976, page 35.

Gannon, Devin. 2019. "City Has Repaired Just Two of 35 NYCHA Developments Damaged by Hurricane Sandy." *6sqft*. www.6sqft.com/city-has-repaired-just-two-of-35-nycha-developments-damaged-by-hurricane-sandy/.

Gelman, Andrew, Jeffrey Fagan, and Alex Kiss. 2007. "An analysis of the New York City Police Department's 'Stop-and-Frisk' Policy in the Context of Claims of Racial Bias." *Journal of the American Statistical Association* 102: 479. www.stat.columbia.edu/~gelman/research/published/frisk9.pdf.

Harvey, David. 1973. *Social Justice and the City*. Oxford: Blackwell.

Knapp Commission. 1973. *The Knapp Commission Report on Police Corruption*. New York: George Braziller.

Krase, Jerome. 1982. *Self and Community in the City*. Washington, DC: University Press of America.

Krase, Jerome, and Timothy Shortell. 2016. "Seeing Residential Im/mobilities in New York City, in (Im)Mobilities of Dwelling: Places and Practices." Special Issue of *Cultural Studies* 30 (3) edited by Sybille Frank and Lars Meier: 376–400.

Krase, Jerome, and Judith N. DeSena. 2016. *Race, Class, and Gentrification in Brooklyn: A View from the Street*. Lanham, MD: Lexington Books.

Lewis, Kristen, and Sarah Burd-Sharps. 2019. "The Pace, and Face, of Gentrification: Population Change in Five Brooklyn Neighborhoods." *Urban Matters*, The Center for New York City Affairs, The New School, March 13. www.centernyc.org/the-pace-and-face-of-gentrification.

Malanga, Steven. 2008. "The Death and Life of Bushwick." *City Journal* (Spring). www.city-journal.org/html/death-and-life-bushwick-13083.html?wallit_nosession=1.

Manoni, Mary. 1973. *Bedford-Stuyvesant: The Anatomy of a Central City Community*. New York: Quadrangle Books.

McGeehan, Patrick. 2011. 'Report Lays Out State's Great Gains and Losses from Globalization." *New York Times*, October 25, 2011. www.nytimes.com/2011/10/26/nyregion/report-details-globalization-effects-in-new-york.html?ref=todayspaper.

Morello-Frosch, Rachel, and Bill M. Jesdale. 2006. "Separate and Unequal: Residential Segregation and Estimated Cancer Risks Associated with Ambient Air Toxics in U.S. Metropolitan Areas." *Environmental Health Perspectives* 114 (3): 386–93.

Moynihan, Daniel Patrick. 1963. *Beyond the Melting Pot*. Cambridge: MIT and Harvard University Press.

Muzzio, Doug. 2021. "Voting Rights Act." *Gotham Gazette*. www.gothamgazette.com/index.php/about/1856-voting-rights-act.

National Advisory Commission on Civil Disorders. 1968. *Report of the National Advisory Commission on Civil Disorders*. New York: Bantam Books.

Park, Robert E., Ernest W. Burgess, and Roderick D. McKensie. 1925. *The City*. Chicago: University of Chicago Press.

Pereira, Sydney, Dobkin, Jake, and Akpan, Nsikan. 2021. "City Reveals White New Yorkers Have Received Lion's Share of COVID-19 Vaccine Doses." *The Gothamist*, January 31, 2021. https://gothamist.com/news/white-new-yorkers-triple-nyc-covid-19-vaccine-doses.

Piven, Frances F., and Richard A. Cloward. 1971. *Regulating the Poor*. New York: Pantheon Books. Pritchett 2002.

Rawls, John. 1971. *A Theory of Justice*. Cambridge, MA: Harvard University Press.

Rawls, John, and Erin I. Kelly. 2003. *Justice as Fairness: A Reinstatement*. Cambridge, MA: Harvard University Press.

Sassen, Saskia. 2009. "When Local Housing becomes an Electronic Instrument: The Global Circulation of Mortgages—A Research Note." *International Journal of Urban and Regional Research* 33 (2): 441–26.

Schwartz, Joel. 1993. *The New York Approach: Robert Moses, Urban Liberals, and Redevelopment of the Innercity*. Columbus: Ohio State University Press.

Sheller, Mimi, and John Urry. 2006. "The New Mobilities Paradigm." *Environment and Planning* 38 (2): 207–26.

Shepard, Peggy. 2016. "Building Justice: NYC's Sacrifice Zones and the Environmental Legacy of Racial Injustice." *City Limits*. https://citylimits.org/2016/10/10/building-justice-nycs-sacrifice-zones-and-the-environmental-legacy-of-racial-injustice/.

Smith, Neil. 1998. "Giuliani Time: The Revanchist 1990s." *Social Text* 57: 1–20.

Taeuber, Carl E., and Alma F. Taeuber. 1969. *Negroes in Cities*. New York: Atheneum.

Tesfahuney, Mekonnen. 1998. "Mobility, Racism and Geopolitics." *Political Geography* 17 (5): 499–515.

Thabit, Walter. 2003. *How East New York became a Ghetto*. New York: New York University Press.

von Nostitz, Glenn, and Jonathan Bowles. 2011. *New York in the World*. New York: Center for an Urban Future and SUNY Levin Institute. https://nycfuture.org/pdf/New_York_in_the_World.pdf.

Wilder, Craig Steven. 2000. *A Covenant with Color: Race and Social Power in Brooklyn*. New York: Columbia University Press.

15

PARTICIPATORY ACTION RESEARCH IN RIO DE JANEIRO'S SUSTAINABLE *FAVELA* NETWORK

From Self-Sufficiency and International Collaboration to Participatory Regenerative Infrastructures

Leonel Lima Ponce

Introduction

For over a century, global urbanization and climate change, as well as subsequent stresses to natural ecosystems, political systems, and physical and social infrastructure, have resulted in a proliferation of informal urban settlements, exacerbating trends toward unsustainable living conditions for billions around the world. Most of the global population now resides in cities, with especially rapid urbanization occurring in Majority World countries. The urban population share of Brazil grew from an estimated 36% in 1950 to 87% by 2018, with over 55 million urban inhabitants (United Nations 2019, 59). Internally displaced people, motivated by economic opportunity, often settle informally in available, affordable land. In Rio de Janeiro city, *favelas* have historically received this population. As of 2019, over 22% of Rio's residents lived in *favelas* (Municipality of Rio de Janeiro 2021, 121). In 2019, an estimated 453,571 (19.28%) of residential units in the municipality were informal settlements (IBGE 2021, 9).

This influx strains aging public infrastructure, as informal housing is constructed by residents outside of formal infrastructure systems and their functional capacities. Though infrastructure upgrade projects have been constructed in many *favelas*, municipal urbanization programs have not provided continuity of basic services to all residents. As of 2019, approximately one-third of Rio's sanitary sewage was left untreated (Oliveira et al. 2021, 119). Through infrastructural, social, environmental, and health deficits, environmental racism, and state violence, *favela* residents have autonomously responded by pursuing sustainable projects that generate ecosystem services, build economic capacity, and enhance social resilience.

In support of these responses, Catalytic Communities (CatComm), a *favela* reporting and technical assistance NGO that has long promoted grassroots

DOI: 10.4324/9781003217442-17

sustainable development initiatives and projects, formed the Sustainable Favela Network (SFN) in 2018. This coalition of community organizers, researchers, and technical experts strives to recognize, connect, and build upon grassroots socioenvironmental initiatives in Rio's *favelas*, fostering knowledge exchanges, capacity building, and impact expansion (Catalytic Communities 2020, 2). Established activities by the SFN's Environmental Education working group are sharing and devising models for nonextractive collaboration to support sustainable, *favela* asset-based community development. These can be adapted with environmental justice principles to guide just, international regenerative infrastructure projects and processes.

Out of Necessity, Grassroots Development: Leveraging *Favela* Assets Toward Community-Led Sustainability and Regeneration

Favelas play an intrinsic role in the culture and heritage of Rio de Janeiro. These vibrant communities, well-known for traditional *samba* and *carnaval* and for contemporary *baile funk* music, are also portrayed internationally as host to drug trafficking kingpins and violent battles between police and gangs (Meirelles 2002). Rio's *favelas* have been perennially stigmatized and pushed to the outskirts of society. The first *favela*—Providência, established in 1897—was founded by soldiers who were promised land ownership but who were forced by local government to squat on hillside parcels (Perlman 2009, 26). From the beginning, *favelas* were viewed by the government as temporary solutions to housing crises, their population deemed necessary but expendable. Rio de Janeiro's public sector continues underserving *favelas* by both inaction and active contempt—through defamatory public narrative and policy, displacement, public infrastructure disinvestment, and violent policing.

The cumulative outcome of public actions addressing Rio de Janeiro's *favela* residents has been a patchwork of inadequate projects and continuous marginalization of residents by the government and general public (Williamson 2017, 58–59). The collective failure of urbanization programs and policies to fundamentally improve living conditions for a significant portion of Rio's population (Williamson 2012) can be directly correlated with this stigma, leaving a sustainable development vacuum.

Though subject to violence, disinvestment, and vulnerability, it is a mistake to categorize Rio de Janeiro's *favelas* as squalid "slums" or to describe their residents as passive victims of oppression. Residents have embraced autonomous development and implemented hyperlocal sustainable urbanist solutions with tangible social resilience and environmental sustainability cobenefits. *Favelas* offer desirable land, housing, community, and resources to their residents, in the absence of outside or government support—providing valuable assets to the city writ large (Williamson 2012). While physical precarity, risks, and vulnerabilities exist, these qualities are not universal and do not define *favelas*.

Self-determined and responsive by nature, Rio de Janeiro's *favela* communities and organizations maximize their assets efficiently, providing a basis for organic, sustainable, asset-based community development (ABCD) that must be respected and integrated when considering urban development and climate adaptation interventions in these territories.

Asset-based community development eschews prototypical, top-down, deficit and needs-driven planning processes, instead focusing on the capacity of community members and resources to effect change. A community-led ABCD process complements existing efforts from local individuals, associations, institutions, and their connections in a sustained manner (Kretzman and McKnight 1993, 1–11). In Rio de Janeiro's *favelas*, the ABCD approach organically emerges as the operational model for community-based organizations (CBOs) to plan and implement sustainable development projects, overcome environmental hazards and infrastructural deficits, and achieve dignified housing and existence. Relying on lived experience and professional expertise, *favela* organizers maximize value and efficiently leverage their proximity to environmental resources, infrastructural informality and flexibility, entrepreneurship and creativity, and social cohesion and adaptive capacity to devise community-led projects that address immediate needs and broader, systemic fractures.

Rio's *favelas* are often driven to settle on public land at interstitial undeveloped zones between the built environment and sensitive tropical ecosystems, encroaching on rainforest, mangroves, shorelines, and riparian zones—terrains prone to environmental hazards (e.g., flooding and mudslides). This proximity also nurtures connections to land and nature and environmental stewardship expertise. Waterfront communities like Vila Autódromo form out of subsistence fishing traditions (Williamson 2017, 61–62), whereas *favelas* encircled by lush rainforest develop agroforestry projects, like Verdejar in Alemão (Depies 2018). A spiritual connection to nature is forged through projects like the open-air Horto Museum, commemorating a freed slave community's historic relationship with the rainforest (Patel 2018). Proximity to Rio's natural resources cultivate socioenvironmental education and management organizations. Alfazendo, in Cidade de Deus, integrates Paulo Freire's popular and mutual education theories into its EcoRede youth education program, building collective knowledge and sustainable resource management between students and informal waste pickers (SFN 2020a). *Favela* community stewardship and socioenvironmental education contributes to environmental hazard reduction through awareness building, reduction of degradation, and risk mitigation (Catalytic Communities 2018, 36).

When deprived of basic water and electricity services, *favela* residents improvise tapped connections to the formal electrical and water grids, circumventing high public service costs. Though ingenious, taps are subject to service outages. Off-grid, decentralized infrastructures like rooftop solar power (e.g., InSolar in Santa Marta) and vegetated roofs (e.g., Teto Verde Favela in Parque Arará) have proliferated to meet demand. Installations utilize existing rooftops, reduce energy use, promote energy sovereignty, reduce heat island effect, and improve air quality (Andrew

2019; Silva 2021). System implementation is facilitated by local, generational construction expertise. Vulnerable residents use organic waste, stormwater, and sanitary sewage, typically dumped directly into water bodies, as valuable resources. In Vale Encantado, nestled in the Tijuca Rainforest, residents have constructed two anaerobic biodigesters: one treats residential sewage, the other collects the community kitchen's food waste. Biogas is used for clean cooking, and liquid biofertilizer nourishes the community farm (Suarez 2021).

Culturally vibrant and socially and economically vigorous, Rio de Janeiro's *favelas* exhibit more street activity than formal neighborhoods, providing opportunities for creative entrepreneurship to take hold. Colocation of homes and stores present a model for live–work spaces, reducing strains on transportation systems. Constant social interactions lead to solidarity and collective sweat equity action; neighbors assist one another on home renovations and improvised public infrastructure repairs (Williamson 2012). Local informal and solidarity economies generate considerable value. In Maré, the DEVAS women's artisan collective provides sewing classes, networking, and access to dignified working conditions (Pizzimenti 2017). Informal waste picker cooperatives such as Transvida in the Penha Complex complement material sorting skills with business administration training to support local resource recovery economies (Bazin 2020).

Perhaps the most critical assets of *favela* residents are social cohesion and adaptive capacity, traits embedded in the identity of informal communities responding to constant change. Grassroots organizations harness these qualities to fill public policy vacuums and to provide social services, environmental education, and dynamic climate change adaptation. Leaders in Morro dos Prazeres support alternating priorities and programs, working with youth to reclaim public space, upcycle plastics, and reduce hillside erosion through Reciclação, while also providing students with digital literacy and entry points into emerging job markets through Vai Na Web (Gomes 2019). Sustainable *favela* organizations operate from an inclusive lens; participatory community plans such as the Vila Autódromo Popular Plan function as collective advocacy agendas, prepared and updated to reflect evolving local realities and preferences (Williamson 2017, 69–79).

Community-based approaches to sustainable development reveal *favelas* as positive references for formal urban development (Catalytic Communities 2020). However, community development projects cannot overcome severe cumulative injustices on their own. Institutional support is needed to sustain projects, build capacity, scale up solutions, and affect policy. These gaps are filled through partnerships with technical support organizations and researchers providing legal assistance, policy development, geotechnical engineering, sustainable design, and other services. Critical to such collaborations is a researcher's approach. Rather than perpetuating the "parachute researcher" stereotype of project-based, extractive unilateral consultation, where consultants drop in to complete a study or installation and leave without operating plans or policy pathways (SFN 2020a), inclusive researchers pursue participatory action research and codesign approaches, developing the work in close coordination with residents. Researchers and residents exchange knowledge about natural processes and technological systems that leverage a community's understanding of climate, terrain, and socioeconomic systems to inform implementation of environmentally just projects.

Participatory, asset-based research takes time, trust, and resources and is most readily pursued by community organizations and local researchers invested in long-term relationships. Continuous, asset-based engagement between universities and community partners is a process replete with institutional, logistical, and relational challenges, and long-term collaboration depends on commitments from all the parties involved, highlighting the importance of collective structures that support capacity and impact expansion.

Sustainable Favela Network: Aggregating, Sharing, and Expanding Capacity, Knowledge and Impact

The nongovernmental organization, Catalytic Communities (CatComm), has helped advance *favelas* as solution generators for Rio de Janeiro's—and the world's—sustainable, equitable development for over 20 years. CatComm provides strategic training to *favela* leaders, facilitates partnerships with global civil society, academics, and researchers, and supports sustainable, community-responsive, asset-based community development in these territories. Three core programs are administered by CatComm. The RioOnWatch website reports on current *favela* news. The Favela Community Land Trust project pools alternative land tenure and ownership resources. The Sustainable Favela Network (SFN) connects *favela* organizations and partners working on sustainable development projects, providing a grassroots-organizing forum for community development across Rio's *favelas* (Williamson 2019).

Formed in 2018, the SFN is a coalition led by community-based designers and organizers and is supported by academic researchers, technical experts, and nonprofits. Its mission is "to realize the potential of *favelas* as sustainable communities through a network of knowledge sharing, mutual support, and development of joint efforts between community initiatives working on environmental sustainability and social resilience, and technical allies." In this emergent development model, *favelas* grow sustainably, organically evolving into resilient, creative, solutions-based places. Historic neglect is more readily overcome, and the importance of *favela* assets to the city is recognized (Catalytic Communities 2020, 2).

To achieve its mission, the Sustainable Favela Network works to (1) recognize extant socioenvironmental sustainability and resilience initiatives in Rio's *favelas*, (2) provide visibility and accessibility for these initiatives, (3) create knowledge exchange networks among partners, (4) provide training and development opportunities, and (5) extend the Network's impact internationally (Catalytic Communities 2018, 7). Members meet regularly to discuss emerging priorities, struggles, projects, and opportunities, with the administrative support of CatComm. The Network is structured into seven working groups: Environmental Education, Gardens and Reforestation, Income Generation, Memory and Culture, Solar Energy,

Solid Waste Management, and Water and Sewage. The Environmental Education working group is committed to strengthening community mobilization for environmental education in *favelas*, bringing together environmental educators and community organizers to generate content and advocacy for just and inclusive socioenvironmental education methodologies (SFN n.d.). To date, the group has (1) hosted a discussion on research methods and ethics of *favela*–academia relationships, (2) written an open letter requiring sustainable *favela* policy and action commitments from local election candidates, and (3) drafted a manifesto to guide local *favela*–academia relations, just research methods, and environmental education.

1 The public panel, "Dialogue Favela and Academy: Necessary Pathways for Environmental Education" in October 2020 discussed the relationship between *favela* communities and academia. Eight experts from grassroots environmental education organizations and local universities examined these relationships, interweaving formal and informal pedagogical expertise with lessons borne out of personal experience. Speakers noted that most academic research based in *favelas* is not inclusive and is still rooted in a savior complex that reduces residents to research subjects. Researchers restrict access and use of documents by community educators, limiting tangible outcomes for residents. Barriers to entry into higher education for the poor, Black population of *favelas* perpetuate this dynamic. Constrained by institutional bureaucracies that limit experiential methodologies, student researchers from *favela* backgrounds must work outside academic structures to generate practical demonstration projects within their communities. These participatory methodologies are critical to local sustainable development, and yield priorities not always aligned with research mandates. Panelists agreed that academics invited into a community must survey its members, understand local priorities, and collaborate on appropriate and inclusive interventions. The most successful approaches and projects foster a diversity of understandings through mutual exchanges, with the primary goal of improving residents' living conditions. Solutions may include supporting community researchers within grassroots organizations—as exemplified through Raízes em Movimento's CEPEDOCA research center in Alemão (SFN 2020a).

2 The COVID-19 pandemic caused disproportionate health impacts for residents of *favelas* in the eve of local elections. The SFN published an open letter to Rio's mayoral and city council candidates on October 16, 2020, highlighting inequitable burdens, charting just and sustainable policies, and requesting that candidates express their support for the Network's recommendations. In its recommendations, the Environmental Education working group stressed the potential for environmental education to promote a healthy society and to empower Rio's *favela* youth to integrate environmental considerations into their lives. They requested candidate commitments to public policies and actions related to municipal schools serving *favela* children: the deepening and expansion of environmental education activities inside schools, the

promotion of programs and partnerships in areas surrounding schools in *favelas*, and the promotion of environmental education activities with *favela* youth in the broader city (SFN 2020b). The letter was submitted to candidates, and informed a subsequent debate with mayoral hopefuls, hosted by the SFN.

3 The Favela-Academia Manifesto was collectively drafted during the working group's October 2020 meeting, containing a set of guidelines for ethical, just, inclusive, and impactful *favela*-based research. The Manifesto, currently undergoing revision by members of the working group, builds upon the panel discussion and letter and on the collective experience of the group. It outlines common principles to guide academics seeking to engage in research within *favelas*. The document holds researchers accountable to the communities with which they work and invites them to participate in an inclusive, mutual, and transparent dialogue to cocreate participatory action research approaches centered on *favela* residents, priorities, and assets.

The SFN Environmental Education working group's activities and statements frame public programs and *favela*–academia collaborations as potential contributors to the environmental consciousness, educational experience, and sustainable development of Rio de Janeiro's *favelas*. These practices must be democratized to better respond to community assets and evolving infrastructural issues, center the work on the lived experiences and priorities of *favela* residents—particularly its youth—and provide pathways for socioenvironmental and economic just transitions for these underserved communities.

International *Favela*–Academia Sustainable Development: Toward Inclusive, Sustained, and Just Collaborations

Actions proposed by SFN's Environmental Education working group, particularly the Favela-Academia Manifesto, are being deliberated at the time of this writing and are thus not fully vetted through the Network. It would be premature and unjust to discuss final priorities and the desires of community leaders concerning local *favela*-based research approaches. However, based on existing documents and ongoing discussions, draft criteria can be extrapolated for international *favela*–academia partnerships, with a focus on ABCD approaches, just and inclusive pedagogies, sustainable development, and environmental education. These criteria for inclusive, sustained, and just international knowledge exchange and technical transfer between *favela* communities and international researchers also reference the work of United States–based environmental justice and social movements and their partners.

While grassroots *favela* socioenvironmental organizations draw their approaches and core values from Brazilian social movements and popular education theorists, their struggles have parallels in the United States–based environmental justice (EJ) movement. Borne out of the social and racial justice activism of the 1960s, the movement builds on the historic work of Black, indigenous, and people of color

(BIPOC) and low-income communities and organizations at the front lines and fence lines of polluting infrastructure, environmental hazards and health inequalities, racial and socioeconomic injustices, and climate change impacts. To pursue sustainable development, EJ organizations rely on connections to their environment, collective understandings of place and heritage, creativity, social cohesion and adaptive capacity, and intergenerational transfer of knowledge. EJ groups across the country have formed coalitions, from the local to international levels, inclusive of academics and technical experts in environmental sciences, public health, planning, economics, and sustainable design. These partners reinforce grassroots campaigns, experiences, and expertise, resulting in significant sustainable and regenerative community development victories. Challenges inherent in such partnerships include partnership power imbalances, mission and scope misalignment, and partner capacity and logistical limits. In response, EJ movement leaders have developed various principles promoting values, goals, and methods for just and impactful community engagement. At the core of these documents, as outlined in the Jemez Principles of Democratic Organizing, are the shared values of (1) inclusion and a plurality of voices, (2) bottom-up organizing, (3) self-determination, (4) solidarity and mutuality, (5) just relationships, and (6) commitment to self-transformation (SNEEJ 1996).

Although community-based groups have charted pathways to facilitate inclusive academic engagement toward environmental justice, it is not their responsibility to integrate and operationalize these tenets within the infrastructures of academia. Researchers must actively work to learn from previous projects and approaches, foster long-term relationships that center community priorities, and work toward new institutional models and structures of community support that transcend dominant academic frameworks. SFN's organizational guidelines set clear ethical expectations and responsibilities for technical allies, which include guaranteeing that local leaders' demands and proposals guide SFN activities, providing structure and support for grassroots actions, building technical awareness and rigor into grassroots initiatives, providing mutual knowledge exchanges cultivating *favela* capacity and leadership, and fostering self-reflection within supporting roles (SFN 2021). Adapting the structure of the Jemez Principles, we can establish preliminary principles for international academic support to grassroots, sustainable *favela* development organizations:

1 *Forge international bridges across inclusive, transdisciplinary networks.* Typically, international academic engagements include short-term excursions, field research projects, and the establishment of satellite campuses beholden to host governments. These structures, while limited, provide opportunities for further inclusion and redistribution of resources. Researchers can access international networks and institutional capacity to support connections *between* international environmental justice organizations or provide pathways for a broader "ecology of understandings" by connecting international grassroots actors with like-minded transdisciplinary research and technical experts. The

first Sustainable Favela Network International Exchange in August 2021, organized by the SFN and CatComm with assistance from Pratt Institute's Graduate Center for Planning and the Environment (Pratt GCPE) in Brooklyn, New York, brought together SFN members and community-based socioenvironmental organizations from the United States, Kenya, Nigeria, and South Africa to share projects, advocacy strategies, and cultural programming and to discuss lessons and collaboration opportunities.

2 *Support existing and emerging favela collaborators, on their terms.* Academic research goals and technical support approaches are often driven by project-specific funding and legal and ethical policies, which center university and consultant priorities and benefits. To shift this power dynamic and center *favela* partners as clients, academic researchers, students, and administrators must work to create flexible programs and projects that respond to the ongoing and emerging needs of community partners. Programs such as Pratt GCPE's Environmental Justice Graduate Fellowship, in partnership with NYC Environmental Justice Alliance, place students in yearlong technical assistance positions with grassroots organizations in support of daily operations, education programs, organizing campaigns, development projects, and public policy advocacy. The scope of work is adapted to emerging organizational priorities.

3 *Amplify and follow community voices.* Centering research work on the lived experiences of *favela* residents, with tangible impacts on their quality of life, requires the development of long-term bonds and mutual trust. This process begins by broadening the platforms of *favela* community voices. Once mutual trust is established, academic partners may be called upon to provide expertise through advocacy planning and development work to resolve emergent and structural struggles faced by a community. Initiatives such as Recovery, Adaptation, Mitigation, and Planning (RAMP), a community recovery planning initiative after Superstorm Sandy led by New York City EJ organizations and Pratt Institute, are built upon decades of advocacy planning work and trust building. The initiative brought together diverse stakeholders through open courses, public workshops, panels, and training programs, producing projects reflective of community recovery needs and development priorities, enhancing grassroots climate adaptation capacity, and increasing frontline community visibility and agency.

4 *Align project goals and processes in solidarity and mutuality with favela organizations.* The traditional model of values-neutral technology transfer and research between academics and research "subjects" fails to fully account for structural inequalities and power imbalances faced daily by *favela* residents. True solidarity and mutuality require intent: a clear alignment of research goals with community organization values, integrating environmental justice, democratic organizing, and ethical research principles, supported by clear communication and shared resources. This may begin with the cocreation of common project goals, the drafting of terms of engagement, roles, and responsibilities,

the establishment of appropriate research methodologies, and the codesign of projects. Continuous assessment by the team helps ensure the achievement of common projected outcomes. Train-the-trainer programs like those implemented by Solar CITIES and Taboá in Vale Encantado (see p. 180) apply the mutual exchange of knowledge between consultants and community to the codesign, construction, and operations and management of food-water-energy sustainable systems that reflect existing local assets and priorities.

5 *Build and expand models for just favela-academia sustainable development collaborations and relationships*. Once successful projects have been completed, partners can learn from their successes and build sustainable models of engagement. Participatory, transparent and accountable processes, if established and supported through ongoing efforts between *favela* communities and academics, can foster long-term collaborations and bonds. Such processes require commitment and institutional support. While institutions typically do not provide financial and logistical support for initiatives beyond academic calendars and terms (e.g., dedicated grant funding periods), some academic departments have built in community engagement infrastructures to support such relationships. For example, as a core component of its community-based curricula, Pratt GCPE requires students to participate in practice-based, transdisciplinary studio courses, responding to briefs developed by faculty and partner organizations. Grassroots clients are usually involved from conception through semester's end, and final documents support ongoing advocacy, planning, and development initiatives. The relationship can extend beyond one semester through independent student projects, technical support of CBO grant applications, student internships, and organizing and advocacy campaign support.

6 *Chart pathways for institutional transformation in support of grassroots struggles*. Much like the community-led initiatives that they support, successful long-term, participatory community-based academic projects and programs—particularly those outside of dominant institutional structures and processes—face constant strains and are vulnerable to shocks such as the COVID-19 pandemic. Faculty and administrators must balance community priorities with curricular and accreditation requirements, organizational capacity, and other structural obstacles. Parallel and alternative structures can be formed to circumvent these challenges and institutionalize technical assistance networks for sustainable community development. Community-focused research and development centers, for example, can operate on longer timelines and broader agendas than academic departments. Founded to provide professional planning and policy support to NYC-based community development groups, the Pratt Center for Community Development (Pratt Center) sometimes builds on Pratt GCPE's studio-based projects. The New School's Tishman Environment and Design Center's Environmental Justice Movement Fellowship program provides financial and technical support for community EJ leaders to pursue research projects.

In response to climate and racial justice movements that have swelled and stepped to the forefront of public consciousness since the beginning of the COVID-19 pandemic, as well as growing demand from students, faculty, and society at large, US-based academic institutions are beginning to focus on more holistic models of engagement, inclusion, and mutual education in the climate classroom. This movement harkens back to activist campuses of the 1960s but is contextualized in interconnected global social, economic, and environmental systems. Universities are more international than ever, and it is incumbent upon researchers to reflect local best practices for community engagement in their transnational work. Logistical and structural obstacles can limit international implementation: reciprocity of context, limitations on access and travel, and funding limits are a few challenges to long-term, international *favela*–academia relationships. While the recommended criteria do not inherently overcome structural obstacles, they reveal core strategies that international sustainable asset-based community development researchers can adopt and adapt to strengthen their partnerships.

Conclusion

The city is an ecosystem, one suffering from centuries of unsustainable pressures and power imbalances maintained by the same power structures that control higher education institutions. But the city is not a laboratory to be explored by academics at their own whim. Vulnerable social, economic, cultural, and environmental urban systems, threatened by the intertwined crises of climate change, racial and socioeconomic injustice, and public health, provide ample opportunities for community-based academic research. However, just approaches must be undertaken to center on the community, combine and share resources, and transform how so-called experts engage local leadership. In Rio de Janeiro, the hard work of grassroots organizing and sustainable development, in addition to the consequences of failed initiatives, have for too long been borne by *favela* residents. The Sustainable Favela Network provides the infrastructure for nascent international collaborations on necessary and urgent sustainable development projects, plans, and policies in Rio's informal communities. Academics looking to participate must step down from insular ivory towers and shift narratives, approaches, and structures of community engagement to provide real value. When contributing to the work from a place of humility, respect, solidarity, and mutuality, academics can follow *favela* leadership toward transformative, inclusive, nonextractive, and just models of development that can support a collective vision of regenerative, sustainable infrastructures and communities.

Acknowledgments

This work references decades of grassroots struggles, responds to centuries of unjust power structures, and builds upon grassroots principles for academic engagement. Although this chapter does not involve coauthors, it is critical to acknowledge this community-led work and open avenues for future grassroots collaborations. Special acknowledgment must be made to members of the SFN Environmental Education working group and members, New York City–based environmental justice leaders and youth, and faculty and students from Pratt Institute GCPE.

References

Andrew, Colin. 2019. "Insolar: Democratizing Access to Solar Energy in Santa Marta and Beyond." *RioOnWatch*, November 13, 2019. www.rioonwatch.org/?p=56067.

Bazin, Sofia. 2020. "The Transvida Recycling Cooperative and Kids' Home of Dreams: Transformed Lives in Rio de Janeiro's Favela of Vila Cruzeiro." *RioOnWatch*, April 14, 2020. www.rioonwatch.org/?p=56950.

Catalytic Communities. 2018. *Sustainable Favela Network Map (2017) Report*. Rio de Janeiro, Brazil. https://catcomm.org/wp-content/uploads/2018/02/Sustainable-Favela-Network-Map-2017-Final-Report.pdf.

———. 2020. *2019 Sustainable Favela Network Annual Report*. Rio de Janeiro, Brazil. https://catcomm.org/wp-content/uploads/2019/12/2019-annual-report-sustainable-favela-network.pdf.

Depies, Jessica. 2018. "3rd Sustainable Favela Network Exchange: Agroforestry in Rio's Densely Populated North Zone." *RioOnWatch*, December 5, 2018. www.rioonwatch.org/?p=49338.

Gomes, Zoraide. 2019. "Sustainability is the Quality of Relationships with Oneself, with Others, and with the Planet." Presentation at Planner's Network 2019 Conference, Pratt Institute, Brooklyn, NY, June 21, 2019.

IBGE—Instituto Brasileiro de Geografia e Estatística, Ministério da Economia, Diretoria de Geociências, Coordenação de Geografia. 2021. *Aglomerados Subnormais 2019: Classificação Preliminar e informações de saúde para o enfrentamento à COVID-19—Notas Técnicas*. Rio de Janeiro, Brazil. https://biblioteca.ibge.gov.br/visualizacao/livros/liv101717_notas_tecnicas.pdf.

Kretzmann, John P., and John L. McKnight. 1993. *Building Communities from the Inside Out: A Path Toward Finding and Mobilizing a Community's Assets*, 1–11. Evanston, IL: Institute for Policy Research.

Meirelles, Fernando, and Kátia Lund, dir. 2002. *Cidade de Deus = City of God*. Rio de Janeiro, Brazil: O2 Filmes/Lumiére Pictures.

Municipality of Rio de Janeiro. 2021. *Plano de Desenvolvimento Sustentável e Ação Climática da Cidade do Rio de Janeiro*. Rio de Janeiro, Brazil. www.rio.rj.gov.br/web/planejamento/pds.

Oliveira, Gesner et al. 2021. *Ranking Do Saneamento: Instituto Trata Brasil 2021*. São Paulo, Brazil: GO Associados. www.tratabrasil.com.br/images/estudos/Ranking_saneamento_2021/Relat%C3%B3rio_-_Ranking_Trata_Brasil_2021_v2.pdf.

Patel, Gitanjali. 2018. "Museums of Counternarratives and Resistance, Part 4: Horto Florestal." *RioOnWatch*, May 17, 2018. www.rioonwatch.org/?p=43431.

Perlman, Janice. 2009. *Favela: Four Decades of Living on the Edge in Rio De Janeiro*. New York: Oxford University Press. https://search-ebscohost-com.ezproxy.pratt.edu/login.aspx?direct=true&db=nlebk&AN=319799&site=eds-live&scope=site.

Pizzimenti, Sophie. 2017. "Devas—Maré's Women Artisans." *RioOnWatch*, December 18, 2017. www.rioonwatch.org/?p=39950.

Silva, Luis Cassiano. 2021. "Green Roofs or Air Conditioning in Parque Arará Favela?" *RioOnWatch*, February 24, 2021. www.rioonwatch.org/?p=64226.

Southwest Network for Environmental and Economic Justice (SNEEJ). 1996. *Jemez Principles for Democratic Organizing*. Jemez, New Mexico. www.ejnet.org/ej/jemez.pdf.

Suarez, Jaqueline. 2021. "Energy That Comes from Waste: Biodigester Turns Sewage into Biogas in Vale Encantado Favela." *RioOnWatch*, August 1, 2021. www.rioonwatch.org/?p=66297.

Sustainable Favela Network. 2020a. "Diálogo Favela e Academic: Caminhos Necessários Para Educação Ambiental." October 14, 2020. Panel discussion. www.youtube.com/watch?v=fzUCowACGQM.

———. 2020b. *Letter of Commitment presented by the Sustainable Favela Network to Rio de Janeiro's Mayoral and City Council Candidates, Municipal Elections—Rio de Janeiro 2020*. Rio de Janeiro, October 30, 2020. https://catcomm.org/wp-content/uploads/2021/02/Sustainable_Favela_Network_Commitment_Letter_2020.pdf.

———. 2021. "Princípios e Diretrizes da Rede Favela Sustentável." Accessed August 22, 2021. https://favelasustentavel.org/?s=diretrizes.

———. n.d. "GT Educação Ambiental." Accessed August 8, 2021. https://favelasustentavel.org/gt-educacao-ambiental/.

United Nations, Department of Economic and Social Affairs, Population Division. 2019. *World Urbanization Prospects: The 2018 Revision (ST/ESA/SER.A/420)*. New York: United Nations. https://population.un.org/wup/Publications/Files/WUP2018-Report.pdf.

Williamson, Theresa. 2012. "Favelas at the Vanguard: Rethinking Our Assumptions in Sustainable Development." Lecture at Pratt Institute, Brooklyn, NY, October 24, 2012.

———. 2017. "Not Everyone Has a Price: How the Small Favela of Vila Autódromo's Fight Opened a Path to Olympic Resistance." In *Rio 2016: Olympic Myths, Hard Realities*, edited by Andrew Zimbalist, 57–96. Washington, DC: Brookings Institution Press. https://search-ebscohost-com.ezproxy.pratt.edu/login.aspx?direct=true&db=nlebk&AN=1655929&site=eds-live&scope=site.

———. 2019. "What's in a Name? Introducing 'Favela'." Presentation at Planner's Network 2019 Conference, Pratt Institute, Brooklyn, NY, June 21, 2019.

PART III
The City as Classroom

Part III addresses the application of sustainability theory in (in)formal learning environments. These chapters shed light on theory–practice connections and provide insights into the kind of instruction that motivates action-taking and behavior changes, insights that include the importance of interdisciplinarity and the resulting consideration of the objectives of sustainability education. Contexts involve the K–16 classroom, include partnerships with city leadership and nonprofit groups, and cover a range of subject matter including global citizenship, art, ecology, and critical literacy. The authors of these chapters bring fresh perspective to these topics because of their focus on urban connections, a relatively new focus within sustainability education (Russ and Krasny 2017).

Part III begins with Jay M. Shuttleworth's "Teaching About Sustainable Living in Urban Settings," which explores theory–practice connections of K–12 teachers. Shuttleworth describes instructional practices centered around the social issues of sustainable living and climate change. Using several case studies, he examines how teachers have used the local urban landscape to overcome resistance to thinking beyond the boundaries of one's community and nation-state

In "An Environmental Justice Lens on Indianapolis's Urban Ecosystem," Elizabeth Kryder-Reid, Laura M. Holzman, Aghliah Nadaraj, and Leah Humphrey discuss the process and pedagogy of a community-engaged collaboration that partnered Indiana University–Purdue University students and faculty and staff at the nonprofit Kheprw Institute (KI) with community groups. The project was designed to engage public audiences in the history and legacies of environmental injustices in Central Indiana and amplify the voices of Indianapolis's affected communities for a more sustainable city.

Henry Way and Rob Alexander, in "Urban Rewilding Through a Modified Placemaking Lens," describe a senior-level class that focused on efforts to "rewild" two riparian sites in the city of Harrisonburg, Virginia. Approaching urban

DOI: 10.4324/9781003217442-18

sustainability with a geographical lens and emphasizing the value of communication and storytelling allowed for a range of pedagogical opportunities to better develop students' urban environmental understanding.

Similarly, Loren Byrne's chapter, "Using the Prepositional Framework for Urban Environmental Education," uses that prepositional approach—ecology *in*, *of*, *for*, and *with* the city—to structure an undergraduate urban ecology course. Byrne presents examples of activities to help students develop the interdisciplinary knowledge and skills necessary to navigate the complex issues in urban ecology.

Mary Steggles's "Manitoba Hydro's Mega-Dams and the Ethical Teaching of Ceramics and Sustainability" chronicles an interdisciplinary course, focusing on ceramics practice, sustainability, and the social injustice of electricity, that questions industry messaging about hydropower. The author notes that the course "aims to separate facts from myths in Manitoba Hydro's message, especially with respect to indigenous communities, while challenging students to conduct a critical assessment of evidence and arguments regarding kiln fuels."

This section ends on an optimistic note with Alice S. Horning's "Resilience for Sustainability Everywhere: Critical Literacy for Students and Everyone Else." Horning draws on critical literacy across and beyond the disciplines, illustrating the need for critical literacy skills to create a resilient population with the ability to understand and evaluate information to address urban crises and create a more sustainable city.

References

Russ, Alex, and Marianne E. Krasny, eds. 2017. *Urban Environmental Education Review*. Ithaca, NY: Comstock Publishing Associates.

16

TEACHING ABOUT SUSTAINABLE LIVING IN URBAN SETTINGS

Global Citizenship and Overcoming the Arcadian Myth

Jay M. Shuttleworth

Introduction

The need to teach youth about sustainable living has never been more acute. As the world becomes more interconnected and interdependent, young people need to be prepared to solve global issues like climate change, pollution, and habitat loss. Instruction through a global citizenship education (GCE) lens is a suitable approach for K–12 teachers, and it is a broad way of framing instruction around the notion that people are members of a world community and have responsibilities beyond the boundaries of their nation-state (Gaudelli 2003, 2009, 2016).

Environmental scientists agree on the pressing nature of climate change and humans' outsized impact upon it (Cook et al. 2013; Skuce et al. 2017). Yet teaching about the need to live sustainably faces obstacles at the K–12 level. The nation-centric and anthropocentric curriculum provides challenges about expanding responsibilities for environmental issues that transcend political borders (Rappoport 2020; Kissling and Bell 2020). Teachers face pressures to teach topics like climate change as controversies, and, even then, many teachers wonder where to include it in an already highly standardized curriculum (Ho and Seow 2015; Prowse and Forsythe 2018). As a result, a GCE approach to sustainable living is rarely utilized in US classrooms (Myers 2020).

Teachers report a gap between what students know about sustainable living and what they are willing to do about it (Pope and Patterson 2012; Shuttleworth 2015; Shuttleworth and Marri 2014). This gap is particularly significant at the elementary level, even though recent studies indicate that young children are ready to discuss complex topics and are able to think empathically about people they may not know (Halvorsen 2017; Shuttleworth 2021). As a result, GCE researchers have prioritized studies of classroom instruction for their ability to bridge theory and practice (Grossman 2017; Pashby 2018), affirming the necessity of educating K–12

DOI: 10.4324/9781003217442-19

youth about sustainable living (Hayes and Magraw 2020). These objectives inform this chapter, which investigates how this sort of instruction looks in urban settings at the elementary, middle, and secondary education levels.

Theoretical Framework

Sustainable living is the logical premise of making decisions to meet one's needs without impinging on the ability of other (future) living things to do the same (Nolet 2009). The decision to live sustainably often comes from a philosophical perspective and does not have to follow what others are doing (or not doing) (Shuttleworth 2017). Living sustainably in an age of innumerable consumer choices may come out of a moral commitment to care for self and for others. For example, Leopold's (1949/1986) "land ethic" and Snyder's (1990) "natural contract" discuss ways to establish an almost aesthetic interaction with the world, where one might derive satisfaction from living a life that respects one's self and others.

GCE is helpful in guiding students to rethink their role and responsibilities in the world, particularly when it comes to issues that go beyond artificial borders (Gaudelli and Schmidt 2018). GCE focuses on instruction that asks participants to rethink the idea of community and their responsibilities to it. It draws heavily from what scholars term "indigenous environmental knowledge" (IEK) or "traditional ecological knowledge" (TEK) (Longboat, Kulnieks, and Young 2008; Porter 2014). While all people degrade environments, IEK/TEK notes that indigenous peoples cultivated lands for centuries while European colonizers caused those same ecosystems to collapse in one or two generations. Inherent in an IEK/TEK perspective is the idea that one has a duty to conserve resources, just as groups like the Iroquois laid out rules to consider the needs of future generations (Lyons 1980).

GCE borrows this kind of thinking when it asserts that its purpose is for students to rethink their responsibilities beyond the borders of one's nation-state (Soong 2018). No longer beholden only to self or one's immediate community, they are challenged to look beyond imposed hierarchies and toward an understanding of the world's interdependence and interconnectivity. GCE posits that humans are morally obligated to ensure that meeting one's needs does not disrupt the ability of (future) others to do the same. Determining a global understanding of morality is tricky (Dill 2013), but GCE asserts that our social contract to one another is a duty to consider the needs of all people (Banks 2004). These case studies focus on how students practiced empathy, especially for those beyond their circle of friends and family.

Case Studies

This section explores ways three urban K–12 teachers bridge theories of sustainable living through a GCE perspective. The first case is of a secondary teacher who engaged students with discussions about where "nature" begins and a city ends.

The second case study is of a middle school teacher who discussed her own experiences as she transitioned from a rural setting to an urban community. The final case study is of an elementary teacher who utilized GCE to study the idea of both human and animal living "shelters" as a way to understand interconnectivity and interdependence. All teacher names in this section are pseudonyms.

Secondary School Classroom

James teaches secondary social studies at an urban public high school that caters to recently immigrated students. Prior to this appointment, he was a Peace Corps volunteer in Micronesia for three years and primarily taught English.

Each year, James teaches a unit on climate change in his global history class toward the end of the second semester. Thematically, he contextualizes climate change as an extension of imperialism, with Pacific Islander nations being the most vulnerable. He uses a GCE perspective to frame instruction about knowledge and values. Regarding values, he said:

> I think the values I was trying to impart to students is equality of all humans and the interconnectedness of humanity at the current moment. But also trying to see themselves within a larger world in which they are a part . . . and how small actions when combined can lead to larger actions.

He similarly framed knowledge as being able to see one's self as part of a global community. Emphasizing the need for students to activate their agency:

> I want them to see themselves in a bigger picture . . . [of] climate change and the history of what's going on and how it impacts people. I want them to wade through the murkiness of that knowledge and know that what they consider knowledge can be contested, but also the knowledge that they can criticize and see climate change all around them and how it impacts them.

One of James's introductory lessons asked students to challenge dualism: the often mistaken strategy of categorizing things into opposing or contrasting groups. James began the lesson by asking students what the word "nature" meant to them. Student responses included answers like "animals," "trees," and "plants." He also asked them, "Where do you see 'nature' near the school?" Sample responses included, "there isn't any" and "the garden [across the street from the school]."

James shifted the class discussion by asking more philosophical questions about "nature" and humans. Students were less sure how to respond when he asked, "Where does [nature] begin? End?" Shortly thereafter, James asked, "Are humans a part of that nature? Why or why not?" A few said yes, but most students said no. By breaking down assumptions of human's relationship with nature, James was moving students toward a transformational way of thinking about their own relationship with the world. Where they once saw humans as separate from nature, students were introduced to the idea that they were members of a broader community. In a reflection afterward, James talked about his GCE objectives for this and other introductory lessons:

> The idea of being a global citizen, to me, that's kind of a mind-frame. The fact that so much of global citizenship, when people think of it, it implies learning a new language and traveling to a new place, but these are often unidirectional, as some are able to travel more so than others. We don't need to travel to have the mindset of a global citizen. That's kind of what I hope the students take away from it, that there's a sort of mindset, and wherever you're living, even if not traveling a lot, you can be a global citizen.

As a part of the assessment for this unit, James asked students to deliberate about countries' responsibilities for climate change and those displaced by it. He guided this activity with questions: "What is a sustainable citizen?"; "How can countries that create pollution help countries affected by climate change?"; "Are people obligated to decrease their pollution?" and, "How should countries respond to people who become refugees because of climate change?" He used these questions to engage students with two real-world events: the Syrian Civil War and the Micronesian climate refugee crisis.

After two days of instruction learning about how climate change affects Micronesians, James divided the class into four interest groups: Hawai'ian taxpayers, persons living on a Micronesian atoll, climate scientists, and American politicians. James involved Hawai'ian perspectives because their state receives the largest percentage of Micronesians displaced by climate change.

During the class dialogue about each of the interest groups' responsibilities, James posed follow-up questions once they each shared their stance on how to address the needs of climate change refugees. After students representing the Hawai'ian taxpayer group said that they would not recommend accepting Micronesian refugees, James asked why. A student from that group said, "Because the homeless rate in Hawai'i has increased 41%, and maybe they will come and there will be more homeless. Maybe they should go to [another state]." After a few other student responses, James asked, "You said as a Micronesian, they should move; Henry said it's not fair to the people who live there and can't move. How do we solve this problem?" With this question, James started another dialogue about responsibilities for Micronesians; many had lost their livelihood and needed to move, but who was responsible for resettling them was debatable.

James revisited the question of US responsibility for creating a lot of the pollution that causes climate change. He asked, "Is the US obligated to help Micronesia because the US creates so much pollution?" One student said, "Yes, and other countries, too, because they have created the pollution." Another student said, "Blame can't be put on one country. No, they shouldn't help." Much of the remaining dialogue involved how to divide the responsibility so that one country wasn't shouldering too much of the obligation.

After the lesson, James reflected that the division among students' responses might have come from their limited knowledge about CO_2 emissions. He said: "I wasn't convinced of their reasons for the division. It's not their fault; they're not entirely informed about the percentage of CO_2 that's produced by other countries. They didn't have that information [and] I could go deeper." He explained that he wanted to spend more time talking about the history of industrial emissions and felt that if students knew more about the United States' long-term role in its productions, they might make more informed decisions.

James said that he hoped students would understand their responsibilities to assist others who have been disadvantaged by US capitalism. In an interview, he said:

> Why should I care, right? I think a lot of countries react in this way. And some don't, which is important to recognize. It's less tangible than Syria. . . . [I]t's not in your face like their refugee crisis. . . . The influx of immigrants might not bankrupt the US, but the US is responsible for creating the conditions that these people are [experiencing].

James wanted students to connect America's long-standing role as an industrial polluter to CO_2 atmospheric increases. Without more in-class time spent on understanding this context, he reasoned that students might continue their reluctance to extend responsibilities of nations like the United States to people displaced by climate change.

Middle School Classroom

Samantha teaches sixth grade social studies at an urban, public K–8 school. Prior to this appointment, she was an interdisciplinary educator in an outdoor-based school in a national park in the Western United States. Using what she described as a "park-as-learning lab" approach, she taught geography, natural science, and sustainable living. Instruction catered to students who lived all over the country, and the students returned to their communities after taking courses with Samantha.

Her first challenge in teaching sustainable living occurred when her former students reached out to her after they had returned to their hometowns:

> I was living in [a national park]. So, the environment is stupendous. And we would teach the students about the environment, about sustainability, about geography, about the natural world. . . . But I got a lot of emails from students once they go back to their "real lives," and they didn't know what to do with this information. They weren't sure how to use it. I had one girl who was with me for a month backpacking, and it was a total transformation, and then she went back to [her home state] to this really urban area, and she emailed me saying she feels like she should be hiking or outside, but I don't know what to do, I don't know how to do it.

By situating this kind of instruction in a national park, Samantha inadvertently set up an understanding that sustainable living was applicable only in rural, wilderness-type places. In reflecting on this experience, Samantha said, "And one of the big things I think I learned in [this Western state] is, if things aren't relevant, and things aren't transferable to students' own lives, then it's really not going to go anywhere."

Samantha clarified that effective sustainable living instruction relied on *relatability* to students' lives and that it is not based on a particular place. This observation helped to dispel the idealization of rural wilderness over urban spaces. She said that basing sustainable living instruction too closely to the appreciation of wilderness areas was ineffective because many of her students did not live in comparable areas.

Once Samantha began her teaching career in an urban setting, she refocused her sustainable living instruction to be transferable to any setting. First, she established a local-as-global perspective: interacting with one's local environment could be a microcosm of the world. Second, she emphasized the importance of bridging theory and real-life experiences according to *student interest*. Third, she argued that activating student agency was essential for sustainable living instruction; students should see themselves as agents of change.

Samantha utilized a culturally relevant approach to accessing a broader conversation about personal decision making and sustainable living. In what she and her colleagues termed a "kickoff" event designed to generate student enthusiasm at the start of a unit, Samantha offered students choices of food for their lunch. See Table 16.1.

Predictably, student choices heavily favored pizza, cake with whipped cream, and fruit punch. They were divided over apples versus oranges and the two types of chips. Samantha explained her objectives:

> Our objective is to get them to think about why they eat certain foods. In their journal, if you open it up, there's [sic] two different choices. So, for the dinner, it's either Big Mac or potato. And it's [asking] why are you making that choice? One of our guiding questions is, "What is good food?" So, we're going to start exploring that today, and hopefully, this idea of "good food" changes as students learn more and more. . . . The objective is for them to record their choices, just so we have a sense of why they are making the choices they make.

TABLE 16.1 Food choices at kickoff event.

Apples	Oranges
Pizza	Turkey sandwich on whole wheat bread
Doritos chips	Frito's corn chips
Big Mac	Baked potato
Chocolate cake with whipped cream	Chocolate cake without whipped cream
Fruit punch	Water

This activity was in response to a larger essential question, "What is good food?" but it was not meant to push students toward making healthier (and more sustainable) food choices. She was more interested in having a discussion about the gap between what students knew about healthy (and more sustainable) food choices and the decisions they were willing to make with that information. This conversation served as an introduction to the transformation of responsibilities inherent in GCE. Students would soon learn that their individual choices affected a broader network of living things and that they had a responsibility to at least be aware of those connections.

In another way to make personal connections to the content, Samantha also used the "local as global" approach to learning about where food comes from. As an introduction to learning about the interdependence of global food systems, Samantha took students to a nearby park to learn about identifying which plants were edible (e.g., birch leaves, clover, and mustard seeds) and which ones were poisonous (snakeroot). She also took students to a farmer's market and, in conjunction with her math colleagues, had them calculate "food miles"—how far each of the items traveled to get from farm to market. Students compared the relatively short distances food traveled to get to the farmer's market with those from the nearby grocery store (many of which had traveled thousands of miles by ship and/or truck).

Rethinking responsibilities for sustainable living was a persistent theme for Samantha and her students. To get beyond their relatively small worldview as pre-adolescents, she creatively empowered the students to become agents of change for sustainable living. As part of the unit's summative activities, students had to propose dynamic, transformative ways to rethink how their city could have a more sustainable food supply. Proposals tended more toward creative than realistic, but the activity sought to empower students to practice taking more authentic action later in their lives. Samantha said, "I think having this information will be great, so when they become mature enough to actually digest it and use it, they can do it."

In one activity, the guiding question was, "How can we make our school lunch more sustainable?" Samantha said that the goal of the activity was "empowerment and opportunity." She also said, "I'm hoping they see themselves as active advocates for themselves and for others in order to create a sustainable school community." Inherent in this activity was for students to pursue more sustainable alternatives to their existing lunch, which was prepared offsite and served on a disposable Styrofoam tray.

An underlying goal of this activity was for students to practice empathy with their peers. By practicing how their decisions affected someone they knew, they were preparing to think about empathy more abstractly. Samantha suggested that it was difficult for students later in life to be empathic about climate change, which involves so many people they will never meet, if they hadn't first practiced it with people they know.

Elementary School Classroom

Lee is an experienced third grade instructor who teaches at a K–8 elementary school in an urban setting. The unit of study was on living shelters and their associated resources. Her learning objectives included understanding the interconnected nature of resources, practicing empathy, and accounting for where resources might come from in a fully sustainable household and were a way of establishing developmentally-appropriate opportunities to rethink responsibilities for others. She said:

> So, when we start to investigate how people live, we really start to look at what are our basics, what are our fundamentals, and that comes right back into our resources—how families interact with resources, how do families and children and people in our city and in their neighborhood satisfy their human needs? Their biological needs? And the needs for community? And so that gets right back to how we use our natural resources. It also highlights the sense of interconnectedness and interdependence. And it comes back in a very social science–based way that is studying communities and how communities function, and how social systems function.

Lee was also doing a self-study of how she taught systems-thinking, the subject of her thesis for her master's degree in sustainability education. She said her research question was, "How does using outdoor urban spaces impact student learning outcomes for students?" She said that her goal included linking natural systems with social systems. Of these goals, she said, "I hope they will see humans as part of the natural world. . . . I'm aiming for them to see humans as part of the global system."

In one session in a nearby park, Lee designed a lesson to get students to think about the interconnected nature of water resources. She led the students to a large pond, and they walked along the banks of a creek that flowed from it, through a wooded area, and then to a metal grate in which the creek emptied. Lee led students in a conversation about where they thought the water went. After some discussion, the students agreed that the water didn't "go" anywhere, ecologically-speaking. It had joined some other body of water, and it might come back to the park in the form of rain or irrigation to water the plants.

The point of this lesson was to introduce students to systems-thinking and interconnectedness, a key component of GCE. Students returned to this experience in later conversations, like when it came time to clean paintbrushes in the classroom. Rinsing paint brushes in the sink would only cause problems because the paint wouldn't disappear. It would simply join a different water body and cause problems there.

Later instruction involved the analysis of shelters that animals built for themselves in the park. Most of the studied shelters were birds' nests. Students were able to discuss the sustainable nature of these nests because its resources came from the park. Students also discussed the importance of reusing and repurposing materials so that there would be enough resources for other birds (and for other animals).

Studying the sustainability of animal shelters was a prelude to students designing their own idealized dwelling. The purpose of this activity was not to ask why humans built such unsustainable communities while animals did not. Instead, students were tasked with designing their ideal home (using recycled materials from class activities), and they had to account for where all of its resources would come from—electricity, water, food. Students would also have to explain where their waste would "go."

Discussion

These vignettes from three different teachers affirm the bridge between GCE theory and practice and underline the idea that these teachers view instruction about sustainable living as a responsibility to the profession (e.g., Saylan and Blumstein 2011). The teachers' objectives also supported the idea that this kind of instruction emphasized rethinking one's responsibility for present and future living (Fenton 1967).

Insights from these three teachers also reveal that teaching about this subject in urban settings faces an additional challenge, owing to a misconception that city living is "unnatural" and that rural and suburban spaces are more hospitable to sustainable living. While this so-called Arcadian Myth is applied narrowly to teaching about sustainable living, (Schmitt 1990), its context is clear: students and teachers alike have to overcome long-standing emphases of past environmental movements, which focused on conserving wilderness spaces in lieu of more social justice–oriented topics like health and interdependence (Adams and Alinder 1996; Alinder 1988; Wolfe 1945/2003).

Deconstructing the Arcadian Myth is similar to GCE because of its goals of rethinking assumptions about living spaces and humans' place within them. The "myth" of Arcadia sets out two assumptions. First, it places humans firmly outside "nature," which creates problems of human exceptionalism. Second, it distinguishes human concentrations in cities as almost evil compared to the mythical rural spaces that harmonize with nature (Schmitt 1990). Each of these problems needs to be deconstructed to properly apply sustainable living in urban settings.

Placing humans outside nature creates a false assumption that all things created by humans are "unnatural" and confuses the objectives of sustainable living. For example, if humans are not a part of nature, then *what* is worth sustaining: distinctively "unnatural" human cities or bucolic settings? This false dichotomy is untenable because humans inevitably influence all spaces—from cities to wilderness preserves. By properly placing humans within ecosystems governing all living things, it contextualizes the human experience as one governed by resource limits and not exceptionalism.

Similarly, if cities are "unnatural," then teaching about sustainable living in urban settings is unworkable. Cities *are* natural, but the more important matter is their degree of sustainability. Additionally, if cities cannot be studied for their sustainability, urban students learn about sustainable living only in the abstract. Breaking down the Arcadian Myth assists students in properly placing humans, ecologically, in context; it doesn't single out human behavior as separate from that of other living things, nor does it favor some kinds of human living spaces as more "natural" than another (Schmitt 1990).

James's instruction had students challenge the Arcadian Myth in the first few days of his unit on climate change. His goal was to break down assumptions that humans were separate from the rest of the world's living things. In doing so, he helped set the stage for students to also transform their assumptions about what their "community" encompassed.

Samantha grappled early in her teaching career with the Arcadian Myth, especially when it came to making sustainable living instruction relevant outside of national park–type settings. Once she moved to an urban setting, she realized that sustainable living tenets were universal and not limited by bioregion. Utilizing spaces within the community were more than adequate in getting students to think about the interconnectivity of food systems.

Lee set up instruction so that students could build model shelters using methods that animals had used. They used recycled materials, and all of their resources (like energy, food, and water) had to be accounted for. The single-family dwellings that most students made were not comparable to a bird's nest. But by emphasizing that the students could make sustainable choices *like animals* was an important step for conceptualizing humans as not living in a different world from that of animals.

These three teachers' instructions suggest that sustainability education excels when the urban environment is treated no differently than any other setting. Part of this process will involve getting students to discuss *what* is worth sustaining. Instead of scaring them about what students might lose, each of these teachers focused on how they might sustain what they already have. Breaking down assumptions about humans' place in the world is a helpful first step in facilitating this conversation.

References

Adams, Ansel, and Mary Street Alinder. 1996. *Ansel Adams: An Autobiography*. New York: Little, Brown, and Company.

Alinder, Mary Street, ed. 1988. *Ansel Adams: Letters 1916–1984*. New York: Little, Brown, and Company.

Banks, James. 2004. "Teaching for Justice, Diversity, and Citizenship in a Global World." *The Educational Forum* 68 (4): 296–304. http://docplayer.net/23915555-Teaching-for-social-justice-diversity-and-citizenship-in-a-global-world.html.

Cook, John, Dana Nuccitelli, Sarah Green, Mark Richardson, Bärbel Winkler, Rob Painting, Robert Way, Peter Jacobs, and Andrew Skuce. 2013. "Quantifying the Consensus on Anthropogenic Global Warming in the Scientific Literature." *Environmental Research Letters* 8 (2): 1–7. https://iopscience.iop.org/article/10.1088/1748-9326/8/2/024024/pdf.

Dill, Jeffrey. 2013. *The Longings and Limits of Global Citizenship Education: The Moral Pedagogy of Schooling in a Cosmopolitan Age*. New York: Routledge.

Fenton, Edwin. 1967. *The New Social Studies*. New York: Holt, Rinehart, & Wilson.

Gaudelli, William. 2003. *World Class: Teaching and Learning in Global Times*. Mahwah: Lawrence Erlbaum.

———. 2009. "Heuristics of Global Citizenship Discourses Towards Curriculum Enhancement." *Journal of Curriculum Theorizing* 25 (1): 1–4. https://journal.jctonline.org/index.php/jct/article/view/GAUDHEU.

———. 2016. *Global Citizenship Education: Everyday Transcendence*. New York: Routledge.

Gaudelli, William, and Sandra Schmidt. 2018. "Global Citizenship Education and Geography." In *The Palgrave Handbook of Global Citizenship Education*, edited by I. Davies, L.-C. Ho, D. Kiwan, C. L. Peck, A. Peterson, E. Sant, and Y. Waghid, 473–88. London: Springer Nature.

Grossman, David. 2017. "Global Education." In *The Wiley Handbook of Social Studies Research*, edited by M. M. Manfra and C. M. Bolick, 519–54. Hoboken: John Wiley and Sons.

Halvorsen, Anne-lise. 2017. "Children's Learning and Understanding in Their Social World." In *The Wiley Handbook of Social Studies Research*, edited by M. M. Manfra and C. M. Bolick, 385–413. Hoboken: John Wiley and Sons.

Hayes, Dennis, and Daniel Magraw. 2020. "Earth Day at 50: A Time to Engage." *Social Education* 84 (1): 18–23. 10.1080/00933104.2019.1673267.

Ho, Li-Ching, and Tricia Seow. 2015. "Teaching Controversial Issues in Geography: Climate Chagne Education in Singaporean Schools." *Theory and Research in Social Education* 43 (1): 314–344. 10.1080/00933104.2015.1064842.

Kissling, Mark, and Jonathan Bell. 2020. "Teaching Social Studies Amid Ecological Crisis." *Theory and Research in Social Education* 48 (1): 1–31. 10/1080/00933104.2019.1673267.

Leopold, Aldo. 1986. *A Sand County Almanac with Essays on Conservation from Round River*. New York: Ballantine.

Longboat, Dan Roronhiakewen, Andrejs Kulnieks, and Kelly Young. 2008. "Beyond Dualism: Toward a Transdisciplinary Indigenous Environmental Studies Model of Environmental Education Curricula." In *Contemporary Studies in Environmental and Indigenous Pedagogies: A Curricula of Stories and Place*, edited by A. Kulnieks, D. R. Longboat, and K. Young, 9–19. Boston: Sense Publishers.

Lyons, Oren. 1980. "An Iroquois Perspective." In *American Indian Environments: Ecological Issues in Native American History*, edited by C. Vescey and R. W. Venables, 171–74. Syracuse: Syracuse University Press.

Myers, John. 2020. "Understanding the Landscape of Teaching Issues." In *Research on Teaching Global Issues: Pedagogy for Global Citizenship*, edited by J. P. Myers, 1–12. Charlotte: Information Age Publishers.

Nolet, Victor. 2009. "Preparing Sustainability-Literate Teachers." *Teachers College Record* 111 (2): 409–442.

Pashby, Karen. 2018. "Identity, Belonging, and Diversity Education for Global Citizenship: Multiplying, Intersecting, Transforming, and Engaging Lived Realities." In *The Palgrave Handbook of Global Citizenship and Education*, edited by I. Davies, L.-C. Ho, D. Kiwan, C. L. Peck, A. Peterson, E. Sant, and Y. Waghid, 363–76. New York: Macmillan Publishers.

Porter, Joy. 2014. *Native American Environmentalism: Land Spirit, and the Idea of Wilderness*. Lincoln: University of Nebraska Press.

Pope, Alexander, and Timothy Patterson. 2012. "Two Sides of the Megalopolis: Educating for Sustainable Citizenship." *Journal of Social Studies Education Research* 3 (2): 1–20.

Prowse, Alicia, and Rachel M. Forsythe. 2018. "Global Citizenship Education: Assessing the Unassessible? In *The Palgrave Handbook of Global Citizenship and Education*, edited by I. Davies, L.-C. Ho, D. Kiwan, C. L. Peck, A. Peterson, E. Sant, and Y. Waghid, 607–623. New York: Macmillan Publishers.

Rapoport, Anatoli. 2020. "The C3 Framework and Representation of Global Citizenship in State Standards." In *Research on Teaching Global Issues: Pedagogy for Global Citizenship education*, edited by J. P. Myers, 51–68. Charlotte: Information Age Publishers.

Saylan, Charles, and Daniel Blumstein. 2011. *The Failure of Environmental Education (And How We Can Fix It)*. Berkeley: University of California Press.

Schmitt, Peter. 1990. *Back to Nature: The Arcadian Myth in Urban America*. Baltimore: Johns Hopkins University Press.

Shuttleworth, Jay. 2015. "Teaching the Social Issues of a Sustainable Food Supply." *The Social Studies* 106 (4): 159–69. www.tandfonline.com/doi/pdf/10.1080/00377996.2015.1043602.

———. 2017. "Leopold's Third Step: A Framework for Teaching his Ethical Consideration of Land Use." *Selected Proceedings from the Twentieth Annual Conference of the Association for Core Texts and Courses. The Association of Core Texts and Courses* 20: 59–61. www.coretexts.org/wp-content/uploads/2019/01/20th-Annual-Conference-Proceedings_2014.pdf#page=69.

———. 2021. "Sustainable Living, Global Citizenship, and Teaching About Community." *Oregon Journal of Social Studies* 9 (1): 19–29. https://drive.google.com/file/d/1UCf_t-QOB7hC0NDreYSPVy4C33bmIhIK/view.

Shuttleworth, Jay, and Anand Marri. 2014. "Teaching Sustainability as a Social Issue: Learning from Dialogue in the Social Studies Classroom." In *Cases on Pedagogical Innovations for Sustainable Development*, edited by H. E. Muga and K. D. Thomas, 328–47. Hershey: IGI Global.

Skuce, Andrew, John Cook, Mark Richardson, Bärbel Winkler, Ken Rice, Sarah Green, Peter Jacobs, and Dana Nuccitelli. 2017. "Does it Matter if the Consensus on Anthropogenic Global Warming is 97% or 99.99%?" *Bulletin of Science, Technology, & Society* 36 (3): 150–56. https://doi.org/10.1177/0270467617702781.

Snyder, Gary. 1990. *The Practice of the Wild*. Berkeley: North Point.

Soong, Hannah. 2018. "Transnationalism in Education: Theoretical Discussions and the Implications for Teaching Global Citizenship Education." In *The Palgrave Handbook of Global Citizenship and Education*, edited by I. Davies, L.-C. Ho, D. Kiwan, C. L. Peck, A. Peterson, E. Sant, and Y. Waghid, 165–77. New York: Macmillan Publishers.

Wolfe, Linnie. 1945/2003. *Son of the Wilderness: The Life of John Muir*. Madison: The University of Wisconsin Press.

17

AN ENVIRONMENTAL JUSTICE LENS ON INDIANAPOLIS'S URBAN ECOSYSTEM

Collaborative Community Curation

Elizabeth Kryder-Reid, Laura M. Holzman, Aghilah Nadaraj, and Leah Humphrey

Introduction

As representatives from a large urban university (Indiana University Purdue University Indianapolis [IUPUI]) and a small nonprofit organization dedicated to youth empowerment and community wealth building (the Kheprw Institute [KI]), we had a remarkable opportunity to collaborate on a multiyear project focused on environmental justice in Indianapolis's urban core. The project, entitled "Legacies of Environmental Justice in Indianapolis," investigated the history of the city's public waterways to develop a range of environmental humanities projects that would engage broad publics in the history of pollution and its contemporary consequences for affected communities. As such, the work was situated at the intersections of critical heritage studies, critical pedagogy, and community empowerment. It was also part of a larger international collaborative initiative, Climates of Inequality: Stories of Environmental Justice (COI), coordinated by the Humanities Action Lab (HAL), which is a "coalition of universities, issue organizations, and public spaces that collaborate to produce community-curated public humanities projects on urgent social issues" (Humanities Action Lab). Indianapolis was one of 22 localities participating in the HAL project, each with a team of faculty, students, and community partners focusing on a local issue related to environmental and climate justice and contributing to a traveling exhibit that toured internationally. Together, IUPUI students, faculty, and KI members contributed to the traveling exhibit and created related exhibits, digital humanities projects, and public programs (Figure 17.1). But beneath the apparent successes of the final projects lies a more complex story of the messy, rich work of environmental justice through community collaboration, particularly the understanding of what a sustainable ecosystem means to affected communities. This chapter critically examines our experiences with community-engaged environmental justice work and how it illuminates understandings of the city as ecosystem.

DOI: 10.4324/9781003217442-20

FIGURE 17.1 "Parachuting into Environmental Justice," an interactive art-making activity designed to spur conversation about environmental justice as a lived experience.

Source: Photograph by Elizabeth Kryder-Reid.

Introducing the Partnership

Kheprw Institute (KI) is a nonprofit organization focused on empowering youth and building community wealth in Indianapolis. Since 2003, KI has worked to create a more just, equitable, human-centered world by nurturing youth and young adults to be leaders, critical thinkers, and doers who see the people in any community as the most valuable assets and are committed to working with disinvested communities to bring about change that leads to empowered, self-reliant, and self-determining communities. Mutual trust and respect are critical foundations for KI's work. Before formally entering any partnership, KI staff ask several questions. Do we like the people we are collaborating with? Do they feel genuine? Are they trying to do something different and coming in with good intentions? As an organization and community dedicated to the radical act of centering people and relationships, these questions are the primary filters KI uses to assess potential projects. KI had collaborated with IUPUI before through a previous HAL project, States of Incarceration. Members of KI's Environmental Justice team, including youth leaders passionate about their communities and the planet, also met with an IUPUI faculty member to discuss the new project. As is their practice, KI used a collective decision-making process to commit to the project, agreeing that it was an opportunity to shape narratives about the neighborhood and to deepen relationships and connections with the university partners. It also was a space for the youth leaders who are passionate about environmental justice work to gain experience and develop leadership skills.

IUPUI is Indiana's largest urban public research university. The faculty and students involved in the project were primarily from Museum Studies and Public History, and the project was coordinated by the Cultural Heritage Research Center. The IUPUI Museum Studies faculty who led the project are committed to public scholarship (creating knowledge with and for communities that address relevant issues), activist scholarship that addresses social inequalities, and community-engaged teaching in which students learn the skills and principles of community-focused work for their future careers by doing projects with community partners. This work aligns with important shifts in the cultural heritage field to recognize shared authority, democratize knowledge production, include more diverse voices and points of view, and embrace critical perspectives on intersections between heritage and social inequalities (Adair, Filene, and Koloski 2011; Gentry and Smith 2019; Harrison 2013; Little and Shackel 2014). We approach the cultural arts and heritage sector with a critical lens, and we often collaborate with partners outside of the university to define project goals and work toward mutually beneficial outcomes together. This approach aligns with higher education's embrace of community engagement and, more particularly, with our own campus's commitment to teaching and research that increases the well-being of our city, country, and world (Alivizatou 2022; Bennett and Gadlin 2012; Boyer 2015; Dostilio 2017; Dumlao 2018; Holzman et al. 2014; Holzman 2021; IUPUI Office of Community Engagement; Jay 2012; Mitchell 2014; Shah 2020).

It is important to point out some of the implications of the partners' respective histories, missions, and scale—particularly the legacy of harm in the creation and ongoing expansion of the urban university. IUPUI was created in 1969, and the present-day campus was formed through the systematic purchase of properties on the city's near west side, razing a nearly 300-acre long-standing community and displacing the predominantly African-American residents (Grey 2003; Mullins 2006; Mullins and White 2010; Paschall 2020). The displacement has continued as the campus has grown over the past 50 years, and the threat of gentrification persists with continued expansion on the campus's western boundary. The university partners brought with them organizational resources and institutional capacity that dwarfed KI in terms of staff and budgets, but KI had deep expertise in environmental activism, was involved in national environmental justice networks including the Climate Justice Alliance, and had deep relationships with affected communities. KI team members were predominantly people of color whereas students in the COI classes were predominantly white.

In consultation with KI, we focused on Indianapolis's urban waterways (Fall Creek, White River, Central Canal, and a Superfund-eligible groundwater contamination site designated as Site 0153). The sites were chosen because the three waterways run through Indianapolis's urban core, and each offered opportunities to explore different dimensions of the structural causes and community impact of pollution. Specifically, Fall Creek has a number of combined sewer overflow (CSO) outlets and has been the focus of redevelopment plans such as Destination Fall Creek. The White River, which has CSO outlets as well, also has a long history of pollution related to industrial waste both on the shores of the river in the city and upstream contamination from manufacturers and agricultural runoff to the north. The Central Canal is a critical part of Indianapolis's public water infrastructure, and it also demonstrated the differential investment in recreational amenities in communities of color. Site 0153 was not initially on our prospective site list, but KI felt that the recent controversy about whether to designate it as a federally recognized Superfund site raised important political issues about the management of contamination and would also raise awareness of the relatively invisible groundwater contamination.

The university curriculum associated with the project was designed as a three-semester sequence: a research seminar (Fall 2018), an "Exhibit Planning and Design" course that incorporated the exhibit as an applied project (Spring 2019), and an applied research course to develop public programs (Fall 2019). KI sent someone to every class in the first semester and periodically in the subsequent courses (Figure 17.2). KI members provided feedback, shared insights, and offered

FIGURE 17.2 IUPUI students, faculty, and KI leaders meet at KI in Fall 2018 for a conversation about KI's environmental justice work and the principles of just transition.

Source: Photograph by Elizabeth Kryder-Reid.

FIGURE 17.3 Students in the research seminar tour Fall Creek with Mapleton Fall Creek Development Corporation CEO Leigh Riley Evans and Destination Fall Creek advocate Doug Day (Fall 2018).

Source: Photograph by Elizabeth Kryder-Reid.

instruction on the principles of environmental justice work, particularly Just Transition, "a vision-led, unifying and place-based set of principles, processes, and practices that build economic and political power to shift from an extractive economy to a regenerative economy" (Climate Justice Alliance). Over the three semesters, student teams researched archival sources, secondary literature, government (especially IDEM and EPA) reports, and conducted oral history interviews and went on walking tours with community members (Figure 17.3). The teams produced an annotated bibliography (on Zotero), research briefs, identified primary sources (images, texts, etc.) for an exhibit, and developed a contribution to the COI traveling exhibit and two companion exhibits. Working individually or in small groups, the students also created public-facing projects, primarily in digital formats and public programs, to engage audiences with the history of environmental justice in Indianapolis (Indy Environmental Justice History 2018; Toxic Heritage-Collaborative Research n.d.).

We hosted the COI traveling exhibit in January and February 2020 at the Central Library in downtown Indianapolis (Figure 17.4), along with two companion exhibits the students curated: "Fall Creek: A Look at Art and the Environment," featuring the work of photographer Tim Faris, and an exhibit showcasing

FIGURE 17.4 Climates of Inequality exhibit, Indianapolis Public Library – Central Library (January 2020).

Source: Photograph by Elizabeth Kryder-Reid.

four community organizations' environmental advocacy work. Students developed programs including a workshop on monitoring water quality at home, an interactive art-making project that invited people to respond to prompts about environmental justice by drawing or writing on a parachute, a talk about pollution at a state park followed by a cleanup hike, a book group discussion of Octavia Butler's *Parable of the Sower*, and an EJ writing workshop facilitated by two environmental reporters from the *Indianapolis Star*. Fortunately, all but one of the programs were completed before the COVID-19 pandemic disrupted plans.

Indianapolis's Environmental Justice Ecosystem

As can be expected of a project of this scope and complexity, there were substantial challenges, significant outputs, and positive outcomes. For the purposes of this volume, we focus on four ways the project informed our understanding of the city as ecosystem. First, the project illuminated the importance for affected communities of a holistic sense of sustainable ecosystems including not only air, water, and soil, but also issues of policing, food access, and personal safety. Second, the students came to understand the history of Indianapolis through an environmental justice lens, recognizing the enduring structural inequalities and environmental racism that have informed urban design and policy for a century and a half. Third, participants in the project came to new understandings of their own lived experience and positionality within the city's ecosystem. Fourth, the project highlighted the intertwined social, political, and intellectual capital that is central to environmental justice and sustainability workers and movements. It demonstrated that as much as infrastructure and ecology are integral to the city's ecosystem, so too are the social relationships of its communities.

1. *Indianapolis's Ecosystem and Holistic Well-Being*

From our first meeting with the KI team, it was clear that their understanding of environmental justice and a healthy ecosystem was far broader than the standard EPA definition as "the fair treatment and meaningful involvement of all people regardless of race, color, national origin, or income, with respect to the development, implementation, and enforcement of environmental laws, regulations, and policies" (Environmental Protection Agency n.d.). Imhotep Adisa, cofounder of the Kheprw Institute, explained to the students that a healthy, safe community isn't just about lead exposure in soil and water or particulate matter in the air. For KI, environmental justice included freedom from overpolicing, access to healthy food, and policies that provided for personal safety such as adequate street lighting and sidewalks. In short, it was a view of environmental justice as "everything related to well-being," and it is ultimately about "inequitable relation to power" (Adisa 2018). The conversation immediately broadened the project participants' understanding of the scope of Indianapolis's waterways research to consider not just pollution and water quality but also the policies and practices that prevented some residents from fully enjoying the benefits of the resources and that led to disparate impacts of environmental harm on some communities. Consequently, when students examined the history of the Central Canal, they did not stop at concerns about the contamination from industry and railroad lines but also considered the differential investment in improvements and the experiences of local residents. For example, to the north of the Riverside neighborhood, the towpath was restored as a popular running-biking trail. In the downtown section to the south, the canal was intensively redeveloped for residential and commercial use, as well as an attraction for tourists and residents to rent paddleboats, gondolas, and Segways. In contrast, the Near Westside's "middle section" of the canal that runs through predominantly low-income and communities of color was largely overgrown, and portions were impassable[1] (Davenport 2018; Markley and Will 2018) (Figure 17.5).

2. *Indianapolis's Ecosystem as Environmental Injustice*

Faculty strive to instill critical thinking and self-reflection in all curricula. This project was a rich opportunity for students not only to gain an appreciation for the complexity of environmental justice issues but also to connect issues of environmental harm to the structural racism and power dynamics that have led to

FIGURE 17.5 Walking tour of the Central Canal in the Riverside neighborhood.

Source: Photograph by Elizabeth Kryder-Reid.

generational disinvestment and marginalization of Indianapolis's communities of color and lower economic status. One student reflected:

> During our meetings with KI, Groundworks Indy, and Reverend Sheldon, it became apparent that these ideas [about environmental justice in the class readings] are privileged and do not account for the more complicated nature of structural racism and classism that are reinforced in systems of power and through practices of false equity. . . . While community members were invited to the table to plan what happened next to the polluted site, the final decision was made by those in power.

Other students gained critical perspectives on the power dynamics in Indianapolis, including the problematic history of their own university and their chosen profession. One museum studies student wrote, "Within this course, I've learned about the deep distrust that communities of color feel toward museums. This distrust was something I was aware of, but I'm grateful that members of KI were very forthcoming in their opinions about museums. There is also a distrust among people of color toward many institutions in Indianapolis, like the *Indy Star* [newspaper]." The introduction to Jacqueline Rooksberry's timeline of Fall Creek exemplifies the students' comprehension of the complex structural racism implicated in the history of polluted waterways.

> Fall Creek was intimately tied to African American residents' access to the basic human right of safe clean spaces for their children to play daily. Not only was the area polluted by human waste and garbage, but the creek became a symbol of the cities [sic] neglect and bureaucratic violence in the fight for equal rights and treatment in Indianapolis.
>
> *(Rooksberry 2018)*

Similarly, Topher Anderson's history of the groundwater contamination Site 0153 and the Riverside neighborhood examined the industrial history of the area and connected it to the history of redlining and blockbusting that contributed to the decades-long decline of property values and generational poverty (Anderson 2018). Students' public programs also demonstrated the connections they made between the particular histories of Indianapolis's waterways and broader issues of structural racism. One student developed a book group discussion of Octavia Butler's *Parable of the Sower* (1993) with guiding questions to support conversations about environmental racism. Another student, in partnership with Young Actors Theater, developed a play called "Let's Take ACTion on Environmental Justice." It was intended to apply Forum Theater principles to engage audience members in the deeper structural causes and consequences of Indianapolis's polluted urban waterways. Regrettably, the pandemic shutdown in March 2020 forced the cancellation of the performance.

3. *Indianapolis's Ecosystem and Positionality*

Inevitably, critical heritage, critical urban studies, and critical pedagogy are disruptive and uncomfortable. They require people not only to grapple with the centuries of structural racism and inequalities that are inextricable from contemporary environmental justice issues, but also to locate themselves in that history. As faculty, we knew that this work, particularly when done in close collaboration with community partners, often creates discomfort for students (Kryder-Reid 2019). Student reflections throughout the project demonstrated how they confronted their own lived experience with the city's environmental history and contemporary ecosystem. Many noted changes in their attitudes, emotions, relationships, and motivations, and some wrestled deeply with their own positionality, whether they perceived it as one of privilege and complicity or of vulnerability and harm. Students reported that they valued becoming more aware of the particular history and issues in Indianapolis. Some talked about changing personal behaviors to reduce waste and minimize their carbon footprint, while others indicated that they planned to continue advocacy work on environmental issues. Many described their experience as a wake-up call, with comments such as noting that the class "has made me more aware and opened my eyes to problems around Indianapolis" and that it "has forced me to become more aware of what is happening around the world." A student who lives within the project area wrote, "This project forced me to confront my ignorance and utilize my skills to learn more about our local history. I am now a more informed citizen, individual, and Indy resident because of this project, and I now understand the responsibility I have to work toward a more environmentally just Indianapolis." Another student reflected:

> I've lived in the Indianapolis community for the past five years and prior to this course I can say I knew nothing of its history and issues. It made me recognize that it's important to really learn about where you reside and call home and to not be a robot who drives by Fall Creek Parkway every Saturday and Sunday but doesn't exert the effort it takes into finding out why it smells so horrible.

Students expressed their discomfort in a variety of ways. Some identified ethical concerns related to their positionality as outsiders. In addition to their institutional affiliation with IUPUI, most were white and not from Indianapolis, and none had grown up in the affected communities. One student observed, "It was challenging to properly articulate racism without sounding like an outsider. Language is so important, and I often felt that my lack of personal experience made my input less valuable. I also would just rather listen to other people who have dedicated their lives to environmental justice, who have fought the battle and made strides in the world." Others were troubled by the premise that in one semester they were supposed to develop authentic, trusting relationships necessary for co-curation. Several framed this discomfort with their positionality and the project's constraints as an issue of being "white saviors." For example, one student wrote:

> I hold collaboration, not consultation, as an ethical standard for my curatorial practice when interpreting a community's experience. However, this ethical standard was difficult to adhere to with our single-semester time constraint and I do not feel that we were able to incorporate our community partner enough in the final, or in my individual project. . . . For me, balancing my own ethics with expectations was the most challenging part of this course. I did not want to approach this project with a "white savior" lens or by dominating the narrative with my own authority and position of power.

Although the faculty had structured the project and partnership to share authority and resources among partners in a mutually beneficial way, the students clearly needed more help to understand and navigate their role in the collaboration. One student expressed the discomfort as both personal and institutional guilt.

> I found it difficult to reconcile with feeling like I was part of an exploitative university process that collaborates with [a] marginalized community and then leaves. I did my best to get to know the KI members as individuals by meeting with them outside of class and attending their Facebook Live events. I often felt guilty during class when people, including myself, expressed uninformed views or made ignorant statements. I struggled to move past the feelings of guilt I had about how ignorant I am to the injustices marginalized communities experience and how I never felt like I had enough time to build deep relationships with community members.

As much as some students struggled with their positions as privileged citizens and participants in an academic-based project, other students found satisfaction and pride in being a part of education and advocacy for environmental justice issues. One noted, "It is extremely rewarding for me to see this program fall in place and have it be a part of the environmental justice conversation." Another student wrote, "It made me feel that through the completion of this project, I have actually done something that has made a difference in the world." Students' reflections indicated that some came to see themselves not just completing a course requirement, but as part of the ongoing environmental justice work of the city. One student reflected, "For me, . . . this project is successful beyond even my initial expectations. Exhibits are temporary, but tools for community empowerment are a long-term impact that will be here long after we are gone." Some student programs developed to complement the exhibit were born of their own connections to environmental issues. For example, one student developed a workshop to teach people how to test the water in their residences after he became concerned about his own home's water supply and started investigating lead in the city's water system. Another student developed a program combining information about the urban watershed with a trash cleanup at a nearby state park that had been a long-standing way for her to connect with nature (Figure 17.6).

4. *Indianapolis's Ecosystem and the Exchange of Capital*

The final aspect of the COI project's significance for understanding the city as an ecosystem is the importance of capital—social, political, and intellectual—that was built and exchanged over the four years. Students connected with community members, advocates, activists, and workers. Faculty built relationships, knowledge, and trust with community partners. Community partners strengthened relationships and benefited from research. KI staff commented specifically on the value of the students' research that helped them to easily access important historical and contemporary data and narratives. They have used this information as they build

FIGURE 17.6 Trash cleanup at Fort Benjamin Harrison State Park following a program called Follow the Fall Creek Flow.

Source: Photograph by Elizabeth Kryder-Reid.

curriculum, provide context of current environmental harm, and develop other storytelling projects. KI staff also valued the relationships they formed with students, some of which they continue to nurture and cultivate. For example, during a brainstorming session with filmmakers about a documentary on Indianapolis waterways, KI youth leaders recalled the COI collaboration and the work students had put in nearly three years prior and immediately connected the filmmakers with the students' research.[2] One student continued to collaborate after the COI project to help build out KI's "Democratizing our Data (DoD) program."[3] The COI project also enabled KI youth interns to go to the HAL Convening, speak about their experiences working for environmental justice, and develop their leadership potential.

Many students similarly valued the relationships they made with community partners, particularly the key KI members who were most involved in the project. They commented on how "we were able to define our expectations and goals, and establish a sense of trust which enabled us to openly go to KI with questions and feedback. . . . I am seeing the relationships I built with KI grow and deepen." Another student agreed that "[b]oth groups came in with an open mind and flexible attitude" and appreciated that both were "passionate about the project and its goals." One noted "I greatly value the relationships founded on mutual respect I have made over the past semester with KI members like Alvin, Keenan, and Aghilah." Another shared that "I learned so much from the experience and have made profound friendships that will shape how I view the world moving forward." Other students expressed appreciation for the opportunity to expand their networks, meeting nonprofit leaders and advocates and, for those who were able to participate in the Humanities Action Lab Convening at Rutgers, the opportunity to meet, as one student put it, "amazing people from around the world." Students built their own social capital, and they came to recognize the importance of personal and professional networks for environmental issues advocacy and the work of advancing urban sustainability. As much as the research project focused on the history of urban infrastructure and physical aspects of environmental harm, students also recognized that ultimately the project was about building relationships and connecting people. As one final reflective essay concluded:

> [T]o see the project go from research to planning to finally execution in this class is exciting for me. The research work done by individuals in the first class are now being shown in the companion exhibits that we have created. . . . The opening event allows us to see the interest that people have in this topic and that we are connecting with people and raising concerns through this work.

The outcomes for the hundreds of people who saw the exhibit or participated in the public programs are difficult to measure. The digital humanities projects developed through the COI project continue to get hits and comments, but we do not know, ultimately, whether the educational resources change the conditions or structural inequities of Indianapolis's polluted waterways. What we do know is that social capital is central to the sustainability of environmental justice workers and movements in the Indianapolis community and beyond. This project demonstrated that as much as infrastructure and ecology are integral to the sustainability of the city's ecosystem, so too are the social relationships among those who live there.

Notes

1 The Canal Path has since been extended to reach much of the Riverside and Near Westside neighborhoods.

2 The project, Water Scouts, follows the story of Girl Scout Troop 2715 and their quest to protect their community, tying into the broader struggle of communities of color and their relationship to land, environment, and the city.

3 DoD is an initiative to train and develop intergenerational grassroots capacity to use public data to inform community projects and public policy advocacy.

References

Adair, Bill, Benjamin Filene, and Laura Koloski. 2011. *Letting Go? Sharing Historical Authority in a User-Generated World*. Philadelphia: The Pew Center for Arts and Heritage.

Adisa, Imhotep. 2018. "Oral History Interview about Environmental Justice." Posted to YouTube by Jacqueline Rooksberry, November 25, 2018. www.youtube.com/watch?v=7hrYf6LGHos.

Alivizatou, Marilena. 2022. *Intangible Heritage and Participation Encounters with Safeguarding Practices*. New York: Routledge.

Anderson, Topher. 2018. "Site 0153: A Blogger's History." Published online November 2018. http://site0153.blogspot.com/.

Bennett, L. Michelle, and Howard Gadlin. 2012. "Collaboration and Team Science: From Theory to Practice." *Journal of Investigative Medicine* 60 (5): 768–75. doi:10.231/JIM.0b013e318250871d.

Boyer, Ernest L. 2015. *Scholarship Reconsidered: Priorities of the Professoriate*. Edited by Drew Moser, Todd C. Ream, and John M. Braxton. Expanded ed. (originally published 1990). San Francisco, CA: Jossey-Bass.

Butler, Octavia. 1993. *Parable of the Sower*. New York: Four Walls Eight Windows.

Climates of Inequality (HAL). https://climatesofinequality.org/.

———. Indianapolis https://climatesofinequality.org/story/inequity-along-the-white-river-local-advocacy-for- change/.

Climate Justice Alliance. "What Do We Mean by Just Transition?" https://climatejusticealliance.org/just-transition/.

Davenport, Alexa. 2018. "Narratives of Development: The Central Canal." https://humanizingdevelopment.weebly.com/.

Dostilio, Lina D., ed. 2017. *The Community Engagement Professional in Higher Education A Competency Model for an Emerging Field*. Boston: Campus Compact.

Dumlao, Rebecca. 2018. *A Guide to Collaborative Communication for Service-Learning and Community Engagement Partners*. Sterling, VA: Stylus.

Environmental Protection Agency. n.d. "Environmental Justice." www.epa.gov/environ mentaljustice.

Gentry, Kynan, and Laurajane Smith. 2019. "Critical Heritage Studies and the Legacies of the Late-Twentieth Century Heritage Canon." *International Journal of Heritage Studies* 25 (11): 1148–68. doi:10.1080/13527258.2019.1570964.

Grey, Ralph D. 2003. *IUPUI-the Making of an Urban University*. Bloomington: Indiana University Press.

Harrison, Rodney. 2013. *Heritage: Critical Approaches*. New York: Routledge.

Holzman Laura M. 2021. "Structuring Academic Jobs for Engaged Art History." In *Socially Engaged Art History and Beyond*, edited by Cindy Persinger and Azar Rejaie, 267–73. Cham: Palgrave Macmillan. https://doi.org/10.1007/978-3-030-43609-4_22.

Holzman, Laura M., Elizabeth Wood, Holly Cusack-McVeigh, Elizabeth Kryder-Reid, Modupe Labode, and Larry J. Zimmerman. 2014. "A Random Walk in Public Scholarship: Exploring our Convergent Paths." *Public: A Journal of Imagining America*. http://public.imaginingamerica.org/blog/article/a-random-walk-to-public-scholarship-exploring-our-convergent-paths/.

Humanities Action Lab (HAL). www.humanitiesactionlab.org/.

Indy Environmental Justice History—2018 Student Projects. https://lizkryderreid.wixsite.com/indyejhistory/iupui-student-projects.

IUPUI Office of Community Engagement. https://engage.iupui.edu/.

Jay, Gregory. 2012. "The Engaged Humanities: Principles and Practices for Public Scholarship and Teaching." *Journal of Community-Engagement and Scholarship* 3 (1): 51–63.

Kryder-Reid, Elizabeth. 2019. "Do the Homeless have Heritage? Archaeology and the Pedagogy of Discomfort. In *Pedagogy and Practice in Heritage Studies*, edited by Susan J. Bender and Phyllis Mauch Messenger, 129–47. Gainesville: University Press of Florida.

Little, Barbara J., and Paul A. Shackel. 2014. *Archaeology, Heritage, and Civic Engagement: Working toward the Public Good*. Walnut Creek: Left Coast Press.

Markley, Laura, and Dan Will. 2018. "The Forgotten Canal." (Published online November 2018). https://uploads.knightlab.com/storymapjs/a93711701ed1a8813ab340106a087319/central-canal-project/index.html.

Mitchell, Tania D. 2014. "How Service-Learning Enacts Social Justice Sensemaking." *Journal of Critical Thought and Praxis* 2 (2). https://doi.org/10.31274/jctp-180810-22.

Mullins Paul R. 2006. "Racializing the Commonplace Landscape: An Archaeology of Urban Renewal Along the Color Line." *World Archaeology* 38 (1): 60–71. https://doi.org/10.1080/00438240500509884.

Mullins, Paul R., and Glenn White. 2010. *The Price of Progress: IUPUI, the Color Line & Urban Displacement*. Indianapolis: IUPUI.

Paschall, Wildstyle. 2020. "Indiana Avenue: The Ethnic Cleansing of Black Indianapolis." *New America*, February 4, 2020. www.newamerica.org/indianapolis/blog/indiana-avenue-ethnic-cleansing-black-indianapolis/.

Rooksberry, Jacqueline. 2018. "Fall Creek." https://cdn.knightlab.com/libs/timeline3/latest/embed/index.html?source=1Tm-0g2TK_5Cx0qxFO4fzdGW5K1x3R1krGKSMZn9Ifc8&font=Default&lang=en&initial_zoom=2&height=650.

Shah, Rachael W. 2020. *Rewriting Partnerships: Community Perspectives on Community-Based Learning*. Louisville, CO: Utah State University Press, an imprint of University Press of Colorado.

Toxic Heritage—Collaborative Research. "Climates of Inequality." https://toxicheritage.com/research-projects/climates-of-inequality/.

18

URBAN REWILDING THROUGH A MODIFIED PLACEMAKING LENS

A Teaching and Learning Journey

Henry Way and Rob Alexander

Introduction

Urban ecosystems, like all ecosystems, have a geographical foundation: they exist somewhere and have geographical form. The teaching and learning experience we explore in this chapter examines these geographical characteristics using the lens of "place" in the pursuit of better outcomes in urban sustainability education and in rewilding and restoration ecology within cities. This "place" lens, because of its necessarily holistic, integrative focus, facilitates the greater incorporation of the human dimension into the urban environmental context and allows for distinctive and productive learning opportunities for students confronted with new settings and diverse disciplines. Likewise, it elevates ecological priorities within the more traditionally human-focused urban planning and design approaches.

We discuss a multidisciplinary undergraduate course that examined and developed opportunities for place-based rewilding in two urban riparian sites. In addition, we demonstrate how such place-based activities are valuable to student learning. As an experimental pedagogical adventure by the authors, this course also revealed the opportunities and challenges for refocusing urban sustainability education along—and beyond—geographical lines. We first examine the central focus of our spatial approach to urban sustainability here: the notion of place.

Convergence: Place

"Place" is one of those seemingly straightforward, but ultimately ambiguous terms we use to navigate our world. What, really, is a "place"? A first observation might be that place denotes something unique, a point or location on the Earth's surface, a piece of geographic space. Place implies a certain distinctiveness (Relph 1976).

DOI: 10.4324/9781003217442-21

Examining the concept more deeply, places have physical characteristics—they have form, substance, extent—but at the same time places also have a certain identity, meaning, or nonphysical character. Places are conceived and perceived. "Space is transformed into place as it acquires definition and meaning," according to Yi-Fu Tuan (1977, 136). This dual nature has been theorized by geographers (and others) and offers a rich—and sometimes foundational—way to understand the world. Defining place as a "meaningful location" captures this hybrid spirit of physicality and subjective expression (Lewicka 2011, 209; Cresswell 2015, 12).

Agnew's (1987, 16) definition adds complexity: a place is a three-layer structure with three simultaneous elements: location (the geographical position of the place), locale (the physical setting and characteristics of the place), and sense of place (the local "structure of feeling", in his words). Beyond Agnew's trilevel model, Casey (1996, 24) discusses how places "gather." Places are seen as having an immense power because of the way that they inherently bring together the material, the social, the emotional—all dimensions to existence. Cresswell (2015, 53), in his survey of the development of meanings of place, points to Dovey's (2010, 16) assertion that "all places are assemblages," a perspective that echoes the body of scholarship around "assemblage theory" (e.g., DeLanda 2006). As discussed later, the more emotional human dimension to place has the potential to be marshalled in valuable ways for urban environmental interventions.

Why, then, is the lens of "place" important for teaching and practicing urban sustainability? First, adopting the lens forces our attention to the distinctiveness of sustainability project *sites*. While it is valuable to think in terms of the broader policy, cultural, and economic contexts in which projects are developed, our urban sustainability projects also constitute actual, unique locations with distinctive "place" characteristics and needs. Turner and Kaplan (2019) write that, pedagogically, using a place lens demands learning experiences sensitive to the particularities of location and recommends using case studies to understand the extent to which sustainability projects pay attention to the distinctiveness of their places. Practitioners of "place-based education" argue that the approach productively connects learners to their immediate community (e.g., Sobel 2004).

Second, understanding that places are assemblages and encompass location, locale, and human meaning is essential for integrating the human with the ecological in urban sustainability. Because urban ecosystems are inherently hybridized between the "natural" and the human, the "places" these habitats manifest should be looked at as both human and ecological. This elevates the position of the environment in urban design (as exemplified well by the "biophilic cities" movement [e.g., Beatley 2010]) and requires, reciprocally, that human dimensions are actively incorporated into restoration ecology projects in urban places. These "human" dimensions are the contextual layers for the site—historical character, regulatory elements (such as zoning codes), immediate economic conditions—as well as the ways in which people find meaning and attachment to that place. Rooting urban sustainability in "place" in this way can lead to more appropriate place-specific and culturally and socioeconomically framed projects, adding value to the human conditions and durability to the ecological prospects. Pedagogically, this puts greater emphasis on ecological and human approaches that are integrated and layered, metaphorically like a GIS, onto the site. It can also prioritize the development of systems thinking approaches.

Cocreation: Sense of Place

Drawing from environmental psychology, a sense of place is defined as having both a cognitive, or "place meaning," element as well as an affective, or "place attachment," component (Semken and Freeman 2008), adding a more specific human, subjective direction. *Place meaning* entails people ascribing symbolic meanings to places that are personal and informed by the individuals' intersection of their sense of self, experience with place, and lessons learned about the place through the storytelling of others (Kudryavtsev, Stedman, and Krasny 2012). *Place attachment* refers to the "emotional bonds between people and a particular place or environment" (Seamon 2014, 11). Going further, Tuan identifies a deep individual human need for an "affective bond between people and place," which he terms "topophilia" (Tuan 1974, 4). Examining the sense of place concept demands a focus on people's subjective understanding and emotional experience of places and considers the ways the structure of places generates particular human responses.

In urban sustainability, the closest field of practice that attends to such concerns might be the "placemaking" approach, promoted in urban design and elsewhere by organizations such as the Project for Public Spaces (see pps.org for examples). Such work, prioritizing community input and the creation of urban places that are contextualized by local conditions and with the goal of inclusive, comfortable physical places, draws upon the rich existing work of a number of designers and theorists (i.e., Whyte 1980; Gehl 1971; Lynch 1960).

The placemaking approach most directly attends to the human dimensions of urban design and sustainability, opening the way to more compact, appealing cities which have positive "environmental" effects. By conceiving of place more holistically here, we consciously included more directly *ecological* dimensions to sustainability into "sense of place" thinking. This is not an untrod path (e.g., Ardoin 2006) but is relatively underdeveloped. By the same token, could such urban restoration ecology projects be more successful if they explicitly integrated how people *feel* about the places or sites in question? Moreover, how does a sensitivity to student emotional understandings of places shape how students engage with teaching and learning about urban sustainability?

Our focus on these more qualitative, subjective elements led us to value *in situ* education, field visits, interviews, and storytelling as ways to augment the traditional "environmental science" perspectives in urban restoration ecology. This meant understanding places with the individuals and communities that care about the spaces in question, while cocreating the class with the students as the learning experience evolved. This approach is consistent with the view of Kudryavtsev, Stedman, and Krasny (2012, 240) that teaching and learning that reinforce the development of "place meaning" and the solidification of "place attachment" provide an opportunity to generate proenvironmental behaviors.

The approach to place just described—incorporating both the integrative understanding of "place" and the deep subjectivity of "sense of place"—coupled with the promise of shifting student mindset, attitudes and behaviors toward those conducive to a more resilient, inclusive, and sustainable rewilding of urban spaces—shaped the development of our course.

Course Design

At our university, no academic unit is focused solely on interdisciplinary environmental or sustainability studies, so students interested in this work are dispersed across many colleges and departments. While some programs have developed concentrations and tracks for students interested in sustainability, many students access a suite of environmental and sustainability minor programs that complement their major courses of study. Three of these minor programs share an upper-level culminating class, "Capstone in Environmental Problem Solving," taught every semester by different pairs of faculty members grounded in distinct disciplinary perspectives. This chapter focuses on our version of this class. Thirteen juniors and seniors representing six different disciplines were enrolled, meeting once a week.

Prior to the start of the semester, the faculty team identified two urban areas in Virginia within an easy day-trip visit of our campus—Roanoke and Richmond—and visited each to walk through possible place-making tours we could use to engage students with "place." We looked for spaces and places where the intersection of diverse human activity, such as housing, transit, industry, and recreation met with diverse ecological/nonhuman activity to set up the challenge of reimagining these spaces as rewilded through the lens of place.

Then, we looked for a comparable case within our own city of Harrisonburg, Virginia. Scaling down but retaining the same human/ecological intersections, we identified two sites linked by both riparian and pedestrian pathways to be used as a "case study" for applying principles of sustainable placemaking and rewilding that we developed through our course content and urban field trips.

We designed the class as modified experiential learning with an initial phase to make desired group norms and behaviors explicit, followed by a series of iterative cycles of content delivery, applied experience, and reflection. The class then evolved into a student-led inquiry and production process that resulted in student-led work products focused on placemaking and reimaging the case studies as exemplars of urban rewilding.

Recognizing the role that technology plays in our placemaking work, particularly as it applies to telling the stories of these places, we inserted a weekly assignment of using and reviewing a range of technological tools relevant to placemaking and/or project management, such as ThingLink, ESRI StoryMaps, Google Tour Creator, Instagram Stories, and Basecamp. We saw this as an opportunity for students to develop their technological skills on a variety of platforms. After learning and performing basic functions in these tools and reviewing examples of place-based storytelling produced by others, students recorded their thoughts on such questions as: "How well did the tool combine mapping and environmental information?" "Did the tool convey a 'story' or narrative? If so, what was the story?" And, "How well did it present information?" Class discussion of these thoughts led to a more critical examination of how each tool transformed the story of place being told. Table 18.1 summarizes the organization of our topics and activities, illustrating how we developed a range of content-based, competency-based, and disciplinary approaches as the course progressed.

TABLE 18.1 Overview of course design broken down by topics, key questions, and technologies reviewed.

Weeks	*General topics*	*Key questions: reflection*	*Technology exploration*
1	• Introduction to "place" • Storytelling • Rewilding • Class covenant	• What do we see when we focus on "place"? • What is "wild"? What is "wilderness" in an urban neighborhood? • Who are we as a group when we are at our best?	
2	• Environmental philosophy	• What is "natural"? What are various natural values?	ESRI StoryMaps
3	• Restoration ecology	• Can cities be seen as "natural"? • What are the prospects for urban "rewilding"? For novel ecosystems?	ThingLink
4	• "Place" and "Placemaking" • Richmond field experience	• What are the key characteristics of successful urban design? How do we interact with the built environment? What informs "sense of place"?	Instagram Stories
5	• Reflections on Richmond	• What makes a good "place"? • How does "place" tie into ideas about restoration ecology and environmental problem solving? • Stepping back: what is this class about?	Google Tour Creator
6	• Introduction to case sites • Harrisonburg exploration	• How are these sites "places"? • What is the potential for placemaking at these sites?	Google Earth

(Continued)

TABLE 18.1 (Continued)

Weeks	*General topics*	*Key questions: reflection*	*Technology exploration*
7	• Biophilic cities; green infrastructure; urban planning	• What is planning and zoning, and what are the regulations shaping our sites? How can biophilic urbanism inform what we develop?	SketchUp Public GIS
8	• Planning for the Roanoke exploration	• How should we design our Roanoke exploration? • What can we do there that will help us start work on our case sites?	
9	• Roanoke field experience	• (Student-generated questions *in situ*)	
10	• Roanoke reflections • Organizing for Harrisonburg case sites	• What stories about self did you hear? • What stories about society did you hear? • To what extent were the places explored examples of "sustainable placemaking"?	360-degree cameras
11–14	• Student exploration, analysis, and projection of Harrisonburg case sites	• How are we now defining "sustainable placemaking"?	Basecamp; website creation
15	• Walkabout presentation of Harrisonburg case sites with key stakeholders	• What is the potential for sustainable placemaking in these sites? • How do we communicate that potential?	

Course Implementation

Recognizing the need for students to shift their mindsets regarding place, urban sustainability, and what learning "looks and feels like," we produced a deliberative "covenant" or set of aspirational behaviors that would maximize our ability to learn differently and with an open mind (e.g., Dukes, Piscolish, and Stephens 2000). This covenant became a tool against which we could check our behaviors and interactions to ensure that we were living up to our own expectations as well as preventing ourselves from sliding back to existing, more restrictive habits of learning.

Central to the class were the field trips to Richmond and Roanoke where we immersed ourselves in place-based experiences. In our Richmond visit, we facilitated walking tours that prompted the students to feel their sense of place emerging. We observed the key tenets of urban design in a variety of streets and spaces, and discussed how the built and natural environment generated emotional responses and feelings. We considered the layers of history inherent to these places while taking time after each stop to share our reflections and appreciate the divergent ways that we, as a learning group, engaged with our own sense of place. For example, we spent time prior to the field trip learning basic built environment histories and the social stories attached to them and then designated key stopping points on our tour where we could share those stories with students just prior to having them silently experiencing that space on their own. One of the key messages for the students was how the urban landscape we experience did not emerge by accident: it was the historical and geographical product of systematic forces around economics, social relations, and political dynamics, as well as responses to the immediate natural environment and determinant regulatory structures. Richmond's economic and social history, along with its riverine location, provided compelling on-the-ground exemplars of these shaping forces. Added to this, at each stop in the exploration, we encouraged critical reflection on how these "layered" places in their meanings and their physical structure generated feelings for us: How did the composition of the built environment generate comfort or unease? How did the integration of natural elements in the urban design soften the cityscape or undermine legibility? How did each of us as individuals have different responses to elements in the built environment, and where did each come from?

We focused the second urban visit in Roanoke on experiencing place through the lenses of those who were embedded within them. We focused this trip, therefore, on student-led "walk and talks" with key actors—a city planner, a greenway manager, a community activist—listening to their placemaking storytelling. In Roanoke, we also introduced questions of social equity, particularly around the historical displacement of the Black and African American residents and how that might impact the sense of place for these communities.

In both field experiences, we raised awareness of exemplars of topics and themes we had covered in our readings and on-campus work, including "good/bad" urban design, environmental remediation, and restoration ecology. Students subsequently engaged in reflective writing exercises to reinforce the interconnectivity between instruction and experience.

After journeying with students through sense-of-place from the personal to the urban levels, we shifted ownership of the course to the students with the charge of proposing "a rewilding of our small city field sites that maximizes placemaking potential and engenders urban sustainability." The students engaged in place-based observation of and interviews with place "owners" and users to establish a narrative of the "place that is" in order to develop a vision of "place that could be," articulating both human and ecological elements. Students also explored historical archives to learn the evolution of these places.

Case Study Details

We selected two sites in Harrisonburg linked by Black's Run, the primary water body running through our city, and a way-finding walking path consisting of sidewalks, street crossings, and a small walking bridge across the stream. Both sites, about a third of a mile apart, consisted of mostly neglected stream banks utilized by a variety of urban wildlife. One site (one-third acre) to the north sits adjacent to a rapidly transitioning edge of our small downtown but is otherwise overlooked, constrained by a busy road intersection and billboard advertising, and containing numerous invasive species. Owned by a billboard operator willing to entertain opportunities to improve its condition, the site was earmarked as a site for a pollinator garden to be prepared and maintained by the city, which had the potential to enhance the quality of the environment and attractiveness in the context of the postindustrial redevelopment in the immediate area. The smaller (0.10 acre) southern site is in the heart of the downtown and is more intensively used by people, as it was connected by a small bridge to a parking deck on the other side of the stream, and it was coterminous with the rear patios and entrances of a popular set of stores and restaurants. The owners of one of these stores—a bike shop—also owned this site and were keen to find ways to improve the environmental condition of the streambank, the aesthetic attractiveness of the adjacent patio, and the general use of the site by customers and visitors. Linking these sites as our "case study" allowed us to explore a greater variety of contextual and site-specific needs and opportunities. Curating the connected sites also allowed us to emphasize the value of more systems-oriented thinking and connectedness in both effective human urban design and ecosystem dynamics.

Utilizing the tools and content of our class, students accumulated observations, interviews, and historical records of the spaces in and around these sites. As this work ensued, a narrative emerged of heavy past industrial use (for example, a tannery once sat adjacent to the southern site) upon which new layers of the built environment were placed and against which current human and ecological uses occurred. Human use was largely of the "passing through" nature rather than "sit and stay." People would often pause on the bridge across Black's Run to view the occasional ducks and water snakes, and dog walkers from a nearby condominium complex would utilize the northern site as a dog relief area, but few, if any people were observed inhabiting either of these spaces for more than a few minutes at a time. When the students themselves spent a class period at either site, integrating their historical, economic, and regulatory understanding with their emotional engagement with the locations, they observed in their reflection journals how it took some time for their personal attachment to these places to "warm up" and "activate" to develop any sort of sense of place, raising interesting questions about how placemaking techniques might be applied to these sites. In the script prepared by students for their final presentation where they walked, participants from the northern site to the southern site wrote that engaging in sustainable placemaking relies upon the relationship between human and nonhuman entities, which takes both time as well as a place where people are comfortable to stop, rest, observe, and reflect. Therefore, one of the key recommendations that students had for improving these sites was to introduce improved seating that would allow human users to connect to wildlife and habitat without impinging upon them. Otherwise, as one

FIGURE 18.1 Excerpt from student-generated walking tour brochure.

student commented, "[W]ithout the human entity we don't think the site would become 'important' and would end up not being successful."

As students developed the historical story of these sites and their connection to them to prepare ideas for reshaping their future narratives in a more sustainable manner, they incorporated a number of the digital tools we explored during the class. Images of both sites were recorded with a 360-degree camera as well as smartphone cameras, and ESRI StoryMaps and ThingLink were considered as places to embed and annotate these images. Students then developed a website where they shared the human and ecological story of these sites with a 360 tour, slide shows, past/present swipes, and static images of human and ecological site users in and around the sites to make their case what they termed "sustainable placemaking." Figure 18.1 is an excerpt from the hard copy brochure produced in parallel with the website that includes their definition of sustainable placemaking.

Ultimately, students used their walking tour plus the website to make the case for sustainable placemaking in our community. As the "About Us" portion of the website describes:

> We are a group of students attempting to rewild Harrisonburg through the lens of sustainable placemaking, in which we define as the planning and development of dynamic places that sustain long-term transformative human–nature relationships. These places convey local, community-based

> stories, are embedded within a broader system of place, and cultivate a diverse group of community members who care for their natural environments and the other living things within them. The practice of placemaking looks different for every local and individual, which is key for sustainability. This concept heavily influenced our perceptions and ideas for the two sites in downtown Harrisonburg.

While actual rewilding did not occur as part of our work within the class, we feel assured that the students embraced the ideas and opportunities for leveraging sense of place in human actors to improve human–nature relationships in our built environments.

Discussion

Since delivering this course, our reflections on the discussions, writings, and reactions we shared with the students have converged around several key points that we feel are useful to the urban sustainability educator.

Repositioning the Meaning of "Interdisciplinary"

As this class evolved, we observed the "integrating" power of place for reshaping how we might think about the interdisciplinary approach to urban sustainability education. Our modified placemaking methods required that students initially suspend their disciplinary mindsets to connect with their inner sense of place. The holistic conception of place meant that no one discipline was considered more important than the others; the approach necessitated a genuine engagement across all disciplines. This allowed them to then arrive at a more equal footing in the group when contributing their disciplinary perspectives and skills. The field trips to city spaces previously unknown to all the students reinforced the shedding of disciplines to collectively share a new experience together.

The "definition" of sustainable placemaking developed by the students nicely illustrates both the conceptual value of integrating the human and environmental dimensions, the transdisciplinary power of this placemaking approach, and by its generation, the cocreative, inclusive opportunities for place-focused learning.

Elevating the Ecological in Placemaking

As we hoped and intended, this place-focused learning experience elevated and centered ecology in urban planning and design for the students. Immersing ourselves in the ecological through our field experiences, our case sites and our telling and retelling of ecology stories deepened understanding of the broader value of urban restoration ecology. Focusing on the particular needs of case sites led to a richer, more contextualized engagement with restoration ecology, making it more meaningful for students and community members alike.

In addition, as we and the students reflected on the interplay between the human character or "meaning" of these places and the ecological needs of the sites, opportunities for more sustainable and resilient rewilding became clearer. This builds upon the insights of the "biophilic cities" movement and illustrates the relationship between human place attachment and ecological restoration. Understanding the social, emotional, and economic characteristics of a place, along with its built environment, produces a more site-based ecological intervention. When developed with human place attachment needs in mind, rewilding contributes to place meaning by creating an ecological space that helps people find experiential meaning. In essence, not only is restoration ecology supported, but a richer sense of place is developed, greater care emerges for this new ecological site, and longer-term resilience to the usual urban pressures or potential neglect manifests. Students, in their submitted reflections, effectively connected these insights.

Telling the Stories of Sustainable Places

In planning this class, we quickly realized that we needed a simple yet effective means for students to make sense of the rather complex theoretical discussion of making sense of place. Building upon the ideas that places are socially as well as physically conceived and that they evoke feelings of both meaning and attachment, we selected storytelling as a way for students to understand these complex layers of place both from inside themselves as well as through the eyes of others, past and present.

Informed by Leineweaver (2017), we coconstructed a storytelling framework with the students to guide how we would articulate our placemaking work to ourselves, to one another, and to the world writ large. Key elements of this framework included "stories do not need purpose," "stories exist at the individual, societal, and mythic levels," and "these levels interact and inform one another." Practicing the telling of our individual stories of place enabled us to better convey how places hold societal and mythic level stories. The storytelling lens also enabled a greater empathy for how all individual-level stories are of equal validity, especially as we strive to influence the societal and mythic storytelling levels.

Infusing the technology review exercises into our coursework and framing them as "storytelling tools" also created a useful hook for the students to see themselves as digital curators of place-based stories. When students constructed the site stories—past, present, and future—with their chosen digital tools, they demonstrated a growing awareness of how their own personal and disciplinary lenses shaped the story that emerged on the other end. This awareness seemed to deepen their critical thinking of the place stories presented by planners and developers, launching a desire to know the "whole" story through ecological, humanistic, justice, and economic lenses.

Conclusions

This course demonstrated the great potential value of framing urban sustainability education around place and an enhanced approach to placemaking. Our approach allowed us to effectively convene multiple disciplines, elevate the ecological within the human, and leverage the tools of storytelling to understand the critical value of sense of place for effective urban sustainability planning. Students developed a greater appreciation for the complexity of urban sustainability initiatives, for the importance of understanding the development of place meaning by place users, and for the ways in which technology can improve how we understand city places and communicate their planning. The students realized how to pursue urban rewilding and restoration ecology with a greater sensitivity to the human context.

However, we also note several avenues for improving how we teach urban sustainability. Most broadly, a class like this takes time, to an extent that we did not realize at the outset. We thought ambitiously that we could develop the culture of integrative learning, embed the place-oriented understanding, learn from other locations, develop the technical competencies, and cogenerate and communicate the rewilding plans in one semester. As we learned, it takes time to develop a useful transdisciplinary approach that fully connects the human and environmental dimensions.

A two-semester model would work most effectively here: the first semester to develop the thoughtful, place-oriented, integrative mindset and fully understand the places in question; the second to fully develop the ecologically based place-specific plans and connect the communicative and community integrated pieces. By the end of our class, we saw excellent potential for effective rewilding but ran out of time to dive deeply into the actual ecological plans for the sites and fully build the ecological capacities of our students. With the time limits placed on this course, the final result was more human oriented in analysis than fully exploring the ecological implications of the plans.

Having more time to deliver this course would also help better connect the end products with the initial theories and ideas that shaped its codevelopment. For instance, while students embraced the value of storytelling for both building and communicating their ideas, they ultimately lacked the time needed to fully evaluate the efficacy of their storytelling. We finished the course with an "exit interview" discussion with each student to assess the storytelling theory and practice connections, and we discovered that these connections were only just being formed.

The placemaking focus of this learning experience offers an innovative and integrative way for enriching urban sustainability learning, and for pointing the way toward enhancing the prospects for restoration ecology in cities. Although our experience was circumscribed in some ways by time constraints and available resources, it proved a valuable test case for this approach and illustrated the utility of a place-based orientation for transdisciplinary learning and more inclusively connecting with the urban communities in which sustainable placemaking and restoration ecology are undertaken. Pursued with the human understanding of place in mind, urban rewilding can be a valuable fusion of the social and ecological and an intrinsically multidisciplinary and engaging learning experience.

References

Agnew, John. 1987. *Place and Politics: The Geographical Mediation of State and Society*. Boston: Allen & Unwin.

Ardoin, Nicole M. 2006. "Toward an Interdisciplinary Understanding of Place: Lessons for Environmental Education." *Canadian Journal of Environmental Education* 11 (1): 112–26. https://eric.ed.gov/?id=EJ881798.

Beatley, Timothy. 2010. *Biophilic Cities: Integrating Nature into Urban Design and Planning*. Washington: Island Press.

Casey, Edward. 1996. "How to Get from Space to Place in a Fairly Short Stretch of Time." In *Senses of Place*, edited by Steven Feld and Keith Baso, 14–51. Santa Fe: School of American Research.

Cresswell, Tim. 2015. *Place: An Introduction*. Oxford: Wiley-Blackwell.

DeLanda, Manuel. 2006. *A New Philosophy of Society: Assemblage Theory and Social Complexity*. London: Continuum.

Dovey, Kim. 2010. *Becoming Places: Urbanism/Architecture/Identity/Power*. London: Routledge.

Dukes, E. Franklin, Marina A. Piscolish, and John B. Stephens. 2000. *Reaching for Higher Ground: Creating Purpose-driven, Principled and Powerful Groups*. Charleston: BookSurge Publishing.

Gehl, Jan. 1971. *Life Between Buildings: Using Public Space*. Washington: Island Press.

Kudryavtsev, Alex, Richard Stedman, and Marianne Krasny. 2012. "Sense of Place in Environmental Education." *Environmental Education Research* 18 (2): 229–50. https://doi.org/10.1080/13504622.2011.609615.

Leinaweaver, Jeff. 2017. *Storytelling for Sustainability: Deepening the Case for Change*. Oxford: Routledge.

Lewicka, Maria. 2011. "Place Attachment: How Far Have We Come in the Last 40 Years?" *Journal of Environmental Psychology* 31 (3): 207–30. https://doi.org/10.1016/j.jenvp.2010.10.001.

Lynch, Kevin. 1960. *The Image of the City*. Cambridge: MIT Press.

Relph, Edward. 1976. *Place and Placelessness*. London: Plion.

Seamon, David. 2014. "Place Attachment and Phenomenology: The Synergistic Dynamism of Place." In *Place Attachment: Advances in Theory, Methods and Applications*, edited by Lynne Manzo and Patrick Devine-Wright, 11–22. Oxford: Routledge.

Semken, Steven, and Carol Butler Freeman. 2008. "Sense of Place in the Practice and Assessment of Place-Based Science Teaching." *Science Education* 92 (6): 1042–57. https://doi.org/10.1002/sce.20279.

Sobel, David. 2004. *Place-based Education: Connection Classrooms and Communities*. Massachusetts: Orion Society.

Tuan, Yi-Fu. 1974. *Topophilia: A Study of Environmental Perception, Attitudes, and Values*. New York: Columbia University Press.

———. 1977. *Space and Place: The Perspective of Experience*. Minneapolis: University of Minnesota Press.

Turner, V. Kelly, and David H. Kaplan. 2019. "Geographic Perspectives on Urban Sustainability: Past, Current, and Future Research Trajectories." *Urban Geography* 40 (3): 267–78. https://doi.org/10.1080/02723638.2018.1475545.

Whyte, William H. 1980. *The Social Life of Small Urban Spaces*. New York: Project for Public Spaces.

19

USING THE PREPOSITIONAL FRAMEWORK FOR URBAN ENVIRONMENTAL EDUCATION

Teaching and Learning About Ecology *in*, *of*, *for*, and *with* Cities

Loren Byrne

Introduction

A central goal of teaching is to help students understand and further investigate patterns and processes of the "world" broadly defined. Achieving this aim is not always straightforward because many patterns and processes are not observable at spatiotemporal scales perceived by human senses nor are they obvious without relevant knowledge. Further, many characteristics of the world are best understood using abstract representations created by humans, including uncommon words (jargon), metaphors, theories, and mathematical equations (Borghi et al. 2018; Desai, Reilly, and van Dam 2018). Among such abstractions, conceptual frameworks help organize ideas to show interrelationships among variables and terms (e.g., Wald and Daniel 2020). Such frameworks establish common languages that can catalyze shared understanding, effective communication and collaboration among diverse people (including teachers and students). Developing and comparing different frameworks can also reveal divergent views and facilitate finding common ground. Frameworks are especially important in environmental sciences that focus on interacting variables that exist over many spatiotemporal scales (e.g., Groffman, Driscoll et al. 2004; Scherer, Holder, and Herbert 2017).

Among the most complex environmental systems are those in which humans are dominant drivers of patterns and processes. This includes urban areas, conceptualized as social-ecological-technological systems (SETS), for which many frameworks have been developed that provide excellent starting points for teaching and learning about them (reviewed in Zhou, Pickett, and McPhearson 2021). For example, near the beginning of my undergraduate urban ecology course, I have students compare four different foundational conceptual frameworks that integrate natural and social sciences (Grimm et al. 2000, 2008; Alberti et al. 2003; Pickett and Cadenasso 2006). In small groups, students identify the frameworks' strengths and limitations to understand the

DOI: 10.4324/9781003217442-22

complexity of urban SETS and need for interdisciplinary approaches to effectively manage and improve urban systems' biodiversity, ecosystem services, capacities to promote human health, and overall sustainability (McPhearson et al. 2016).

Such interdisciplinary teaching about urban ecology has received increasing attention (e.g., Berkowitz, Nilon, and Hollweg 2003; Barnett et al. 2011; Tidball and Krasny 2011; Biberhofer and Rammel 2017), and the scientific foundations to support courses about urban SETS is well established. However, educators may find the breadth of topics overwhelming to adapt into courses, syllabi and lessons. Likewise, students may find the study of urban SETS challenging because it requires skills for abstract systems thinking. Indeed, teaching and learning about complex urban systems are inherently difficult for people of all levels and backgrounds: Where should one begin and how is best to proceed? (For more discussion about this question and example learning activities, see Schmitt-Harsh and Harsh [2017].)

A prominent way that the study of urban SETS has been organized is using four main topical areas described by the "prepositional framework" (Byrne 2022). This approach categorizes research and knowledge about urban areas as "ecology *in, of, for*, and *with* cities." Pickett et al. (1997) first articulated the *in* and *of* phrasing with further description and expansion by others (e.g., Byrne 2022; Childers et al. 2015; Grimm et al. 2000; Grove et al. 2016; McPhearson et al. 2016; Pickett et al. 2001, 2016).

The goal of this chapter is to describe the use of the prepositional framework to organize an interdisciplinary urban ecology course using examples from my advanced undergraduate course, which was developed for students majoring and minoring in environmental science, biology, and sustainability studies. Brief descriptions of how each preposition relates to the study and management of urban systems are provided in the following sections, alongside example topics and teaching resources. My intent is not to be prescriptive or to offer a fully formed course outline and lessons; instead, I aim to be inspirational by offering a general structure and illustrative activities that instructors can adopt and develop based on their course contexts and objectives. Though my experience is in higher education, I expect that the following framework and some of the ideas are relevant to K–12 teaching, for which others have reported urban environmental education initiatives (e.g., Abu Baker et al. 2015; Aloisio et al. 2018; Barnett et al. 2011; Oliveira et al 2021; Russ 2015; also see Berkowitz, Nilon, and Hollweg 2003 and resources at https://academics.lmu.edu/cures/partners/k12teachers/urbanecolab/). At all levels, more educational efforts about urban SETS are needed, given the increasing rates of urbanization and their relevance to pursuing a more sustainable future for life on Earth. The urban prepositional framework serves as a logical starting point for helping students think about and more effectively engage in diverse SETS around them.

Ecology *in* Cities

Most early urban ecology research investigated traditional ecological variables using standard methods to describe basic patterns and processes. This approach, labeled ecology *in* cities, focuses on abiotic conditions (soils, temperature, water, etc.), nonhuman organisms (their behaviors, resource use, survival, reproduction, etc.), populations (abundance, distribution, growth), communities (composition, interspecific interactions, food webs), and ecosystem processes (biogeochemical cycles, energy flow) of urban-impacted, outdoor places (Pickett et al. 2001, 2016). Though data about these variables can be collected independently, a main goal of urban ecologists is to integrate the data to understand the interacting drivers, including anthropogenic ones, of larger-scale patterns of urban biodiversity and ecosystem functions (e.g., Aronson et al. 2014; Kaushal, McDowell, and Wollheim 2014). Further, comparing urban and nonurban places helps elucidate differences between them to deepen understanding of how urbanization impacts ecological variables (McPhearson et al. 2016). To this end, the basic ecology of uniquely urban habitats like lawns, gardens, and vacant lots is of particular interest since they represent much of the "nature" of urban systems (e.g., Byrne, Bruns, and Kim 2008; Rega-Brodsky and Nilon 2016).

To introduce students to ecology *in* cities, attention-grabbing, unique stories and imagery can be used, such as a rat stuck in a manhole cover (Held 2019), birds nesting in odd places (BoredPanda 2015), and otherwise-arboreal koalas moving on the ground through urban landscapes leading to unexpected interactions (Siossian 2018) and deadly outcomes (Buttigieg 2019). The scientific literature is extensive enough that urban examples can be found to reflect the full range of topics within a general ecology textbook (e.g., McPhearson et al. 2016). An example lesson from my urban ecology course is the student-led "birds & bugs search, share & synthesis" (Byrne in press), for which each student chooses a peer-reviewed article about urban arthropods or birds, presents it to the class (with one or two slides), and then works with a small group to write a synthetic summary and follow-up research questions.

Studying basic ecological patterns in urbanized environments may not seem exciting or relevant to some students. However, it is foundational to solving problems relating to biodiversity conservation, environmental quality, human health, and sustainability. This can be emphasized through ecology-*in*-cities lessons that support connecting basic and applied ecology issues, e.g., managing endangered, invasive, pest and disease-causing species. For example, to mirror a larger societal controversy, students can evaluate studies about the impacts of domestic cats on urban biodiversity and debate their accuracy, interpretation and relevance to decisions about controlling outdoor cats (e.g., Loss et al. 2018). Such examinations help open students' eyes to the need for reliable biological and environmental knowledge to guide informed management decisions (which are central to the *for* and *with* approaches). In addition, basic variables included in the study of ecology *in* cities are required to understand the ecology *of* cities.

Ecology *of* Cities

Ecology *in* cities is a reductionist approach that, while generating valuable and necessary knowledge, is insufficient to fully understand urbanized places. This has led to more holistic studies that examine urban SETS as single, integrated entities with emergent properties (i.e., the urban place is the system, rather than a set of independent variables and systems) (Pickett et al. 2001, 2016; Zhu, Pickett, and McPhearson 2021). From this ecology-*of*-cities view, several key insights emerge that are now accepted as central to urban systems science. First, a fuller range of social, technological and infrastructural (e.g., sewers, roads) variables must be considered as drivers of ecological variables, which can in turn affect the former through interconnected feedback relationships. This contrasts with the *in* view, which places ecological variables at the center of study (i.e., the dependent variables impacted by humans and urbanization), while largely ignoring how they affect other parts of the system.

Second, the *of* perspective underscores the need to examine urban systems at larger scales including landscapes, watersheds, and regions that have emergent properties arising from interactions among their subsystems over long timeframes. Such expanded spatiotemporal analyses support integrated measurement of urban systems' metabolism (energy and nutrient flow through environmental, social and infrastructure components), ecological footprints, resilience to disturbances, relationships with surrounding ecosystems (e.g., food imports from rural areas), and impacts on larger-scale biodiversity patterns (e.g., regional species loss), and ecosystem processes (e.g., cities' greenhouse gas emissions) (Pickett et al. 2001).

Third, and highly relevant to the *for* and *with* perspectives, this more holistic outlook leads to explicit consideration of human lives and social variables (ethics, economics, aesthetics, etc.) within the scope of urban ecology. In turn, this has spurred scientists to study how ecological variables are translated into urban ecosystem services: the benefits that humans receive from ecosystems, including material goods, and processes and conditions that create healthy urban living conditions (reviewed in Tan et al. 2020; Ziter 2016). Altogether, these and other insights build on those from the ecology-*in* view to generate a more systemic understanding of how diverse social, technological, and biophysical patterns and processes interact to form a variety of complex, dynamic urban systems, from neighborhoods through towns, cities, metropolitan regions, and even the entire urban-impacted Earth system.

Studying the ecology *of* cities is likely to be more challenging for students because it requires interdisciplinary, synthetic, systems-level thinking skills. However, I have effectively helped students develop their comprehension with a few lessons. First, I broaden their perspective about relationships between ecological and sociocultural variables using the case study of lawns and gardens (described in Byrne 2015; the instructor guide can be obtained by contacting the author). As part of this lesson, I share research revealing that household socioeconomic variables (e.g., income) help explain plant abundance and species richness in ornamental gardens, a pattern known as "the luxury effect" (Leong, Dunn, and Trautwein 2018) and "an ecology of prestige" (Grove, Locke, and O'Neil-Dunne 2014). Similarly, people's diverse aesthetic preferences are important drivers of urban biodiversity; having students discuss the perspectives of Nassauer (1988) and Byrne and Grewal (2008) facilitates reflection about why people want urban environments to look a certain way and, in turn, how their appearance can feed back to social systems by affecting ecosystem services and choices about how lawns and gardens *should* look to improve their sustainability. In this way, lawns and gardens are an excellent focus for developing insights about the value of the ecology of cities' holistic, multiscale approach.

To help students think about urban systems across different spatiotemporal scales, I have them present to the class one of five papers that use different subjects, methods, and scales to exemplify how scientists study the ecology *of* cities (households: Fissore et al. 2011; city: Baker et al. 2001; Moore, Kissinger, and Rees 2013; watersheds: Groffman, Law et al. 2004; continental: Cubino et al. 2019). Using Socratic questions about each paper alone and then for the whole set, I help students develop their own mental models about how social, ecological and technological variables can be integrated to generate holistic understanding of urban SETS and their spatiotemporal dynamics (e.g., as in the lesson presented by Morin [2016]). Each of these papers can also be discussed in the context of ecosystem services and sustainability to link basic and applied concepts. For that, I have used a teaching activity that asks students to critically evaluate research and rhetoric about the effectiveness of green roofs to provide ecosystem services and improve the social-environmental quality of urban systems (Crispo 2016). Such lessons nicely illustrate the ecology-*of*-cities view while also providing a helpful bridge to the next preposition in the framework: ecological knowledge needed *for* improving the quality and sustainability of cities and people's lives.

Ecology *for* Cities

Reflecting broader trends in science, scholars have increasingly conducted studies directly relevant to decisions about how to better design, manage, and live in urban SETS. Such applied, use-inspired, actionable research and knowledge are encompassed by the recently described perspective of ecology *for* cities (e.g., Du Toit et al. 2014; Grove et al. 2016; Zhou et al. 2019). Integrating and extending the *in* and *of* views, ecology *for* cities is an explicitly goal-driven, interdisciplinary approach that adopts insights and methods from urban planning, engineering, governance, ethics, and other fields relevant to decision making and communicating with stakeholders (Childers et al. 2015; Grove et al. 2016; McPhearson et al. 2016; Pickett et al. 2016). This view recognizes that urban residents, policy makers, planners, managers and others can and should influence the types of questions that urban scientists investigate, inspiring them to more thoughtfully consider the types of information needed to address the practical, everyday challenges faced by urban communities (e.g., environmental justice, managing biodiversity, restoring ecosystem services, reducing pollution, mitigating effects of climate change). At the same time, the scientific integrity of such projects should be maintained with rigorous, objective methods and data analyses (Pickett et al. 2016). Though the phrase ecology *for* cities is relatively new, exemplary applied research and perspectives have a long history, providing plenty of material for urban ecology courses.

Ecology-*for*-cities lessons are valuable because they support broader educational goals such as developing skills for critical thinking, evaluating evidence, and communicating effectively, especially related to translating scientific ideas for general audiences. Focusing on evidence-based decision making in "real world" contexts lends itself well to helping students consider how science can be used *for* addressing urban social-environmental and sustainability issues. For example, as part of lawn-and-garden ecology lessons, I have introduced the topic of conservation biological control using data about how landscape management affects populations of predatory arthropods that consume harmful pests (which doubles as a lesson on urban food webs; e.g., Ellis et al. 2005; Kunkel, Held, and Potter 1999; Shrewsbury and Raupp 2006). Students are then asked to prepare a memo for campus landscape managers about actions to help conserve beneficial biodiversity, enhance the ecosystem service of pest consumption, and reduce pesticide use. Also related to biodiversity management, Hane and Korfmacher (2022a, 2022b) describe a pair of activities in which students actively engage in ecology-*for*-cities practices by designing "insect hotels" and campus outdoor spaces to achieve conservation and restoration goals.

Another exemplary topic related to *for* cities is urban greening programs, especially in the context of using basic ecological science for informing tree-planting decisions. An exemplary activity I have used begins with students examining data about the role of different tree species in releasing volatile organic compounds (VOCs) that contribute to formation of ground-level ozone pollution (Crumrine 2016). Then students read Churkina et al. (2015) to learn about tree species that form fewer or no VOCs and discuss how such information relates to the social and ecological trade-offs surrounding people's planting choices (e.g., some low-VOC-emitting trees may have negative effects on other variables; some high emitters are more suitable for certain locations). Similarly, students can be introduced to the growing body of research about where new trees should be planted to help reduce the urban heat island effect and its negative health outcomes (e.g., Grabmeier 2021). Such topics can help students develop deeper understanding of how the *in*, *of*, and *for* perspectives inform one another (e.g., the goals of cleaning and cooling cities' air has led to research about abiotic conditions, tree physiology and where to install new trees) by exemplifying how basic science is relevant to social concerns like environmental racism and justice (marginalized, vulnerable communities often have fewer trees and thus hotter conditions; e.g., see Popovich and Flavelle 2019 and the activity by Long 2016). An emerging tenet of such applied research—and the community-oriented projects it informs—is that working *with* stakeholders is integral to successfully identifying local problems and implementing solutions. This insight points to the fourth dimension of the urban ecology prepositional framework.

Ecology *with* Cities

Alongside increasing applied research *for* societal goals, another trend in science is to directly collaborate *with* people in the processes of data collection, learning, and evidence-based decision making. Such methods challenge scientists to consider different research and teaching methods because just generating information *for* people (and giving it to them) will not always catalyze the desired changes. For urban contexts, the phrase ecology *with* cities was proposed by Byrne (2022) to encompass the many ways of engaging and communicating with students, citizens, policy makers, managers, planners, and others (also see Grove et al. 2016). These ways include citizen/community science, participatory action research, stakeholder engagement, democratization of science, coproduction of knowledge, urban designed experiments, civic ecology, and translational ecology (see references in Byrne 2022). In education, pedagogies that have students work *with* information and one another to develop their own knowledge through active engagement (rather than passively listening to lectures) is known as learner-centered teaching (Byrne 2016); this approach reflects ecology *with* cities because it emphasizes the learner's role in collaborating *with* the teacher to achieve learning outcomes, including knowledge about how students can generate positive changes in the urban SETS in which they live and work.

Students can be introduced to the *with* perspective by examining and evaluating case studies of community gardening, tree planting and restoration projects that involve local people in their design and implementation (e.g., Moran, Perreault, and Smardon 2019; Sipilä and Tyrväinen 2005). Urban citizen science projects can be analyzed to learn about methods for engaging people in studying their local environments (e.g., Taylor et al. 2021). For deeper engagement, students can be given opportunities to collect data for existing projects (e.g., for birds: https://celebrateurbanbirds.org/cub/instructions; for fireflies: www.massaudubon.org/get-involved/community-science/firefly-watch) or partner with a local organization to complete a service-learning project (Tidball and Krasny 2011). In my class, some students helped a local conservation committee by investigating how communities around the United States reduced impacts on water quality from lawn fertilizer and making recommendations for effective actions. For another project, sustainability majors worked with campus landscape managers to research, design and build a nesting platform for osprey, a coastal bird-of-prey. In such projects, a key goal is for students to learn about the need to communicate and collaborate effectively with stakeholders *and* how to do so (see Chapters 5 and 18, this volume). This, combined with discussions of other ecology-*with*-cities examples, will help prepare them to become more effective urban stewards and leaders who engage with others to improve the sustainability of urban SETS.

Conclusion

Urban ecology's prepositional framework organizes the field's diverse interdisciplinary themes into a simple, accessible outline. Thus, it provides an effective approach to teaching and learning about the complex and often abstract nature of urban SETS because it can help students organize and synthesize their knowledge. Further, the four prepositional categories overlap and have feedback relationships. For example, data about ecology *in* cities informs ecology *of* and *for* cities;

goals developed from an ecology-*for*-cities view can influence *in* and *of* research agendas. This can help facilitate development of students' systems thinking skills for integrating insights from all four perspectives. In addition, an urban ecology course that is designed using the prepositional framework emphasizes the value of generating and using scientific evidence while working *with* diverse stakeholders to study and manage complex urban social-ecological challenges. To that end, I hope this chapter will inspire others to develop rich, learner-centered lessons that help students understand the many basic and applied dimensions of urban ecology and how to help transform urban SETS into healthier, more sustainable places for everyone.

References

Abu Baker, Mohammed A., Sara E. Emerson, and Joel S. Brown. 2015. "Studying Behavioral Ecology on High School & College Campuses: A Practical Guide to Measuring Foraging Behavior Using Urban Wildlife." *The American Biology Teacher* 77 (6): 432–438.

Alberti, Marina, John M. Marzluff, Eric Shulenberger, Gordon Bradley, Clare Ryan, and Craig Zumbrunnen. 2003. "Integrating Humans into Ecology: Opportunities and Challenges for Studying Urban Ecosystems." *BioScience* 53 (12): 1169–79.

Aloisio, Jason M., Brian Johnson, James D. Lewis, J. Alan Clark, Jason Munshi-South, Su-Jen Roberts, Deborah Wasserman, Joseph Heimlich, and Karen Tingley. 2018. "Pre-college Urban Ecology Research Mentoring: Promoting Broader Participation in the Field of Ecology for an Urban Future." *Journal of Urban Ecology* 4 (1): juy023.

Aronson, Myla F. J., Frank A. La Sorte, Charles H. Nilon, Madhusudan Katti, Mark A. Goddard, Christopher A. Lepczyk, Paige S. Warren, et al. 2014. "A Global Analysis of The Impacts of Urbanization on Bird and Plant Diversity Reveals Key Anthropogenic Drivers." *Proceedings of the Royal Society B* 281 (1780): 20133330.

Baker, Lawrence A., Diane Hope, Ying Xu, Jennifer Edmonds, and Lisa Lauver. 2001. "Nitrogen Balance for the Central Arizona—Phoenix (Cap) Ecosystem." *Ecosystems* 4 (6): 582–602.

Barnett, Michael, Meredith Houle Vaughn, Eric Strauss, and Lindsey Cotter. 2011. "Urban Environmental Education: Leveraging Technology and Ecology to Engage Students in Studying the Environment." *International Research in Geographical and Environmental Education* 20 (3): 199–214.

Berkowitz, Alan R., Charles H. Nilon, and Karen S. Hollweg, eds. 2003. *Understanding Urban Ecosystems: A New Frontier for Science and Education*. New York: Springer Science & Business Media.

Biberhofer, Petra, and Christian Rammel. 2017. "Transdisciplinary Learning and Teaching as Answers to Urban Sustainability Challenges." *International Journal of Sustainability in Higher Education* 18 (1): 63–83.

BoredPanda. 2015. "59 Unusual Bird Nests Built in The Weirdest Places." Accessed July 9, 2021. www.boredpanda.com/unusual-bird-nests/.

Borghi, Anna M., Laura Barca, Ferdinand Binkofski, and Luca Tummolini. 2018. "Varieties of Abstract Concepts: Development, Use and Representation in the Brain." *Philosophical Transactions of the Royal Society B* 373 (1752): 20170121.

Buttigieg, Melissa. 2019. "'It Will Haunt Me': Sad Story Behind Photo of Dead Koala 'Cradling' Her Joey." Accessed July 9, 2021. https://au.news.yahoo.com/sunshine-coast-koala-cradling-joey-road-death-093616940.html.

Byrne, Loren B. 2015. "Exploring Lawns and Gardens as Complex Socio-Ecological Systems." Accessed April 20, 2022.http://faculty.rwu.edu/lbyrne/papers/Byrne_STIRS_Student_Case.docx.
———. 2016. "Learner-Centered Teaching for Environmental and Sustainability Studies." In *Learner-Centered Teaching Activities for Environmental and Sustainability Studies*, edited by Loren B. Byrne, 1–28. New York: Springer.
———. 2022. "Ecology with Cities." *Urban Ecosystems* 25 (3): 835–837.
———. In press. "Synthesizing Science." In *The Pocket Instructor: Writing*, edited by A. I. Wilkins and K. Shaw. Princeton: Princeton University Press.
Byrne, Loren B., Mary Ann Bruns, and Ke Chung Kim. 2008. "Ecosystem Properties of Urban Land Covers at the Aboveground—Belowground Interface." *Ecosystems* 11 (7): 1065–77.
Byrne, Loren B., and Parwinder Grewal. 2008. "Introduction to Ecological Landscaping: A Holistic Description and Framework to Guide the Study and Management of Urban Landscape Parcels." *Cities and the Environment* 1 (2): 20.
Childers, Daniel L., Mary L. Cadenasso, J. Morgan Grove, Victoria Marshall, Brian McGrath, and Steward T. A. Pickett. 2015. "An Ecology for Cities: A Transformational Nexus of Design and Ecology to Advance Climate Change Resilience and Urban Sustainability." *Sustainability* 7 (4): 3774–91.
Churkina, Galina, Rüdiger Grote, Tim M. Butler, and M. Lawrence. 2015. "Natural Selection? Picking the Right Trees for Urban Greening." *Environmental Science & Policy* 47: 12–17.
Crispo, E. 2016. "Evaluating the Effectiveness of Green Roofs: A Case Study for Literature Research and Critical Thinking." In *Learner-Centered Teaching Activities for Environmental and Sustainability Studies*, edited by Loren Byrne, 253–59. New York: Springer.
Crumrine, P. W. 2016. "Don't Blame the Trees: Using Data to Examine How Trees Contribute to Air Pollution." In *Learner-Centered Teaching Activities for Environmental and Sustainability Studies*, edited by Loren Byrne, 245–50. New York: Springer.
Cubino, Josep Padullés, Jeannine Cavender-Bares, Sarah E. Hobbie, Diane E. Pataki, Meghan L. Avolio, Lindsay E. Darling, Kelli L. Larson et al. 2019. "Drivers of Plant Species Richness and Phylogenetic Composition in Urban Yards at the Continental Scale." *Landscape Ecology* 34 (1): 63–77.
Desai, Rutvik H., Megan Reilly, and Wessel van Dam. 2018. "The Multifaceted Abstract Brain." *Philosophical Transactions of the Royal Society B* 373 (1752): 20170122.
Du Toit, Marie, Juanee Cilliers, Francois Retief, and Ernst Drewes. 2014. "Actionable Urban Ecology in China and the World: Integrating Ecology and Planning for Sustainable Cities." *Landscape and Urban Planning* 125: 207–8.
Ellis, Jodie A., A. D. Walter, John F. Tooker, Matthew D. Ginzel, Peter F. Reagel, Emerson S. Lacey, Ashley B. Bennett, Erik M. Grossman, and Lawrence M. Hanks. 2005. "Conservation Biological Control in Urban Landscapes: Manipulating Parasitoids of Bagworm (Lepidoptera: Psychidae) with Flowering Forbs." *Biological Control* 34 (1): 99–107.
Fissore, C., L. A. Baker, S. E. Hobbie, J. Y. King, J. P. McFadden, K. C. Nelson, and I. Jakobsdottir. 2011. "Carbon, Nitrogen, and Phosphorus Fluxes in Household Ecosystems in the Minneapolis-Saint Paul, Minnesota, Urban Region." *Ecological Applications* 21 (3): 619–39.
Grabmeier, Jeff. 2021. "Throwing Shade: Measuring How Much Trees, Buildings Cool Cities." Accessed July 9, 2021. https://news.osu.edu/throwing-shade-measuring-how-much-trees-buildings-cool-cities/.
Grimm, Nancy B., Stanley H. Faeth, Nancy E. Golubiewski, Charles L. Redman, Jianguo Wu, Xuemei Bai, and John M. Briggs. 2008. "Global Change and the Ecology of Cities." *Science* 319 (5864): 756–60.

Grimm, Nancy B., J. Grove, Steward T. A. Pickett, and Charles L. Redman. 2000. "Integrated Approaches to Long-Term Studies of Urban Ecological Systems." *BioScience* 50 (7): 571–84.

Groffman, Peter M., Charles T. Driscoll, Gene E. Likens, Timothy J. Fahey, Richard T. Holmes, Christopher Eagar, and John D. Aber. 2004. "Nor Gloom of Night: A New Conceptual Model for the Hubbard Brook Ecosystem Study." *BioScience* 54 (2): 139–48.

Groffman, Peter M., Neely L. Law, Kenneth T. Belt, Lawrence E. Band, and Gary T. Fisher. 2004. "Nitrogen Fluxes and Retention in Urban Watershed Ecosystems." *Ecosystems* 7 (4): 393–403.

Grove, J. Morgan, Daniel L. Childers, Michael Galvin, Sarah Hines, Tischa Muñoz-Erickson, and Erika S. Svendsen. 2016. "Linking Science and Decision Making to Promote an Ecology for the City: Practices and Opportunities." *Ecosystem Health and Sustainability* 2 (9): e01239.

Grove, J. Morgan, Dexter H. Locke, and Jarlath PM O'Neil-Dunne. 2014. "An Ecology of Prestige in New York City: Examining the Relationships among Population Density, Socio-Economic Status, Group Identity, and Residential Canopy Cover." *Environmental Management* 54 (3): 402–19.

Hane, Elizabeth, and Karl Korfmacher. 2022a. "Insect 'Bee&Bees' and Pollinator Penthouses: Teaching Students about Pollinators and Their Services in an Urban Environment." *Urban Ecosystems*.

———. 2022b. "Engaging Students in Redesigning a Local Urban Space to Improve Ecosystem Services." *Urban Ecosystems* 25 (3): 719–724. https://doi.org/10.1007/s11252-021-01184-6.

Held, Amy. 2019. "German Firefighters Work to Free Rotund Rat Stuck in Manhole Cover." Accessed July 9, 2021. www.npr.org/2019/02/27/698547607/watch-german-firefighters-work-to-free-rotund-rat-stuck-in-manhole-cover.

Kaushal, Sujay S., William H. McDowell, and Wilfred M. Wollheim. 2014. "Tracking Evolution of Urban Biogeochemical Cycles: Past, Present, and Future." *Biogeochemistry* 121 (1): 1–21.

Kunkel, Brian A., David W. Held, and Daniel A. Potter. 1999. "Impact of Halofenozide, Imidacloprid, and Bendiocarb on Beneficial Invertebrates and Predatory Activity in Turfgrass." *Journal of Economic Entomology* 92 (4): 922–30.

Leong, Misha, Robert R. Dunn, and Michelle D. Trautwein. 2018. "Biodiversity and Socio-economics in the City: A Review of the Luxury Effect." *Biology Letters* 14 (5): 20180082.

Long, Joshua. 2016. "Helping Students Envision Justice in the Sustainable City." In *Learner-Centered Teaching Activities for Environmental and Sustainability Studies*, edited by Loren B. Byrne, 227–32. New York: Springer.

Loss, Scott R., Tom Will, Travis Longcore, and Peter P. Marra. 2018. "Responding to Misinformation and Criticisms Regarding United States Cat Predation Estimates." *Biological Invasions* 20 (12): 3385–96.

McPhearson, Timon, Steward T. A. Pickett, Nancy B. Grimm, Jari Niemelä, Marina Alberti, Thomas Elmqvist, Christiane Weber, Dagmar Haase, Jürgen Breuste, and Salman Qureshi. 2016. "Advancing Urban Ecology Toward a Science of Cities." *BioScience* 66 (3): 198–212.

Moore, Jennie, Meidad Kissinger, and William E. Rees. 2013. "An Urban Metabolism and Ecological Footprint Assessment of Metro Vancouver." *Journal of Environmental Management* 124: 51–61.

Moran, Sharon, Meredith Perreault, and Richard Smardon. 2019. "Finding Our Way: A Case Study of Urban Waterway Restoration and Participatory Process." *Landscape and Urban Planning* 191: 102982.

Morin, Erica A. 2016. "Location, Location, Location! Analyzing Residential Development in Environmentally-Fragile Areas." In *Learner-Centered Teaching Activities for Environmental and Sustainability Studies*, edited by Loren B. Byrne, 155–62. New York: Springer.

Nassauer, Joan Iverson. 1988. "The Aesthetics of Horticulture: Neatness as a form of Care." *HortScience* 23 (6): 973–77.

Oliveira, Sofia S., Joana Pereira, Paulo Santos, and Ruth Pereira. 2021. "Assessing Soil Quality in Schoolyards: A Hands-On Activity for Middle School Students." *The American Biology Teacher* 83 (5): 331–36.

Pickett, Steward T. A, William R. Burch Jr, Shawn E. Dalton, and Timothy W. Foresman. 1997. "Integrated Urban Ecosystem Research." *Urban Ecosystems* 1 (4): 183–84.

Pickett, Steward T. A., and Mary L. Cadenasso. 2006. "Advancing Urban Ecological Studies: Frameworks, Concepts, and Results from the Baltimore Ecosystem Study." *Austral Ecology* 31 (2): 114–25.

Pickett, Steward TA, Mary L. Cadenasso, Daniel L. Childers, Mark J. McDonnell, and Weiqi Zhou. 2016. "Evolution and Future of Urban Ecological Science: Ecology in, of, and for the City." *Ecosystem Health and Sustainability* 2 (7): e01229.

Pickett, Steward T. A., Mary L. Cadenasso, J. Morgan Grove, Charles H. Nilon, Richard V. Pouyat, Wayne C. Zipperer, and Robert Costanza. 2001. "Urban Ecological Systems: Linking Terrestrial Ecological, Physical, and Socioeconomic Components of Metropolitan Areas." *Annual Review of Ecology and Systematics* 32 (1): 127–57.

Popovich, Nadja, and Christopher Flavelle. 2019. "Summer in the City Is Hot, but Some Neighborhoods Suffer More." *New York Times*, August 9, 2019. www.nytimes.com/interactive/2019/08/09/climate/city-heat-islands.html.

Rega-Brodsky, Christine C., and Charles H. Nilon. 2016. "Vacant Lots as a Habitat Resource: Nesting Success and Body Condition of Songbirds." *Ecosphere* 7 (11): e01578.

Russ, Alex, ed. 2015. *Urban Environmental Education*. Ithaca, NY and Washington, DC: Cornell University Civic Ecology Lab, NAAEE and EECapacity.

Scherer, Hannah H., Lauren Holder, and Bruce Herbert. 2017. "Student Learning of Complex Earth Systems: Conceptual Frameworks of Earth Systems and Instructional Design." *Journal of Geoscience Education* 65 (4): 473–89.

Schmitt-Harsh, Mikaela, and Joseph A. Harsh. 2017. "Engaging Nonscience Majors in Urban Ecology: Recommendations for Course Design." *Journal of Environmental Studies and Sciences* 7 (4): 550–61.

Shrewsbury, Paula M., and Michael J. Raupp. 2006. "Do Top-Down or Bottom-Up Forces Determine Stephanitis Pyrioides Abundance in Urban Landscapes?" *Ecological Applications* 16 (1): 262–72.

Siossian, Emma. 2018. "Knock, Knock, There's A Koala at the Door: It's Breeding Season and the Marsupials Are on the Move." Accessed July 9, 2021. www.abc.net.au/news/2018-07-09/koala-an-unexpected-visitor-at-the-front-door/9931558

Sipilä, Maija, and Liisa Tyrväinen. 2005. "Evaluation of Collaborative Urban Forest Planning in Helsinki, Finland." *Urban Forestry & Urban Greening* 4 (1): 1–12.

Tan, Puay Yok, Jingyuan Zhang, Mahyar Masoudi, Jahson Berhane Alemu, Peter J. Edwards, Adrienne Grêt-Regamey, Daniel R. Richards, et al. 2020. "A Conceptual Framework to Untangle the Concept of Urban Ecosystem Services." *Landscape and Urban Planning* 200: 103837.

Taylor, Mark Patrick, Cynthia F. Isley, Kara L. Fry, Xiaochi Liu, Max M. Gillings, Marek Rouillon, Neda S. Soltani, Damian B. Gore, and Gabriel M. Filippelli. 2021. "A Citizen Science Approach to Identifying Trace Metal Contamination Risks in Urban Gardens." *Environment International* 155: 106582.

Tidball, Keith G., and Marianne E. Krasny. 2011.2010. "Urban Environmental Education from a Social-Ecological Perspective: Conceptual Framework for Civic Ecology Education." *Cities and the Environment* 3 (1): Article 11.

Wald, Navé, and Ben Kei Daniel. 2020. "Enhancing Students' Engagement with Abstract Ideas Through Conceptual and Theoretical Frameworks." *Innovations in Education and Teaching International* 57 (4): 496–505.

Zhou, Weiqi, Brendan Fisher, and Steward T. A. Pickett. 2019. "Cities Are Hungry for Actionable Ecological Knowledge." *Frontiers in Ecology and the Environment* 17: 135.

Zhou, Weiqi, S. T. A. Pickett, and Timon McPhearson. 2021. "Conceptual Frameworks Facilitate Integration for Transdisciplinary Urban Science." *Urban Sustainability* 1 (1): 1–11.

Ziter, Carly. 2016. "The Biodiversity—Ecosystem Service Relationship in Urban Areas: A Quantitative Review." *Oikos* 125 (6): 761–68.

20

MANITOBA HYDRO'S MEGA-DAMS AND THE ETHICAL TEACHING OF CERAMICS AND SUSTAINABILITY

Mary Steggles

Introduction

For one month, ten individuals met online every Tuesday from 19:00 to 20:30 (7:00–8:30 p.m.). Each signed up to be part of a discussion group about Manitoba Hydro, ceramics practice and sustainability, and the social injustice of electricity. The goal of the workshop was threefold: to educate the participants on the impact on humans and nonhumans of building hydroelectric dams in Northern Manitoba; to evaluate electric kilns in terms of their green and clean reputation; and for our participants to determine ways to green their ceramic practice.

Rather than screen the applications for an even distribution of beginning, intermediate, and advanced ceramic practitioners, the first ten applications would be accepted, including nonceramists with a keen interest in the subject. The Zoom workshop was promoted on the Manitoba Craft Council website beginning August 23, 2020, and by word of mouth. The workshop registration was closed on August 27, 2020. Of the ten participants, eight were female and two were male. The age range was 23–70 years. One female was a fourth-year honors student studying ceramics. Three are currently MFA students, two females and one male (two at the University of Manitoba and one at the University of Alberta, Edmonton). One is studying for her PhD in Visual Art at Western University in London, Ontario. Except for a female social worker working with the homeless Indigenous population in the inner city and a male teacher in an inner-city school with mostly Indigenous students, all others were practicing artists with undergraduate degrees in ceramics. Each participant lives in an urban environment as opposed to a rural one. Eight were from Manitoba, one lives in Alberta, and the other lives in Ontario. Three of the participants self-identify as Indigenous.

This Zoom workshop replaced a special topics course offered for credit at the School of Art, University of Manitoba. Due to my retirement and the

DOI: 10.4324/9781003217442-23

COVID pandemic, aspects of the original course had to be altered or dropped altogether. Individuals could not meet in person because of pandemic regulations in Manitoba. Additionally, we were unable to have a group firing and exhibition. These changes did not seem to alter the outcome of the original goals. Instead of being offered by an accredited research university, the workshop was cosponsored by myself and the Manitoba Craft Council (MCC). The workshop was free of charge.

The MCC agreed to provide the promotional materials and accept registrations. The Director of the Manitoba Craft Council, Tammy Sutherland, assisted with the selection of appropriate speakers and provided the funds for their speaker fees. In addition, the MCC provided the Zoom platform for the workshop, and the director of the MCC acted as tech support and cohost for each evening.

During the first evening, all participants introduced themselves. After introductions, members ranked the cleanest fuel for firing kilns out of a choice of five fuels. We discussed the rationale for the ratings. Afterward, we discussed readings that were sent to the group one week prior. The second evening featured Indigenous artist K.C. Adams. Adams talked about her practice, her desire to learn to make traditional Cree pottery, and the impact that traveling to Northern Manitoba had on her work. The third week included a presentation by Will Braun. Braun works for the Interfaith Task Force on Northern Hydro Development and the Manitoba Justice Coalition. Braun gave a presentation on Manitoba's northern dams, focusing on the impact they had on three distinct families. He also addressed the facts in terms of what the damage was to both the environment and the people. Week four was an opportunity for the participants to discuss the presentations and how this new knowledge might impact their practice.

A Brief Introduction to Clay and the Fuel Used to Fire Kilns

Clay is a naturally occurring material found all over the world. Historically, potters dug their clays, removed stones and twigs, and allowed the clay to dry before pounding it into a fine powder. Water was added to the fine clay powder until it was of a consistency that could be used. These naturally occurring clays contain mineral oxides that give them their distinctive color. The hues range from dark chocolates to beautiful reds, yellows, and buff. Porcelain has no mineral oxides and is pure white, often achieving translucency when fired to very high temperatures, such as 1,350°C (Celsius).

In the late 20th century, pottery supply houses began to create commercial clay blends. They added naturally occurring materials to the various clays they extracted to make them easier to form, creating blends better suited for wheel throwing, hand building, and significant architectural work. Mineral oxides create a wide range of colors, while aggregates like grog mean less shrinkage and warpage. The ingredients are mined then sent to a factory where the various materials are blended. They are then packaged and shipped out to clients or pottery supply houses in 20-kilogram or 50-pound cartons.

Whether one digs and processes clay or uses a commercially made product, the clay cannot be reused if it is fired to 573°C. At that point, called quartz conversion, the clay is hard. It will hold water. As a fired object or vessel, it will exist for millennia; even when broken, the shards remain. As an instructor, I cannot emphasize the permanent nature of fired clay enough. Indeed, a sessional instructor declaring that they would fire everything a student made first prompted me to consider the environmental impact that ceramics has on the biosphere.

Ceramic projects are typically fired twice. The first firing is the bisque. At this stage, the clay objects are completely dry. They are baked to no higher than 1060°C. This initial heating converts the unfired clay or greenware into a vessel or object that can hold water. At this point, the clay will no longer return to a state of mud or earth. It is now called bisqueware. The second firing is typically associated with the application of glaze to the surface. A glaze is a suspended glass coating that renders the work food safe and waterproof. Glazes can be transparent, semitransparent, or opaque and contain metal oxides that add color to the surface. Ware is fired at temperatures from 990–1,350°C to achieve this state. The length of this second firing depends entirely on the glaze, the type of kiln used, its efficiency, and the aesthetic desired by the potter. There are specialist applications, including liquid metals, such as gold, that require a third low-temperature firing, 700°C. The fuels used by potters to heat ceramic ware in kilns include electricity, natural gas, liquid propane gas (LPG), coal, animal dung, sawdust, fuel oil, used cooking oil, and wood. Each type of fuel used to heat these kilns has a distinct environmental footprint.

Electric Kilns: Are They Clean and Green?

The focus of this study in sustainability is on the use of electric kilns only. When asked, most potters are quick to say that electric kilns provide clean energy. In a survey conducted over the summer of 2019, 100 individuals working with clay were asked to respond to a single question: Which of the following fuels is the cleanest and greenest source of energy for firing a kiln? The six possible answers were natural gas, wood, liquid petroleum gas (LPG), electricity, fuel oil, or waste oil (such as cooking oil from restaurants). There were 89 responses to the survey. Of those, 87 stated that electricity provided the most environmentally friendly fuel for kilns. The other two said that wood was the best option.

Like the individuals in my survey, eight workshop participants at the beginning of the month-long webinar stated that they believed electricity was a cleaner energy source to fire their kilns. One said wood, and another, the only nonceramist, did not have an opinion.

None of the discussion group members were familiar with the history of the northern dams, nor did they understand the impact of megawatt dams on the local environment. Participants read materials posted online to prepare them for the presentations during weeks two and three. Those materials contained promotional information on the merits of hydroelectricity circulated by Manitoba Hydro and news articles about the impact of the dams on northern residents.

Manitoba's Northern Hydroelectric Dams and Their Impact on Northern Communities

In Manitoba, a Crown-owned public utility produces hydropower for domestic consumption and export. In other words, the energy produced at the mega-dams is for the Southern white towns and cities or for export. This hydroelectric energy originates in the waters of northern Manitoba. It is then carried south to markets beyond provincial boundaries via a vast and intricate transmission network. In the 1960s, the provincial and federal governments built generating stations, powerhouse structures, control dams, and transmission lines on the Churchill and Nelson Rivers and their diversions to produce the electricity Manitobans use and sell. Manitoba Hydro is currently building a new dam 725 kilometers northeast of Winnipeg in the Split Lake Resource Management Area. The Keeyask Generating Station is a 695-megawatt generating station. The current cost of constructing the dam and diverting the river systems is $8.3 billion CAD.

The damming of the rivers in Treaty 5 Territory caused flooding on an unparalleled scale to the homes, the traditional hunting grounds, and the burial sites of Indigenous people. Extensive documentation details the social ramifications of these actions. Ramona Neckoway (2018), a member of the Nisichawayasihk Cree nation impacted by the mega-dam states:

> Manitoba Hydro's vast and impressive network, including the labyrinth of transmission lines, affected and continue to impact entire generations of indigenous peoples in Manitoba. My grandparents' generation, my parents' generation, my generation, my children's generation and their children's generation, have borne witness to and experienced a kind of cultural genocide resulting from Hydro's generation in Manitoba; these experience to varying degrees and severity are akin, in some ways, to the residential school era that devastated many Aboriginal communities, children, parents and grandparents.

The impact of these mega-dam projects is not limited to the province of Manitoba. In November 2019, people from around the world met in Winnipeg, Manitoba, for a conference organized by the Wa Ni Tan, a group trying to stop the building of mega-dams around the world. People came from as far away as Brazil and Panama to mobilize against these hydroelectric projects' social and environmental damage. Underlying their concern were the pronounced changes to their communities since the construction of the dams. According to James Wilt (2019), these include "significant social disorder, the abuse of drugs and alcohol, racial discrimination and the destruction of traditional practices of hunting, trapping, fishing, and gathering." The people who drink the water from the river also eat the fish they caught or the game they killed that relied on the river. Today, all animals, fish, and people have been poisoned by methyl mercury developed in the reservoirs upstream. The result of the mercury poisoning is that many indigenous people living near the mega-dams have had to abandon their fisheries and their traditional diets resulting in an elevated increase of diabetes and other ailments (Austen 2016).

Due to a lack of medical professionals in the North, many of the residents who have developed diabetes or other issues related to the toxins in the land, the water, the fish, and the animals must leave their homes and come to the larger urban centers in the south of the province for treatment. The health problems caused by the mega-dams have a direct impact on the healthcare system in the South, often putting pressure on already crowded hospitals and overworked medical staff and social workers. Many of the patients will spend months in the urban centers getting treatment. The cost to treat, house, and transport the individuals has meant that the taxes of urban dwellers are raised. The care of the Indigenous people of the North often generates resentments between their community and the mostly white Southern settlers whose taxes fund the hospitals within Winnipeg. With the loss of their traditional livelihoods and self-esteem, the men of the Northern communities left their homes to find work in the urban areas. Some succeeded, but most did not. They slipped into a life of desperation, homelessness, and addiction. This lifestyle, in turn, places a heavy burden on social services, not-for-profits helping addicts, the police services, and charities that provide meals and beds for the homeless. Additionally, the violent crime rate has increased in the core urban areas where the Indigenous population lives. The level of crime and people sleeping or taking drugs have created a fear among the non-Indigenous community. That fear has had a direct negative impact on urban development within our city. People and businesses have moved out of the core area of our city causing huge costs in infrastructure and the acquisition of more and more land, which had been used for agriculture, for housing developments.

The Presenters

K.C. Adams is a Winnipeg-based Cree, Ojibway, and British multimedia artist. Adams completed a degree in studio art from Concordia University in Montreal. On her return to Winnipeg, Adams wrestled with her identity as an Indigenous person and the Western techniques she learned during her time at Concordia. In many ways, Adams's return to her home and the land of her ancestors led her to a reawakening of her Indigenous identity in her art practice. Adams spoke about the impact the first of many trips to Northern Manitoba had on her as a person and an artist. She said the first thing she noticed when entering the community of South Indian Lake was the large black-and-white billboards showing portraits of four missing young men. She learned that the destruction of traditional ways of living, by the work on the Churchill and Nelson River dam projects, was the cause of large numbers of suicides and addictions. Other significant issues impacted all the people. One of the biggest was that the hunting and fishing rights that family members had held for generations were taken away. Their water was no longer fit for consumption. Traditional rites of passage, such as the fathers taking their newborn children in the canoe to introduce them to the land of their ancestors, no longer held any meaning. The presence of the workmen from Southern Manitoba and their camps was a source of fear and anxiety. The number of rapes of young women was high. Adams returned many times. Her northern experiences led her to focus her art on the ancestors' knowledge, traditional teachings, language, mythology, and sacred materials.

In 2017, Adams began researching traditional pottery techniques of the Indigenous people of the Lake Agassiz region. She has an extensive collection of shards, and these were, in part, inspiration for an exhibition titled "Shards." Adams also spent many hours in the Manitoba Museums collection. She wanted to recreate the style of the traditional pottery as well as the firing techniques these people used. Adams began to harvest clay from areas around Winnipeg. After collecting the clay, Adams mixes it with sand and crushed granite that she has saved from sacred ceremonies. In this way, spirits empower the clay. Adams makes her tools from natural materials. Each piece is made by hand-used traditional coil and paddle methods. The pieces are then fired in a pit for only 1 hour. When cooled, the vessels can be used for cooking on a pit fire, just as Adams's ancestors had done. Adams is ever mindful that pottery contains the culture, the spirit, the land, and the water of these ancient people, her ancestors. She believes that making a connection to traditional knowledge can be healing for the spirit.

The second presenter was Will Braun, a social activist, a member of the Interchurch Council on HydroPower, and an advocate for several of the bands living in Northern Manitoba who deal with Manitoba Hydro. The Government of Canada uses the term "band" to refer to First Nations Communities. Bands function as municipalities with band leaders and band members. Braun introduced the seminar members to the history of dam building and flooding in Northern Manitoba, as well as its negative impact on the people, the land, and their traditional ways of living. It was clear from Braun's presentation that the mega-dams exacerbated the social injustices of the Indigenous people. Braun also explained how the dams' physical presence undermined the men's ability to perform their traditional roles as providers for their families. Braun showed images of Manitoba Hydro clearcutting around the transmission lines to carry electricity to Southern Manitoba as well as the 836.9 square kilometers of boreal forest lost to flooding at the southern tip of South Indian Lake. A little known fact is that while huge areas were flooded, other water resources that the people depended on were drained. That is, they diminished in area by 39%–76%. The result was devastating to the people. Many lost their inherited trap lines and fishing licenses. The water was clean and clear before the dams. Now it is so laden with mercury that the people cannot drink it or eat the fish living in it. They were told it would take 20 years for the mercury to dissipate. The flooding of the burial grounds of the people brought great emotion and sadness to the community while also increasing their anger at how Manitoba Hydro, a Crown corporation, had negotiated with the communities that would be impacted. Braun compared the earlier agreements at the Churchill-Nelson Dams with the new Keeysak Dam, the last one to be built by Manitoba Hydro. The demand for electricity south of the border has diminished with the new dam.

In contrast to earlier negotiations, where the communities were divided with each arguing for their own interests and not as a group, the leaders and band members whose land was impacted by the Keeysak Dam negotiated with one voice. They insisted on equal partnership and ownership status with Manitoba Hydro. Their demands also included consultations on every aspect of the project, proper environmental studies, and impact reviews. Agreements with Manitoba Hydro and the Province of Manitoba were undertaken for compensation should there be any adverse effects on the communities. The result was much more positive in terms of social, economic, and cultural protections than the initial agreements with the South Indian Lake and Cross Lake people. There, Manitoba Hydro attempted to divide and conquer the communities to get access to the land.

Both of the two presenters brought home the impact of the loss of traditional ways of living on the people. The youth left for Winnipeg because there was nothing for them in the North. This displacement has resulted in persistent homelessness, self-harm, addiction, and violence.

Participant Projects

Michelle Wilson was one of the Zoom Webinar participants. Wilson is a Canadian visual artist currently living in London, Ontario, and a PhD student at the University of Western Ontario (UWO). She creates art that contributes to those relationships in a reciprocal way and prioritizes that relationship over the demands of the gallery. In everything she does, Wilson contemplates the legacy that she leaves as an artist. This workshop led Wilson to reject the use of kilns to fire the clay in her project known as Cleansing the Land, which she created as part of her PhD projects.

Wilson wants to locate herself and her work through a complete understanding of place, not as an Indigenous person but as a settler attempting to naturalize herself by "an understanding, embodiment, science, history, culture, and humility." Wilson understands how important the land and the water are for the Indigenous people of the North. She recalls the ceremony that Will Braun talked about during his presentation to the group—the rite of passage in Northern Manitoba in which the father takes his newborn child in a canoe down the river to show it the land where they were born. It is a place where the Cree have, for generations, hunted and gathered, fished the clean rivers, and lived peacefully. The taking of the land and transforming it with industry was part of the progress of humankind in the minds of colonial powers. Wilson recognizes that humans have transformed the land on which we live to a point where we are all nearing what some call the Sixth Extinction. As an artist, a mother, and an environmentalist, Wilson wants to regenerate the land to the point where it was before colonization. She wants to form a bond with the land where she lives.

Wilson's project focuses on an area of London, Ontario, designated an Environmentally Significant Area or a toxic wasteland. For Wilson, the spot shows promise because of flora and fauna reclaiming the space. At the outset of the project, the artist collected blackberries from the fence line around the old paint factory. Her interest in science and the environment led her to understand that the blackberries take up lead from the soil in a process known as phytoremediation. She gathered samples and had them tested. The results indicated high concentrations of lead; one reading was 1.4 mg/kg. The acceptable level is 0.2 mg/kg. Wilson is collecting more and more of the poisonous berries. She will eventually turn these into a dye to make a textile map of the human and animal trails through the area. She will also obtain the GPS coordinates of all the paths by having groups explore using apps on their phone to map their walks. Already she has begun to mentor three undergraduate students by guiding them through the coves to help with the project and inspire them to create their environmental pieces about the land.

After participating in our online webinar on the impact of Manitoba Hydro's mega-dams on the northern communities of Manitoba, Wilson embraced the notion that ceramics do not have to be fired to have an impact. She used clay that she dug from around the old paint factory to form vessels. She impregnated the clay with the seeds of cattails and native grasses. After they dried, the artist carefully transported them to the site to find the clay and the poisoned berries. "I will return the whole piece to the earth, allowing the clay itself to be recycled for future generations." For Wilson, returning the landscape into the pristine site before it was contaminated by humans is the least she feels she can do as an artist and a human. She hopes the plants will filter the water and the soil to create a healthy environment for future generations to enjoy eventually.

Another workshop participant who was moved by the discussion was Karen Lee Edwards. Edwards moved from the Canadian Prairies, after graduating from the University of Manitoba's School of Art in 2013, to British Columbia. She settled in a small rural desert town on the border between Canada and the United States. After setting up her studio, she was determined to lower the environmental impact of her ceramic work. As an established artist, Edwards knew that the mining of clay and glazes was as problematic as the energy required to turn soft clay into a lasting work of sculpture. To reduce her environmental footprint, Edwards worked at developing a protocol for single firing to 2,336°Fahrenheit (F), while, at the same time, staying true to her desired aesthetic. This new method eliminates the initial bisque firing altogether. At the time, Edwards also began to develop methods for surface decoration that included stains and engobes rather than glazes. These methods allowed her to reduce her energy consumption substantially.

Edwards joined the Zoom webinar to find out other ways of approaching sustainability in ceramics and connecting with other artists to share ideas. She plans to rework many inherited porcelain objects and vessels, along with some functional stoneware and earthenware pieces. Whether or not she will grind the objects to a powder and mix them with her clay or break them up, creating sculptures and wall pieces, is unknown at this time.

During the webinar, Edwards asked one of the visiting speakers, Winnipeg artist K.C. Adams, if she might consider giving a workshop on using natural clays from Treaty 1 Territory and firing them with wood. The term "Treaty Territory" refers to a series of agreements between the Crown, the Government of Canada, and the First Nations of Canada between 1871 and 1921. There were 11 treaties that continue to define the relationship between the Indigenous people and the Government of Canada. These are highly contested and, at the time, allowed for white settlers to take the land of the Indigenous population. Adams agreed to give a workshop for anyone participating in the webinar during the summer of 2021, providing COVID restrictions allowed personal meetings. Adams was excited and said she would instruct them in her fast single firing method using wood. Edwards is also hopeful that Adams will discuss how she incorporates aspects of her Indigenous rituals and beliefs in her ceramic work. The workshop was postponed in 2021 due to COVID restrictions. It will be rescheduled for some time late in 2022.

The participants in the Zoom webinar deemed it to be successful in terms of raising their environmental awareness. Each stated that they were moved to change their practice after learning of the impact that the energy created for them to use had devastated so many Northern people. In addition to two of the participants immediately setting out to lower their use of hydroelectricity, another student is planning on having a community pit firing using local clay. Others discussed using materials for their projects other than clay if they could, while many talked about how to incorporate family heirloom china into their work. The middle school teachers plan to do pit firings with their students as well as taking them on field trips to visit the shard collections at the Manitoba Museum. In addition, the Manitoba Crafts Council has offered six opportunities for funding to local craftspeople seeking to address environmental issues in their work. The project would see the artist being partnered with a scientist and an end-of-program report or exhibition. Clearly, the need to make ceramics more sustainable and to understand the impact that our use of public utilities has on those people in Northern Manitoba has found resonance in the lives of the members of the seminar.

References

Austen, Ian. 2016. "Canada's Big Dams Generate Clean Energy and High Amounts of Mercury." *The New York Times*, November 10, 2016. https://nytimescom/2016/11/11/world/canada/clean-energy-dirty-water-canadas-hydroelectric-dams-have-a-mercury-problem.html.

Neckoway, Ramona. 2018. "A Critical Perspective on Integrated Hydropower in Manitoba: Considering the Manitoba-Minnesota Transmission Project. A Report prepared for Wa Ni Ska Tan: An Alliance of Hydro Impacted Communities." May 4, 2018. https://hydroimpacted.ca/wp-content/uploads/2018/07/A91743-9-WNST-MMTP-Ramona-Neckoway-A-CriticalPerspective-on-Hydro-Production-A6E1C1–1.pdf.

Wilt, James. 2019. "Projects of Death. Impact of Hydro Dams on Environment, Indigenous Communities Highlighted at Winnipeg Conference." *The Narwhal*, November 20, 2019. https://thenarwhal.ca/projects-of-death-impact-of-hydro-dams-on-environment-indigenous-communities-highlighted-at-winnipeg-conference.

21

RESILIENCE FOR SUSTAINABILITY EVERYWHERE

Critical Literacy for Students and Everyone Else

Alice S. Horning

Introduction

The recognition that a city is an ecosystem is the central thesis of this book, building on the variety of ways that living and nonliving things might interact in an urban environment. This includes the process of critical literacy, where people interact with alphabetic and nonalphabetic texts for a wide array of personal and professional purposes. For this discussion, I am defining critical literacy as a lifelong psycholinguistic process of interaction between a visual/audio text and prior knowledge for the purpose of getting or creating meaning. This process includes analysis, synthesis, evaluation (for authority, accuracy, currency, relevancy, appropriateness and bias) and use. It is a cornerstone of essential education for citizens, especially those in cities where most people live, to address the current array of crises and challenges affecting the whole country, including storms, power outages, and fires resulting from climate change. This particular set of environmental crises provides the basis for this chapter, though the tools of critical literacy are key to understanding and responding to any crisis situation. They must play a key role in the environmental education that will support cities in a sustainable way. In every teaching and learning situation, whether in or beyond the formal classroom, online or off, all citizens need instruction in critical literacy as the launch point for their participation in supporting sustainable societies, including in cities. This chapter provides insight into three issues relevant to critical literacy as an element of environmental education to get cities on the path to sustainability: (1) the need for critical literacy to understand important environmental issues among students and the broader population, (2) how critical literacy can lead to more fulfilling lives through empathy and action to build a sustainable environment, and (3) a review of three strategies for improving critical literacy skills.

DOI: 10.4324/9781003217442-24

The Need for Critical Literacy Skills

Several different indicators show that students and others need skills in critical literacy and critical thinking to help build resilience, defined by the American Psychological Association as the ability to bounce back after stress or difficulty (APA 2021). Evidence of the need for these key skills as a component of resilience comes from a set of studies by researchers at Stanford University on middle school, high school, and college students' abilities to assess the credibility of materials found online (Stanford History Education Group 2016, 2019). In untimed exercises, the students drawn from a nationally representative sample were given six different kinds of age-appropriate material (videos, tweets, sites, etc.) and asked to evaluate these items for credibility. Results show that 50%–80% of the students were unable to complete the tasks involved. This finding shows that there is good reason to be concerned about students' skills in judging the information that they find online when given sources to examine on their own. For students and for all adults, critical literacy can provide a foundational set of skills essential to understand and address the challenges of living and working in ways that can sustain urban environments. As this chapter is being written, the impact of Hurricane Ida in the South and Northeast, western forest fires, and other results of climate change are being felt around the nation. Critical literacy can help people to understand why these calamities are happening, to evaluate the scientific evidence available about them, to see the steps they might take both to keep themselves and their families safe, and to contribute to making their communities more able to withstand the impact of these increasingly frequent and intense events.

Results of similar studies on critical literacy appear in reports by college-level librarians who have studied nationally representative samples of students doing research. A key group doing this work is Project Information Literacy, a nonprofit research group that studies how students and others find and use information. Project Director Alison Head and colleagues (2020) have shown that students are overwhelmed by the information available to them for both school and personal research. A full set of critical literacy skills could help students and other citizens explore claims on issues like climate change. The ability to read and understand extended arguments on key issues allows people to evaluate evidence and take reasoned positions on issues like environmental justice, responses to the pandemic, and similar problems that impact the future viability of cities. Critical literacy is thus essential for those who want to work toward healthy environments, both urban and rural, within and beyond schools.

Moreover, even when students have been taught the importance of evaluating materials they are reviewing online and the criteria for doing so, they might understand the criteria but do not know what to do in order to conduct a rigorous evaluation. The essential elements for evaluating sources are illustrated by the widely promoted CRAAP acronym, created by a librarian to capture the key factors in evaluation of sources: currency, relevance, authority, accuracy, and purpose (cf. Breakstone et al. 2018). An additional element of resilience is asking for help (APA 2021). Students do not routinely consult librarians, so developing this strategy as a habit is part of what needs to be taught. But even students who have been taught the CRAAP list do not always know how to use it to conduct a vigorous evaluation of research sources. Having clear goals and routinely asking for help to achieve them can contribute to sustainable resilience that students need and clearly lack, and instruction along these lines is essential. Being able to evaluate information effectively will allow students to become key participants in environmental education as well as in projects to address climate problems.

It isn't only students who need help with these essential critical literacy skills. A study on judgment of fake news, released in June 2021 and widely reported by organizations such as CNN and ABC News, found that about 75% of the US population at large thinks they can sort out fact from fiction in online reports when in fact they cannot do so (Lyons et al. 2021). This finding came from two national surveys of the adult population to examine people's ability to judge fake news as well as their own assessments of their ability. Their actual performance showed that the vast majority are overconfident in their ability to make such judgments. That is, they both cannot judge fake news correctly and think they are right much more often than they are. Sustainable cities require an educated population that can detect true and false information and there is much need to teach relevant critical literacy skills if people are going to participate in effective environmental activities, lest those lacking key skills continue to believe that fire, drought and other climate calamities are part of normal variation and not a by-product of climate change.

Reading as Motivation: Empathy, Resilience and Critical Literacy

There is one further point to be made about the usefulness and relevance of reading for sustainability. An array of different kinds of research shows that reading, especially of fiction, fosters empathy, according to a variety of studies discussed by UCLA reading researcher Maryanne Wolf (2018), and a review of a number of studies of reading, empathy and the brain (Koopman 2018). Wolf (2018) explains why this connection might be important to sustainability in crisis conditions:

> Empathy involves, therefore, both knowledge and feeling. It involves leaving past assumptions behind and deepening our intellectual understanding of another person, another religion, another culture and epoch. . . . It may . . . also be our best bridge to others with whom we need to work together, so as to create a safer world for all its inhabitants. In the very special cognitive space within the reading–brain circuit, pride and prejudice can gradually dissolve through the compassionate understanding of another's mind.
>
> *(53)*

Empathy, then, can provide the motivation or rationale for addressing the problems we face in trying to create sustainable environments. Reading develops skill in understanding others, forming a basis for action and engagement. We can use the empathy that arises from critical reading to help us understand one another as we work together to deal with emergencies and crises. And even for those not in the front lines of a crisis, empathy is an essential element in resilience and sustainability. Empathy allows people to see a situation or an emergency from the point of view of others, essential to understanding and appropriate responses, both to the specific situation and to the wider impact.

Other reviews of recent research, such as the one by science writer Megan Schmidt (2020) show that various studies support the connection of reading to empathy. She writes that people who read the *Harry Potter* novels become more empathic after seeing the various forms of bias that appear in them. She points to studies by psychologists like University of Toronto's Raymond Mar and colleagues who looked at students' improved social abilities in response to reading fiction (e.g., Mar et al. 2006). While no such improvement resulted from reading nonfiction, the critical and analytical skills that can develop from that kind of reading lead to careful analysis of the misinformation and disinformation widely available on a variety of environmental issues. Russ and Krasny, environmental education scholars at Cornell associated with the Environmental Protection Agency, have edited a book on this topic, with contributions from scholars all over the globe. They write on their website about the relationship among education, action and sustainability:

> Environmental education has often been seen as promoting environmental literacy, which encompasses knowledge, affect, and action that benefit the individual and the environment. In cities, this means partnering with organizations addressing public health, justice and equity, community and youth development, and urban planning, among others. In short, environmental education can be one actor among many in addressing sustainability issues.
>
> *(Russ and Krasny 2017)*

Education of this kind that includes critical literacy can draw on empathy that arises from reading as a basis for responses to environmental crises. The resulting resilience can contribute to sustainability.

While resilience is sometimes seen as a contested term because it suggests a potential return to a status quo of many kinds of injustice that have led to or contributed to our present difficulties, other international efforts are underway that can lead to more positive outcomes in terms of sustainability. The Rockefeller Foundation, for instance, has supported the Resilient Cities Network, a global consortium of Resilience Officers in cities around the world working to address the pandemic, climate change, and other key problems with the goal of supporting projects that will contribute to long-term sustainability (https://resilientcitiesnetwork.org/our-story/). This initiative defines resilience specifically for cities this way:

> Urban resilience is the capacity of a city's systems, businesses, institutions, communities, and individuals to survive, adapt, and grow, no matter what chronic stresses and acute shocks they experience.
>
> By strengthening its underlying fabric and deepening its understanding of the risks that threaten its stability, a city can improve its overall trajectory and the well-being of its citizens, prosper in the face of challenges both expected and as yet unimagined.
>
> Urban resilience responds to three converging global megatrends: climate change, urbanization, and globalization.
>
> Urban resilience demands that cities look holistically at their capacities and their risks, including through meaningful engagement with the most vulnerable members of a community. (https://resilientcitiesnetwork.org/what-is-resilience/)

It should be clear that in this work, the focus is on positive results that address the array of crises currently facing city dwellers. The engagement that is needed requires critically literate leaders and citizens who can work toward these goals. Moreover, this enlarged view of resilience reflects the idea of cities as ecosystems requiring the complex interaction of people, organizations and actions for sustainability.

The City Resilience Framework created by the Network offers a number of characteristics and qualities but only touches on critical literacy tangentially (da Silva 2014). Where it does, though, it points to the need for information literacy and management, of which critical literacy is a key part. The Framework also incorporates discussion of a number of the issues and problems currently affecting cities, drawing on such careful sources as *The Conversation*, an independent, transparent news source based in the United States and assorted other countries. One report from *The Conversation* that it draws on, for example, is about the recent bushfires in Australia. Writing under the auspices of the UN Disaster Risk Reduction office, Cook et al. (2021) present a full report on various scientific studies that clearly show the impact of climate change on the number and severity of the fires in 2019 through 2021. While the so-called Black Summer fires did not directly affect the cities of Australia's east coast, they were in that general area. The point is that reports like this one in the context of resilience offer readers research, scientific evidence, discussion of climate skeptics and their response, and extensive information to refute misinformation and disinformation. Thus citizens and leaders who are equipped with the tools of critical literacy can use those tools to read in order to develop not only their empathy but also their ability to evaluate factual information, keys to a sustainable ecosystem in cities and beyond.

A closer look at the Cook et al. (2021) report makes clear how the tools of critical literacy (e.g., the criteria of CRAAP [currency, relevance, authority, accuracy, and purpose]) can be applied to evaluate a document of this kind.

Currency. The report seems quite current with a publication date of November 2021.

Relevance. The report deals with the catastrophe of the fires in Australia to which climate change was a contributing factor. While the areas directly affected by the fires were not the east coast's major cities of Australia, the whole eastern portion of the country was affected, and it is where most of the population is concentrated, so it is certainly pertinent to the overall resilience of the country.

Authority. It is necessary to consider the authors and their credentials. Here again, it is easy to determine the authors' expertise. The research was done under the auspices of the UN, which has an office devoted to Disaster Risk Reduction. Lead author Garry Cook holds a PhD from the University of Queensland, having published over 100 scholarly articles on environmental topics according to his profile on The Conversation website (Garry Cook n.d.). In a 30-year career, this lead author has developed clear expertise on topics pertinent to this published report.

Accuracy. This might be a little harder to assess, but since the report includes extensive citations to maps, data, and other materials, it is reasonable to consider the report accurate.

Purpose. The purpose of this report is to provide proof that the Black Summer fires are the result of climate change. The increasing heat and drought resulting from global warming have a clear connection to the fire catastrophes on the east coast.

This extended application of the CRAAP criteria shows how critical evaluation of source materials can help citizens get appropriate information about environmental issues or anything else. Such information is one key to resilience of a positive kind that can, in combination with empathy, likewise be derived from reading, that contributes to sustainability.

Strategies for Critical Literacy for Sustainable Cities

It's all well and good to say that students should be taught lists like the CRAAP list to help them think about and evaluate materials they read and to show how that process, as illustrated here, might work whether applied to traditional texts or to digital sources of many kinds. Students will also need to understand the three key concepts of urban sustainability as proposed by Deborah Mutnick in Chapter 10 (this volume). Instructors must have Monday morning approaches and specific steps to develop students' skills. Three evidence-based strategies can move directly into almost any classroom and help students become the critical readers and thinkers we need to deal with the emergency situations in which we find ourselves and move toward the goal of sustainability.

The first of these strategies is lateral reading, an approach that appeared in a study by the Stanford History Education Group (Wineburg and McGrew 2017). Investigators compared the evaluation strategies of professional fact-checkers to those of faculty with a PhD in history and Stanford undergraduate students. The main findings of the study are that students and even people who hold a PhDs are easily fooled by misinformation and disinformation from digital sources, chiefly because they rely on vertical reading within a website. By contrast, highly accurate professional fact-checkers read laterally, going to other reputable sites and sources to check the accuracy of information (the analysis of the preceding Australian fire report used this approach in part). The lateral reading approach is defined by the Stanford researchers:

> When reading laterally, one leaves a website and opens new tabs along a horizontal axis in order to use the resources of the Internet to learn more about a site and its claims. Lateral reading contrasts with vertical reading. Reading vertically, our eyes go up and down a screen to evaluate the features of a site. Does it look professional, free of typos and banner ads? Does it quote well-known sources? Are [sic] bias or faulty logic detectable? In contrast, lateral readers paid little attention to such features, leaping off a site after a few seconds and opening new tabs. They investigated a site by leaving it.
>
> Paradoxically, a key feature of lateral reading is *not reading*. . . . It requires knowledge of *sources*, knowledge of how the Internet and searches are *structured*, and knowledge of *strategies* to make searching and navigating effective.
>
> *(38)*

This technique can be taught to students easily enough and can make them much more efficient critical readers of materials found online; the following example shows how this approach might work. And while students and other adults think they are already able to distinguish fact from fiction, Lyons et al. (2021) shows that about 75% of them are not able to do so. To develop the skills needed to understand and address climate change problems happening all around all of us, students need lessons in lateral reading.

In addition, the techniques of mindful reading described by University of Connecticut reading scholar Ellen Carillo (2018) can also be helpful. Mindful reading, as Carillo defines it, is not a particular approach or strategy for reading. Instead, it is a way of thinking about the reading process along with how and why readers engage in it, which Carillo has developed through years of classroom experience and careful research with teaching faculty (see Carillo 2015). The object is for readers to have an array of different strategies at their disposal and to make conscious choices among them depending on the reading situation. As Carillo says, mindful reading is "a particular stance on the part of the reader, one that is open, flexible, and characterized by intentional awareness of and attending to the present moment and the demands it makes on reading" (2018, 66). Among the array of strategies readers might employ are highlighting, annotation, what the text says and does, skimming, rhetorical reading, paraphrasing, reading like a writer, using the CRAAP list, and some others—all discussed in more detail in Carillo's textbook, *A Writer's Guide to Mindful Reading* (2017). Again, these are teachable and learnable strategies that are at the reader's disposal in a mindful framework that takes a reader's purpose and needs into account. Critical readers are mindful readers who can make conscious choices about what and how to read. They can be ready for all kinds of emergencies with the tools to do research and evaluate what they read, making useful and appropriate responses and action possible. Carillo's extensive classroom experience and currently funded research project offer evidence in support of these approaches (personal communication, November 22, 2021).

Finally, a third strategy, suggested by Sullivan (2017), is, like mindful reading, a way of thinking about reading as "a process of inquiry" (143). His coedited volume, *Deep Reading*, earned the Best Book Award from the Conference on College Composition and Communication in 2019. It offers a wide-ranging set of chapters, including some by students, that explore the nature of deep reading as well as offering specific classroom approaches for teaching it. Sullivan's approach is based, like Carillo's, on years of experience in the classroom, and does not have one specific strategy as its focal point. Instead, Sullivan suggests that deep reading includes close reading along with other assorted strategies that "provide opportunities for students to engage in metacognitive thinking about the *process* of learning and to help students assess and reassess their own mental models for understanding the world" (147). Here again, it's clear that giving students the tools for flexible, critical approaches to reading is a fundamental way to help them develop skills that will serve them in and beyond school, across the emergencies we all face.

A brief example shows how these approaches might work. With a search on the Internet, it's easy to find misinformation like this petition (petitionproject.org), developed by members of the Oregon Institute of Science and Medicine (oism.org):

> We urge the United States government to reject the global warming agreement that was written in Kyoto, Japan in December, 1997, and any other similar proposals. The proposed limits on greenhouse gases would harm the environment, hinder the advance of science and technology, and damage the health and welfare of mankind.
>
> There is no convincing scientific evidence that human release of carbon dioxide, methane, or other greenhouse gases is causing or will, in the foreseeable future, cause catastrophic heating of the Earth's atmosphere and disruption of the Earth's climate. Moreover, there is substantial scientific evidence that increases in atmospheric carbon dioxide produce many beneficial effects upon the natural plant and animal environments of the Earth.

The lateral reading strategy would take critical readers immediately beyond the website itself. One Google search takes readers to the petition on the OSS Foundation website: (http://ossfoundation.us/projects/environment/global-warming/myths/31000-scientists-say-no-convincing-evidence), which, at first glance, looks supportive of the petition but actually refutes each point. Readers using mindful strategies would have to think about how to evaluate the claims made. Questions about just what the Kyoto agreement said would be one place to look. Another would be to explore who the scientists are who made the statement or to look at the dubious credentials of the Oregon Institute of Science and Medicine, according to the fact-checking site Source Watch (Center for Media and Democracy 2020). Critical readers can use the references at the Center for Media and Democracy (CMD) website to test the claims made. These strategies can allow readers to evaluate information on the environment, climate change, or any other emergency we currently face.

It should be easy to see how deep reading and the other ideas presented here can offer options for developing students' abilities. There isn't one necessary or essential strategy but rather a set of options accompanied by the awareness of them and when and how best to use each one. The underlying ideas are empathy and resilience, abilities to understand and feel for others as well as to bounce back regardless of the challenge presented by an emergency situation. To be sustainable, cities need people who are empathic and resilient and who are equipped with the tools of critical literacy skills, approaches, frameworks, and so forth. Lateral reading, mindful reading and deep reading offer ways to achieve this goal.

Summary: Resilience and Sustainability Through Critical Literacy

For cities to be sustainable, empathy helps students and all citizens to understand one another's situations and needs. Since the 1950s, the world population has rapidly shifted toward cities, and while some people are now choosing to move away from such close quarters in response to the pandemic and as a by-product of more flexible remote work, the global population at large is increasingly concentrated in cities (United Nations 2018). So, sustainable cities must function as ecosystems. Empathy and resilience are key elements in sustainability, since crises and emergencies are surely going to continue. Public health experts, for instance, say there is no question that there will be more pandemics; it is only a question of when they might occur; similar predictions have been made for hurricanes, flooding, and other climate-related trouble. To prepare for these and other crises as they come, critical literacy offers an essential tool needed to address them. The research on students and the population at large shows the need for much stronger critical literacy skills. Reading develops human empathy as well as the ability to understand and evaluate information and the skills to use that information for a range of purposes. These skills are currently lacking and are urgently needed, especially as we spend more and more of our lives in a digital environment. Not only students but the increasing numbers of city dwellers of all ages can address the forthcoming challenges we all face by more experience with the reading of both fiction and nonfiction. Fiction offers learning empathy, while nonfiction provides the chance to develop skill in evaluation. Developing critical literacy is the work of educators, but it must focus explicitly and directly on hands-on skills and tools like lateral, mindful, and deep reading. Building these critical literacy skills among students and everyone else will encourage the growth of cities that function as sustainable ecosystems.

References

American Psychological Association. 2021. *Building Your Resilience*. Washington, DC: American Psychological Association. www.apa.org/topics/resilience.

Breakstone, Joel, Sarah McGrew, Mark Smith, Teresa Ortega, and Sam Wineburg. 2018. "Why We Need a New Approach to Digital Literacy." *Phi Delta Kappan* 99 (6): 27–32. https://kappanonline.org/breakstone-need-new-approach-teaching-digital-literacy/.

Carillo, Ellen C. 2015. *Securing a Place for Reading in Composition: The Importance of Teaching for Transfer*. Logan: Utah State University Press.

Carillo, Ellen C. 2017. *A Writer's Guide to Mindful Reading*. Fort Collins and Boulder: WAC Clearinghouse and University of Colorado Press.

Carillo, Ellen C. 2018. *Teaching Readers in Post-Truth America*. Logan: Utah State University Press.

Center for Media and Democracy. 2020. "Oregon Institute of Science and Medicine." *Source Watch*. www.sourcewatch.org/index.php/Oregon_Institute_of_Science_and_Medicine.

Cook, Garry, Andrew Dowdy, Juergen Knauer, Mick Meyer, Pep Canadell, and Peter Briggs. 2021. "Australia's Black Summer of Fire Was Not Normal—and We Can Prove It." www.preventionweb.net/news/australias-black-summer-fire-was-not-normal-and-we-can-prove-it#.

da Silva, Jo. 2014. *City Resilience Framework/City Resilience Index*. New York: Arup International Development. https://resilientcitiesnetwork.org/urban_resiliences/city-resilience-framework.

Garry Cook. n.d. "Garry Cook." *The Conversation*. Accessed December 6, 2021. https://theconversation.com/profiles/garry-cook-1293568.

Head, Alison J., Barbara Fister, and Margy MacMillan. 2020. *Information Literacy in the Age of Algorithms*. Project Information Literacy Research Institute. https://projectinfolit.org/publications/algorithm-study.

Koopman, Eva Maria. 2018. "Does Originality Evoke Understanding? The Relation between Literary Reading and Empathy." *Review of General Psychology* 22 (2): 169–77. https://doi.org/10.1037/gpr0000107.

Lyons, Benjamin A., Jacob M. Montgomery, Andrew M. Guess, Brendan Nyhan, and Jason Reifler. 2021. "Overconfidence in News Judgments Is Associated with False News Susceptibility." *Proceedings of the National Academy of Sciences* 118 (23): e2019527118. https://doi.org/10.1073/pnas.2019527118.

Mar, Raymond, Keith Oatley, Jacob Hirsh, Jennifer dela Paz, and Jordan B. Peterson. 2006. "Bookworms Versus Nerds: Exposure to Fiction Versus Non-Fiction, Divergent Associations with Social Ability, and the Simulation of Fictional Social Worlds." *Journal of Research in Personality* 40: 694–712.

Russ, Alex, and Marianne Krasny. 2017. "Environmental Education Generates Urban Sustainability." www.thenatureofcities.com/2017/02/08/environmental-education-generates-urban-sustainability/.

Schmidt, Megan. 2020. "How Reading Fiction Increases Empathy and Encourages Understanding." *Discover*. www.discovermagazine.com/mind/how-reading-fiction-increases-empathy-and-encourages-understanding.

Stanford History Education Group. 2016. *Evaluating Information: The Cornerstone of Civic Online Reasoning*. https://sheg.stanford.edu/upload/V3LessonPlans/Executive%20Summary%2011.21.16.pdf.

Stanford History Education Group and Gibson Consulting. 2019. *Students' Civic Online Reasoning: A National Portrait*. https://stacks.stanford.edu/file/gf151tb4868/Civic%20Online%20Reasoning%20National%20Portrait.pdf.

Sullivan, Patrick. 2017. "'Deep Reading' as a Threshold Concept in Composition Studies." In *Deep Reading: Teaching Reading in the Writing Classroom*, edited by Patrick Sullivan, Howard Tinberg, and Sheridan Blau, 143–71. Urbana: National Council of Teachers of English.

United Nations. 2018. "68% Of the World Population Projected to Live in Urban Areas By 2050, Says UN." www.un.org/development/desa/en/news/population/2018-revision-of-world-urbanization-prospects.html.

Wineburg, Sam, and Sarah McGrew. 2017. *Lateral Reading: Reading Less and Learning More When Evaluating Digital Information*. Stanford History Education Group Working Paper No. 2017-A1. https://papers.ssrn.com/sol3/papers.cfm?abstract_id=3048994. Electronic copy available at: https://ssrn.com/abstract=3048994.

Wolf, Maryanne. 2018. *Reader Come Home: The Reading Brain in a Digital World*. New York: HarperCollins.

22

AFTERWORD

Transdisciplinary Urban Ecosystem Research, Education, and Stewardship

Loren Byrne

In ecology, the concept of the ecosystem has a long history (Golley 1993; Pickett and Cadenasso 2002). Though it has sometimes been contested (e.g., O'Neill 2001), its utility as a multidimensional term to help frame and organize the study of organism–environment interactions gives it staying power (Pickett and Cadenasso 2002). This power is reflected by the existence of similar paradigms among diverse cultures as part of their traditional ecological knowledge (Berkes et al. 1998) and adoption of the word by non-ecologists to describe diverse entities (e.g., media ecosystems, business ecosystems, learning ecosystems, to name just a few that can be found online). Such widespread use risks turning the word into an abused cliché without clear meaning: everything is an ecosystem! On the other hand, when a word like "ecosystem" has such widespread resonance, it has value as a unifying starting point to bring together diverse audiences, perspectives, and bodies of scholarship to examine important issues, especially hyper-complex ones that require transdisciplinary knowledge and problem-solving approaches. In the Anthropocene, few complex issues are as important as creating more sustainable urban ecosystems and societies.

This desire for collaborative examination of the complex topic of urban sustainability is a clear motivation for the present volume. Its breadth of included topics is impressive: history, health, philosophy, biodiversity, economics, ethics, justice, education, sociology, energy infrastructure, and more. All these variables interact to form urban ecosystems, and thus, all must be considered to fully understand urban patterns and dynamics, and resulting levels of sustainability. As such, readers will benefit from reviewing all the chapters rather than just focusing on the ones most closely related to the subjects they already know. Advancing urban sustainability through research and new philosophical perspectives—and their application to urban design and management—requires that we step outside our scholarly and epistemological comfort zones to engage with diverse ways of seeing and

DOI: 10.4324/9781003217442-25

responding to urban ecosystems. Such transdisciplinary thinking is catalyzed by volumes like this one.

In terms of transdisciplinarity, I suggest that, as a collection, the book's diverse chapters point to three emergent reflections. Certainly, these are not the only three, and they are admittedly biased by my personal outlook as an ecologist and educator. Nonetheless, I offer them in this modest afterword to inspire others to consider the broader implications of this volume for the holistic study of urban places, including cities, suburbs, exurbs and other urbanized landscapes.

My first reflection builds on the preceding introductory comments about the word "ecosystem." This book's title is a direct, bold declaration: *The City is an Ecosystem*. However, from an ecologist's view, its contents include lots of topics that do not appear, at first glance, to fit within the traditional purview of ecosystem science's focus on nonhuman organisms, soils, biogeochemical cycles, and other basic ecological variables like sunlight and precipitation. Thus the book's title prompts revisiting basic questions about the definition of ecosystem and whether urban places—for which "city" is often used as a stand-in word—are something other, or more, than ecosystems. In large part, these questions are attributable to the current norms among ecologists for using more descriptive terms for urban places, including human ecosystems, social-ecological systems, and, more recently, social-ecological-technological systems (Zhou et al. 2021). Assumedly, such phrases are preferred because they highlight the anthropogenic variables that comprise a large part of urban systems; this emphasis makes sense given that ecologists intentionally ignored humans for most of the history of the discipline (i.e., humans are not part of ecosystems). Thus is it appropriate to refer to urban places simply as "ecosystems"? Is this perspective missing something, is it misleading, or is it an abuse of ecological science vocabulary? Does the word "ecosystem" alone allow for integrating humans and their actions into the fold?

Two lines of thinking support the assertion that urban places can properly be called ecosystems without sacrificing the fuller transdisciplinary view of such places. The first is arises from an analysis of the writings of Arthur Tansley who coined the word "ecosystem." After reviewing his work, Pickett and Grove (2009) observed that Tansley explicitly noted that humans are ecological variables, and they concluded that "the contemporary definition of the ecosystem can be expanded—without violating either the letter or the spirit of Tansley's precedent—to incorporate humans, their institutions and economies, their buildings and engineered networks of transport and communication" (6). Second, in Eugene Odum's (1971) classic textbook, which is often credited with popularizing the ecosystem concept (Golley 1993), he discussed ecological variables explicitly in the context of cities. Such references indicate that he saw them as ecosystems alongside others like forests. Further, he explicitly called for integration of ecological and social variables to improve urban land use planning. Given these two historical precedents (among others that could surely be found), it is reasonable to conclude that urban places can indeed be referred to simply as ecosystems, thus justifying this book's title. Although the use of the more descriptive phrases just noted will surely continue, perhaps this book points to the value of simplified language and being comfortable with "ecosystem" as a transdisciplinary term that encompasses a fuller range of anthropocentric topics, including all those covered by this volume's diverse chapters.

A second reflection inspired by this book is the need to more explicitly and expansively link the social and ecological dimensions of the study and pursuit of urban sustainability. On the academic side, this means pursuing a fuller integration of variables, methods, and theory from the social and natural sciences, and humanities. On the practical, applied side, this includes wider efforts to help people connect directly with the outdoor ecosystems around them. As emphasized by many chapters, such explicit relationships lead to deeper understanding about the characteristics of urban ecosystems and help people engage directly in the work needed to improve them, such as creating and managing greenspace. In particular, making such linkages means engaging diverse stakeholders in research (e.g., citizen and community science) and its interpretation and translation into decisions and actions. Such work necessitates different types of projects (i.e., transdisciplinary research and community-based, collaborative efforts) from what academic scholars have traditionally pursued but that are common in other contexts (e.g., community organizations, government agencies, architecture and planning firms). Notably, pursuing partnerships and dialogue to promote urban sustainability—especially regarding issues of health, equity, justice, and resilience—is a theme woven throughout the book, both explicitly and implicitly. Such an emphasis echoes that of many others (e.g., Zhou et al. 2021), especially advocates for the views of civic ecology (Krasny and Tidball 2012) and "ecology with cities" (e.g., Byrne 2022 and references therein; also see Chapter 19, this volume). This book advances this conversation with exemplary work about both the academic and applied sides of understanding and working in urban ecosystems.

Though the value of transdisciplinary scholarship and community-based projects is clear, pursuing them is not easy, especially for those in higher education institutions. One of the many challenges to such work is the current structure of educational systems, including curricula and reward criteria for faculty (especially regarding tenure and promotion), which too often hinder holistic, integrated learning and applied projects. Though paradigms have been shifting toward transdisciplinarity and addressing communities' needs, additional and more expansive changes are needed to encourage professors and students to embrace transdisciplinary urban studies and help lead projects aimed at transforming urban ecosystems for improved environmental quality and human well-being. This third reflection is inspired by the insightful chapters in this volume that provide excellent examples of the positive outcomes—for communities, students, and disciplines—that arise when new models of research, teaching, and learning are pursued. As highlighted previously, more intentional linking of ecological and social aspects of urban ecosystems is central to innovative pedagogical models (see Chapter 19). Helping students and community stakeholders explore such connections can facilitate development of more holistic views of complex interrelationships among environments, people, diverse cultural issues, and sustainability. To this end, many chapters in this volume build on a small but expanding body of literature that provides a solid foundation for future work in urban environmental education (e.g., Peterson 2018; Russ 2015; Maharramli, Bredow, and Goodwin 2021).

I readily admit that these three reflections are not new to me; many others—too many to cite in a brief afterword—have written deeper analyses than I can offer here about all these issues. Nonetheless, this book's emphasis on transdisciplinarity—its motivating ethos of synthesizing diverse disciplinary perspectives and methods—points to the continued need to pursue unifying concepts and perspectives that help diverse stakeholders collaborate for integrated research, education, and urban sustainability practices and projects. As noted, "ecosystem" may provide such a transdisciplinary focus because it supports the conceptual and methodological linking of humans, urban infrastructure and ecological variables in a common framework. However, I suggest that the ecosystem view also implies objects of study: ecosystems are entities to be described and evaluated over space and time. These two dimensions of "ecosystem" are a necessary start but feel insufficient; something more is needed to achieve urban sustainability, something that helps people better recognize and embrace their role in that pursuit.

This desire for something else leads me to the idea of stewardship for urban ecosystems. Stewardship has long been used as a theme for examining sustainable human–environment relationships, and it is increasingly being used in the context of urban places (e.g., Andersson et al. 2014; Bennett et al. 2018; Romolini, Bixler, and Grove 2016). In particular, the phrase "social-ecological stewardship" (Cockburn et al. 2018) has been articulated to emphasize that caring for people and the environment are coupled endeavors that cannot be separated. This linkage is especially true in urban places where social and ecological conditions are tightly interconnected: helping people necessitates improving the environment; helping the environment requires helping people, including those who will live in the future. In this context, I propose that urban ecosystem stewardship can serve as a unifying, transdisciplinary focus for urban environmental education, and conversations and collaborations among diverse scholars and community stakeholders. As reflected by this volume, fostering such partnerships and education leads to valuable insights and outcomes that can help make urban ecosystems more sustainable and resilient. Given that increasing numbers of people around the world live in cities and other types of urban ecosystems, the valuable information in this book is sure to have a positive impact on the study and practices of urban sustainability and stewardship for a long time to come.

References

Andersson, Erik, Stephan Barthel, Sara Borgström, Johan Colding, Thomas Elmqvist, Carl Folke, and Åsa Gren. 2014. "Reconnecting Cities to the Biosphere: Stewardship of Green Infrastructure and Urban Ecosystem Services." *Ambio* 43 (4): 445–53.

Bennett, Nathan J., Tara S. Whitty, Elena Finkbeiner, Jeremy Pittman, Hannah Bassett, Stefan Gelcich, and Edward H. Allison. 2018. "Environmental Stewardship: A Conceptual Review and Analytical Framework." *Environmental Management* 61 (4): 597–614.

Berkes, Fikret, Mina Kislalioglu, Carl Folke, and Madhav Gadgil. 1998. "Exploring the Basic Ecological Unit: Ecosystem-like Concepts in Traditional Societies." *Ecosystems* 1 (5): 409–15.

Byrne, Loren B. 2022. "Ecology with Cities." *Urban Ecosystems* 25 (3): 835–837.

Cockburn, Jessica, Georgina Cundill, Sheona Shackleton, and Mathieu Rouget. 2018. "Towards Place-based Research to Support Social—Ecological Stewardship." *Sustainability* 10 (5): 1434.

Golley, Frank B. 1993. *A History of the Ecosystem Concept in Ecology: More than the Sum of the Parts*. New Haven: Yale University Press.

Krasny, Marianne E., and Keith G. Tidball. 2012. "Civic Ecology: A Pathway for Earth Stewardship in Cities." *Frontiers in Ecology and the Environment* 10 (5): 267–73.

Maharramli, Bemmy, Victoria Lowerson Bredow, and Lindsay Goodwin. 2021. "Using Civic Ecology Education to Foster Social-ecological Resilience: A Case Study from Southern California." *The Journal of Environmental Education* 52 (6): 445–62.

Odum, Eugene. 1971. *Fundamentals of Ecology*, 3rd ed. Philadelphia: W.B. Saunders Co.

O'Neill, Robert V. 2001. "Is It Time to Bury the Ecosystem Concept? (With Full Military Honors, of Course!)." *Ecology* 82 (12): 3275–84.

Peterson, Richard B. 2018. "Taking It to the City: Urban-placed Pedagogies in Detroit and Roxbury." *Journal of Environmental Studies and Sciences* 8 (3): 326–42.

Pickett, Steward T. A., and Mary L. Cadenasso. 2002. "The Ecosystem as a Multidimensional Concept: Meaning, Model, and Metaphor." *Ecosystems* 5 (1): 1–10.

Pickett, Steward T. A., and Morgan J. Grove. 2009. "Urban Ecosystems: What Would Tansley Do?" *Urban Ecosystems* 12 (1): 1–8.

Romolini, Michele, R. Patrick Bixler, and J. Morgan Grove. 2016. "A Social-ecological Framework for Urban Stewardship Network Research to Promote Sustainable and Resilient Cities." *Sustainability* 8 (9): 956.

Russ, Alex, ed. 2015. *Urban Environmental Education*. Ithaca and Washington: Cornell University Civic Ecology Lab, NAAEE and EECapacity.

Zhou, Weiqi, Steward T. A. Pickett, and Timon McPhearson. 2021. "Conceptual Frameworks Facilitate Integration for Transdisciplinary Urban Science." *Urban Sustainability* 1 (1): 1–11.

INDEX